Systems for Printed Flexible Sensors

Design and implementation

IOP Series in Sensors and Sensor Systems

The *IOP Series in Sensors and Sensor Systems* includes books on all aspects of the science and technology of sensors and sensor systems. Spanning fundamentals, fabrication, applications and processing, the series aims to provide a library for instrument and measurement scientists, engineers and technologists in universities and industry.

The series seeks (but is not restricted to) publications in the following topics:

- Advanced materials for sensing
- Biosensors
- Chemical sensors
- Industrial applications
- The Internet of Things
- Lab-on-a-chip
- Localization and object tracking
- Manufacturing and packaging
- Mechanisms, modelling and simulations
- MEMS/NEMS
- Micro- and nanosensors
- Non-destructive testing
- Optoelectronic and photonic sensors
- Optomechanical sensors
- Physical sensors
- Remote sensors
- Sensing for health, safety and security
- Sensing principles
- Sensing systems
- Sensor arrays
- Sensor devices
- Sensor networks
- Sensor technology and applications
- Signal processing and data analysis
- Smart sensors and monitoring
- Telemetry

Authors are encouraged to take advantage of electronic publication through the use of colour, animations, video and interactive elements to enhance the reader experience.

A full list of titles published in this series can be found here: https://iopscience.iop.org/bookListInfo/iop-series-in-sensors-and-sensor-systems.

Systems for Printed Flexible Sensors

Design and implementation

Edited by
Tarikul Islam
Jamia Millia Islamia, New Delhi, India

Subhas Chandra Mukhopadhyay
Macquarie University, Sydney, Australia

Boby George
Indian Institute of Technology Madras, Chennai, India

IOP Publishing, Bristol, UK

ISBN 978-0-7503-3935-3 (ebook)
ISBN 978-0-7503-3933-9 (print)
ISBN 978-0-7503-3936-0 (myPrint)
ISBN 978-0-7503-3934-6 (mobi)

DOI 10.1088/978-0-7503-3935-3

Version: 20220901

IOP ebooks

British Library Cataloguing-in-Publication Data: A catalogue record for this book is available from the British Library.

Published by IOP Publishing, wholly owned by The Institute of Physics, London

IOP Publishing, Temple Circus, Temple Way, Bristol, BS1 6HG, UK

US Office: IOP Publishing, Inc., 190 North Independence Mall West, Suite 601, Philadelphia, PA 19106, USA

Dedicated to our parents, the late Md Abul Hossain, the late Anwara Khatun (Tarikul Islam), the late Dyutindra Nath Mukhopadhyay, Mrs Rabi Rani Mukhopadhyay (Subhas Chandra Mukhopadhyay) George, and Maria (Boby George).

Contents

Preface xiii

Acknowledgment xvii

Editors' biographies xviii

List of contributors xx

1 A review on flexible sensors for soft robotics 1-1
Debasrita Kar, Boby George and K Sridharan

1.1 Introduction 1-1
1.2 Important characteristics required for soft robotics sensors 1-2
1.3 Sensing techniques employed to develop soft robotic sensors 1-2
1.3.1 Magnetic sensors 1-2
1.3.2 Inductive sensors 1-4
1.3.3 Capacitive sensors 1-4
1.3.4 Strain gauge/resistance-based sensors 1-6
1.3.5 Eddy current sensors 1-8
1.3.6 Optical sensors 1-9
1.4 Interfacing circuits and associated measurement systems 1-11
1.5 Comparison of different sensing techniques 1-11
1.6 Challenges and future directions 1-12
References 1-12

2 Recent advances in laser-induced graphene (LIG)-based flexible electronic devices 2-1
Khairunnisa Amreen and Sanket Goel

2.1 Introduction 2-2
2.1.1 Characterization techniques 2-4
2.2 LIG fabrication 2-4
2.2.1 LIG chemical modification 2-6
2.3 LIG in electronic devices 2-6
2.3.1 LIG-based miniaturized fuel cells 2-6
2.3.2 LIG-based miniaturized energy-storage devices 2-9
2.3.3 LIG heaters 2-10
2.3.4 LIG sensors 2-11
2.4 LIG in wearable and smart devices 2-21
2.5 Conclusion and future outlook 2-22
References 2-24

3 Printable flexible sensors for hydration monitoring and moisture measurement in concrete structures 3-1

Tarikul Islam and Kazi Jabed Akram

3.1 Introduction 3-1

3.2 Hydration and its monitoring using flexible sensors 3-6

3.2.1 Ultrasonic technique 3-7

3.2.2 Hydration monitoring by temperature measurement 3-10

3.3 Hydration of concrete using transformer principle 3-15

3.4 Determination of the response parameters of concrete 3-16

3.4.1 Fringing field flexible capacitive sensors for hydration monitoring 3-18

3.5 Flexible sensors for moisture measurement in concrete 3-22

3.5.1 Fringing field capacitive sensors for concrete moisture measurement 3-22

3.5.2 Theory of the proposed method 3-22

3.5.3 Modeling of the sensor 3-24

3.5.4 Experimental moisture measurement results in cement slab 3-24

3.6 Conclusion 3-26

References 3-26

4 Resistive sensor interface 4-1

Kaushal Kishore and S A Akbar

4.1 Introduction 4-1

4.2 Auto-balancing interface circuits 4-2

4.2.1 Circuit architecture and operation 4-3

4.2.2 Circuit analysis 4-7

4.2.3 Prototype design and experimental results 4-15

4.2.4 Experimental result with sensor 4-17

4.2.5 Discussion 4-20

4.3 Auto-balancing interface for resistive sensor array 4-20

4.3.1 Circuit architecture and operation 4-21

4.3.2 Circuit analysis 4-24

4.3.3 Prototype design and experimental results 4-27

4.3.4 Experimental measurement with a silicon nanowire array 4-28

4.3.5 Discussion 4-28

4.3.6 Summary 4-30

4.4 Conclusion and industrial aspects 4-31

4.5 Future of sensing technology 4-32

References 4-33

5 Interfacing circuit for capacitive sensors **5-1**
Sougata Kumar Kar and Siddhartha Sen

5.1 Introduction 5-1
5.2 Capacitive sensors: basic principles 5-2
5.2.1 Variable gap type 5-2
5.2.2 Variable area type 5-2
5.2.3 Variable dielectric type 5-3
5.2.4 Differential configuration 5-3
5.3 Sensor interfacing techniques 5-5
5.3.1 Capacitance-to-frequency converter 5-5
5.3.2 Continuous-time interface 5-6
5.3.3 Discrete-time interface 5-8
5.3.4 Chopper stabilization 5-9
5.3.5 Auto-zeroing (AZ) technique 5-12
5.3.6 Correlated double-sampling 5-13
5.4 Capacitance-to-digital converters 5-14
5.4.1 Dual-slope CDC 5-14
5.4.2 Succesive approximation register type 5-15
5.4.3 Sigma delta modulator-type 5-16
5.5 Sensor mismatch cancellation 5-17
5.5.1 Capacitance array 5-17
5.5.2 Voltage-mode charge balance method 5-18
5.5.3 Auto-cancellation of sensor mismatch 5-18
5.6 Case study: integrated capacitance measurement system 5-20
5.6.1 Circuit components 5-20
5.7 Conclusion 5-23
References 5-23

6 Interface electronics and conditioning circuits for triboelectric flexible sensors **6-1**
Chunkai Qiu and Mehmet Rasit Yuce

6.1 Introduction 6-1
6.2 Characteristics of triboelectric sensors 6-3
6.3 Block structure of interface electronics 6-4
6.3.1 Preamplifier 6-4
6.3.2 Filtering 6-5
6.4 Design examples of analog front-end 6-7

6.5 Case study: circuitry for a grating-patterned triboelectric human–machine interface 6-11
6.5.1 Design and working mechanism of a triboelectric sensor 6-12
6.5.2 Signal-conditioning circuit 6-14
6.5.3 Design of the printed circuit board 6-16
6.5.4 Experimental results 6-16
6.6 Summary 6-19
References 6-20

7 Flexible interfacing circuits for wearable sensors and wireless communication 7-1
Moupali Chakraborty, Oliver Ozioko and Ravinder Dahiya
7.1 Introduction 7-1
7.2 Wearable sensors and interfacing circuits 7-2
7.2.1 Non-self-powered sensors and interfacing circuits 7-2
7.2.2 Chemical sensors and interfacing circuits 7-6
7.2.3 Self-powered systems and interfacing circuits 7-8
7.3 Interfacing circuits for wireless communication 7-11
7.4 Conclusion and future perspective 7-13
References 7-15

8 Compact and efficient wireless power and information transfer systems for IoT sensors and implants 8-1
Ramesh K Pokharel and Adel Barakat
8.1 Introduction 8-1
8.2 Examples from the literature 8-2
8.3 Figure-8 inductor 8-3
8.3.1 Structure of a figure-8 inductor 8-3
8.3.2 Self-inductance of the figure-8 inductor 8-4
8.3.3 Losses of the figure-8 inductor 8-4
8.3.4 Mutual inductance between two figure-8 inductors 8-4
8.3.5 Cross-coupling with DGS inductor 8-5
8.4 System design and discussions 8-7
8.4.1 System layout 8-7
8.4.2 Fabrication 8-9
8.4.3 Measurement category 1: S-parameters 8-9
8.4.4 Measurement category 2: misalignment 8-11

8.4.5 Measurement category 3: simultaneous wireless power and information transfer 8-11
8.5 Conclusion 8-16
References 8-16

9 Analysis and design considerations of relaxation-oscillator-based flexible sensor circuits 9-1
Shahid Malik and Laxmeesha Somappa

9.1 Introduction 9-1
9.2 Relaxation-oscillator-based circuits 9-2
9.2.1 Conventional relaxation-oscillator-based interface 9-3
9.2.2 Analysis for time period T 9-3
9.2.3 Conventional relaxation-oscillator-based interface for bridge sensors 9-4
9.2.4 Relaxation oscillator circuits for leaky capacitive sensors 9-5
9.2.5 Relaxation oscillator circuit for differential capacitive sensors 9-8
9.3 Performance analysis with component non-idealities 9-10
9.4 Design criteria 9-13
References 9-14

10 Sensor tags for wireless body-centric communication: challenges and opportunities 10-1
Mohammad Hashmi and Dinesh Rano

10.1 Introduction 10-1
10.1.1 Antenna design 10-3
10.1.2 Conformability of body-worn antenna 10-4
10.1.3 Communication through multiple textile layers 10-5
10.2 Existing challenges 10-5
10.3 Some solutions 10-6
10.3.1 Electromagnetic-band-gap-incorporated antenna design 10-6
10.3.2 Determination of effective dielectric constant 10-6
10.3.3 Conformability analysis of conformal electromagnetic band gap surfaces 10-7
10.4 Compact electromagnetic-band-gap-based antenna for smartwatch applications 10-7
10.4.1 Introduction to electromagnetic band gaps 10-8
10.4.2 Lumped-element circuit model 10-9
10.4.3 Interdigital electrodes unit cell 10-16

10.4.4 Array performance enhancement 10-23
10.4.5 Determination of effective dielectric constant 10-27
10.4.6 Conformal electromagnetic band gap surfaces 10-41
References 10-48

11 The role of capacitive sensors for condition monitoring of transformer insulation **11-1**
Md Manzar Nezami and Shakeb A Khan

11.1 Introduction 11-1
11.2 Literature review 11-3
11.3 Development of a capacitive sensor for condition assessment of an oil-immersed power transformer's insulation 11-4
11.3.1 Materials required for the fabrication of the 2-FAL sensor 11-4
11.3.2 Process of fabrication 11-5
11.3.3 Experimental setup 11-6
11.4 Electrical equivalent model of the developed sensor 11-7
11.4.1 Electrical equivalent circuit of the sensor 11-7
11.4.2 Relation between the sensor's capacitance and the concentration of 2-FAL 11-9
11.5 Performance of the sensor 11-10
11.5.1 Response of the sensor at different frequencies 11-10
11.5.2 Response of the sensor with time 11-11
11.5.3 Calibration of the sensor response 11-11
11.5.4 Statistical analysis 11-13
11.5.5 Repeatability and drift study of the sensor response 11-14
11.6 Modification of the sensor's structure to improve its sensitivity in detecting 2-FAL concentrations in transformer oil 11-16
11.7 Performance evaluation of the comb sensor 11-17
11.7.1 Response of the comb sensor at different frequencies 11-17
11.7.2 Response of the comb sensor with time 11-18
11.7.3 Calibration of the comb sensor response 11-19
11.7.4 Repeatability of the comb sensor response 11-20
11.8 Conclusion 11-22
References 11-22

Preface

In the last decade, printable sensors and electronics on flexible substrates have revolutionized sensing technology for cost-effective production of multi-sensing devices at temperatures that are suitable for different plastic substrates. Some important features of these printable sensors are simple fabrication steps, flexible patterning techniques, low material wastage, and inexpensive fabrication of the planar and non-planar sensors. These flexible sensors can be used for invasive (or destructive)/non-invasive (or non-destructive) detection of various measurement parameters. Another unique feature of the flexible sensors is the fabrication of multiple identical sensors on single large-area substrates. Such flexible sensors can fulfil adverse application needs such as in high corrosivity and relatively high working temperature environments, distributing sensing over large areas with ease of installation on planar and non-planar surfaces. Over the years, printing technologies have matured to pattern a wide range of electronic materials on diverse substrates. The fields of flexible electronics have been extensively explored in the domain of various printing technologies, commonly used substrates, and electronic materials for a variety of electronic devices. Printable technologies on flexible substrates have been applied in the field of sensors, radio frequency tags, thin-film transistors, solar cells, light-emitting diodes, and display devices. For sensors-based electronic systems, there is a need to interface the flexible sensors with hard electronics for data acquisition, signal conditioning, computation, signal processing, analysis, and transmission. Presently, flexible and hard devices are fabricated or assembled on separate substrates, and then interfaced using connectors. This hinders the flexibility of the device and is prone to interconnect issues. By contrast, the integration of soft sensors with hard silicon active devices on the same flexible printed circuit board results in flexible hybrid electronics for system-level implementation.

The fields of printable sensors and electronics are expected to grow continuously in the near future. Significant efforts are being made by the research community to explore new sensing devices, novel materials, and advanced technologies on printable substrates, with a vast literature on technologies, materials, and issues. Efforts are also being undertaken by researchers to integrate sensors and electronic circuits for systems in chip form but limited work has been reported with regards to the electronic circuits to interface the printable sensors. Powering the sensors and integrated systems is another important area of research that has received limited attention, particularly for implantable or wearable devices. This book aims to provide readers with a comprehensive overview to develop electronic systems for flexible sensors. It contains book chapters contributed by experts presenting their work about the latest development on sensing technology, interfacing circuits, managing power, and providing novel applications of the sensors. The main emphasis has been placed on the requirements of interfacing circuits for different types of flexible sensors. To help facilitate an understanding of flexible systems with integration of sensing and electronics circuits for a variety of operations, the book has been carefully organized as described below.

The first chapter, 'A review on flexible sensors for soft robotics', provides an overview of the growth of soft robotic systems on flexible structures due to their environmental adaptability and operational safety. To realize a soft robotic system, smart integration of several subsystems such as sensors, signal conditioning circuits, processors and software, actuators, control and feedback mechanisms are needed. It introduces the different types of sensors needed for soft robotics. Some of these sensors help to realize haptic exploration, collision avoidance, grasping, object identification, and shape recognition. In addition, this chapter emphasizes the important characteristics of these sensors along with their advantages, disadvantages, applications, and different parameters affecting their coherent behavior with the working environment.

The chapter 'Recent advances in laser induced graphene (LIG)-based flexible electronic devices' provides recent advances in the preparation and application of LIG in diverse flexible electronic devices. LIG is basically a 3D porous carbon material, fabricated through direct laser ablation over carbon-based substrates. The LIG technique seems promising for fabricating single or multi-layered different forms of graphene on a variety of flexible substrates with controllable morphological parameters by varying the LIG formation parameters. A broad field of applications utilizing LIG, such as flexible electronic devices, microfluidic devices, sensors, electrocatalysts, biosensing, energy devices, electronics, artificial intelligence, electrochemical sensing, etc, have been reported. The chapter also briefly discusses the future outlook and present research gaps in LIG-based flexible electronics applications.

The chapter 'Printable flexible sensors for hydration monitoring and moisture measurement of concrete structures' gives an overview of some electrical techniques for the health assessment of concrete structures including monitoring of hydration and determination of initial and final setting time, as well as moisture measurement. It highlights some demerits of the conventional and new techniques proposed in the literature. It then explains a fringing field impedance method for hydration monitoring and moisture measurement in concrete using flexible electronics. The performance of these fringing field sensors is compared with commonly used techniques. The impedance sensors are simple in terms of operation and fabrication, and they are cost effective for on-line measurement purposes.

The chapter 'Resistive sensor interface' gives an overview of resistive sensors for industrial, commercial, and scientific applications. The resistive and capacitive methods are the most extensively employed for the fabrication of flexible sensing devices. The diverse applications of such sensors require innovative and appropriate circuits suitable for interfacing the flexible sensors. The chapter focuses on sensor interface design based on auto-balancing circuits and explores the analysis, design methodology, and experimentation with real fabricated resistive sensors to verify their effective working. A novel auto-balancing dual-differential architecture generates a control voltage that balances the interface circuit. A detailed analysis of the same has been presented in the chapter along with applications in gas sensing and temperature sensing. With few circuit modifications, the design is also shown to work for a single-element resistive sensor array. A direct application of the array interface circuit is presented for a silicon nano-wire array.

The chapter 'Interfacing circuit for capacitive sensors' discusses the design, analysis, and implementation of interfacing circuits for various types of single and differential capacitive sensors. The micro- and nanosensors which are gaining popularity day by day impose more constraints on the design of suitable interfacing circuits for the sensor. Reduction of power consumption for battery-operated, sensor-based electronic measurement systems is another major requirement in today's market. Although many interface circuit topologies are available, yet still significant efforts are required to deal with the current requirement of flexible and hard sensors. The main motivation of the present chapter is to introduce some methods to improve the accuracy and resolution of the measurement parameters.

The chapter 'Interface electronics and conditioning circuits for triboelectric flexible sensors' discusses electronic circuits for triboelectric sensors. Triboelectric sensors are inherently low power and susceptible to external noise due to high intrinsic impedance. Such sensors have great potential in sensing environmental, biomedical, chemical sensing, and health parameters. This chapter introduces the analog interface circuits for signal conditioning of triboelectric sensors. The configuration of the sensing circuit of triboelectric-based sensors is described, as well as the issues that should be addressed in the design of the circuit.

The chapter 'Flexible interfacing circuits for wearable sensors and wireless communication' presents some common functional mechanisms of flexible and wearable sensors, including their flexible interfacing and wireless communication circuits. Following the advances in flexible and printed electronics, interfacing circuits for such devices are gradually shifting from conventional rigid versions to the more flexible versions, which will lead to the realization of very comfortable and highly conformal wearable systems. In recent times, interest in the realization of circuits with printing technologies such as screen printing, contact printing, etc, is rapidly increasing. Researchers are now even exploring the use of biodegradable materials including paper for the implementation of disposable circuits. Rapid advances in 3D/4D printing technologies are revolutionizing wearable systems and 3D printed circuits.

The chapter 'Compact and efficient wireless power and information transfer systems for IoT sensors and implants' provides techniques for simultaneous wireless power and information transfer (WPIT) systems for IoT sensors and implants. The importance of selecting a two-channel system over single-channel and dual-band systems is explained. A unique inductance structure in the form of a figure-8 and a defected ground structure inductance for WPIT applications are designed, fabricated, and characterized to evaluate the transfer efficiency and isolation between the power and information channel.

'Analysis and design considerations of relaxation-oscillator-based flexible sensor circuits' are explained in chapter 9. The low hardware complexity, quasi-digital output, wide measurement range, and high resolution make a relaxation-oscillator-based circuit an ideal choice for designing flexible sensor systems for resistive and capacitive sensors. The effects of various component non-idealities such as offset, leakage resistance, and/or parasitic capacitance on the measurement accuracy are analyzed. A prototype of a modified relaxation oscillator circuit for such sensors is fabricated, tested, and analyzed. The performance of the prototype is compared with the design analysis for different components.

The chapter 'Sensor tags for wireless body-centric communication: challenges and opportunities' gives a comprehensive overview of the devices and methods required for interfacing sensor tags for wireless body-centric communication (WBAN). The antenna is an integral part of WBAN devices and plays a vital role in establishing a link between a wearable device and a base communication station situated remotely from the body. The patch antennas used in WBAN applications are often limited by their operational bandwidth, size, reduced efficiency, strong coupling with the body, and high specific absorption rate. This chapter highlights the existing challenges and some solutions for designing an antenna for WBAN applications. An in-depth study of one promising solution related to the design of electromagnetic band-gap surfaces for interfacing the wearable devices and base station is included.

The last chapter, 'The role of capacitive sensors for condition monitoring of transformer insulation', presents an alternative means of diagnosing the health of cellulose insulation by proposing and fabricating a simple, novel, less-expensive capacitive sensor to detect the concentrations of 2-FAL in transformer oil. Capacitive sensors are gaining popularity as they possess many inherent advantages over resistive sensors. These sensors use composite polymers to make target-vapor selective sensing films which exhibit a capacitance change whenever exposed to insulating oil having even traces of 2-FAL. The sensor can be fabricated on a flexible substrate and can be IoT-enabled to measure the degree of deterioration of a transformer's insulation when placed at remote locations.

All the chapters are contributed by authors who are experts in their respective fields of study and who pay special interest to write the chapters in a simple manner. Their valuable time, efforts, and knowledge based on the latest research outcomes in laboratories helped us to produce this high-quality book providing comprehensive knowledge for developing electronic systems using flexible devices. We are very much thankful to the authors for their valuable contributions and willingness to participate in this endeavor, to create what we hope will prove a very useful and interesting book for readers.

Tarikul Islam, New Delhi, India
tislam@jmi.ac.in
Subhas Chandra Mukhopadhyay, Sydney, Australia
subhas.mukhopadhyay@mq.edu.au
Boby George, Chennai, India
boby@ee.iitm.ac.in

Acknowledgment

The Editors would like to thank the authors for their important contribution and the time spent writing the chapters for this book.

Editors' biographies

Tarikul Islam

Tarikul Islam (M'16-SM'18) was born in Murshidabad, West Bengal, India. He gained an M. Tech. in Instrumentation and Control Systems from A. M. U. Aligarh, U.P. in 1997 and a PhD (Engineering) degree from Jadavpur University, Kolkata, India, in 2007. Presently he is working as Professor in the Department of Electrical Engineering, Jamia Millia Islamia (Central University), New Delhi, India. He has over 20 years of teaching and research experience. He has authored/co-authored 10 book chapters, two edited books, filed four patents, and published more than 160 papers in peer-reviewed journals and conferences. His research interests include sensing technologies and electronics instrumentation. He is a Topical Editor of the *IEEE Sensors Journal* and Associate Editor of *IEEE Transactions on Instrumentation and Measurement*.

Subhas Chandra Mukhopadhyay

Subhas Chandra Mukhopadhyay (Fellow, IEEE) holds a BEE (gold medallist), MEE, PhD (India) and Doctor of Engineering (Japan). He has over 31 years of teaching, industrial, and research experience.

Currently, he is working as a Professor of Mechanical/Electronics Engineering at Macquarie University, Australia and is the Discipline Leader of the Mechatronics Engineering Degree Programme. He is also the Director of International Engagement for the School of Engineering of Macquarie University. His fields of interest include smart sensors and sensing technology, instrumentation techniques, wireless sensors and network (WSN), Internet of Things (IoT), wearable sensors, medical devices, healthcare and environmental monitoring. He has supervised over 50 postgraduate students and over 150 honours students. He has examined over 70 postgraduate theses.

He has published over 450 papers in different international journals and conference proceedings, written 10 books and 52 book chapters and edited 18 conference proceedings. He has also edited 35 books with Springer-Verlag and 32 journal special issues. He has been cited so far 14,601 times and has a h-index of 58. He has received various awards, most notably the Australian Research Field Leader in Engineering and Computer Science 2020; Distinguished Lecturer, IEEE Sensors Council 2020–2022; Outstanding Volunteer by IEEE R10, 2019; World Famous Professor by Government of Indonesia, 2018; Certificate of Distinction from IEEE Sensors Council, 2017; IETE R. S. Khandpur Award—India, 2016; and Best Performing Topical Editor of the *IEEE Sensors Journal* from 2013 to 2018, six years consecutively. He has organized over 20 international conferences as either General Chair/co-chairs or Technical Programme Chair. He has delivered 397 presentations including keynote, invited, tutorial and special lectures.

He is a Fellow of IEEE (USA), a Fellow of IET (UK), a Fellow of IETE (India), a Topical Editor of the *IEEE Sensors Journal*, an associate editor of *IEEE Transactions on Instrumentation and Measurements*, and *IEEE Review of Biomedical Engineering*. He is the Editor-in-Chief of the *International Journal on Smart Sensing*. He is a Distinguished Lecturer of the IEEE Sensors Council from 2017 to 2022. He is the Founding Chair of the IEEE Sensors Council, New South Wales chapter.

More details are available at:
https://researchers.mq.edu.au/en/persons/subhas-mukhopadhyay.
https://scholar.google.com.au/citations?hl=en&user=8p-BvWIAAAAJ
https://orcid.org/0000-0002-8600-5907

Boby George

Boby George received MTech and PhD degrees from the Indian Institute of Technology (IIT) Madras, India. He was a Postdoctoral Fellow with the Institute of Electrical Measurement and Measurement Signal Processing, TU Graz, Austria, from 2007 to 2010.

He joined the Department of Electrical Engineering, IIT Madras, in 2010, where he is currently a Professor. He is an Associate Editor for the *IEEE Sensors Journal*, *IEEE Transactions on Instrumentation and Measurement*, and *IEEE Transactions on Industrial Electronics*. His research interests and experience include development of novel sensor systems and electronic instrumentation.

List of contributors

S A Akbar
Chief Scientist and Professor AcSIR(retd.), CEERI, Pilani, India

Adjunct Professor
Department of Electronics and Communication Engineering
SRM Institute of Science and Technology
Chennai
India

Kazi Jabed Akram
Indian Institute of Technology Madras
Chennai
India

Khairunnisa Amreen
Department of Electrical and Electronics Engineering
BITS-Pilani
Hyderabad
India

Adel Barakat
Faculty of Information Science and Electrical Engineering
Kyushu University
Fukuoka
Japan

Moupali Chakraborty
James Watt School of Engineering
University of Glasgow
Glasgow
UK

Ravinder Dahiya
James Watt School of Engineering
University of Glasgow
Glasgow
UK

Boby George
Department of Electrical Engineering
Indian Institute of Technology Madras
Chennai
India

Sanket Goel
Department of Electrical and Electronics Engineering
BITS-Pilani
Hyderabad
India

Mohammad Hashmi
School of Engineering and Digital Sciences
Nazarbayev University
Nur-Sultan
Kazakhstan

Tarikul Islam
Department of Electrical Engineering
Jamia Millia Islamia
Jamia Nagar
New Delhi
India

Debasrita Kar
Department of Electrical Engineering
Indian Institute of Technology Madras
Chennai
India

Sougata Kumar Kar
Department of Electronics and Communication Engineering
NIT Rourkela
India

Shakeb A Khan
Department of Electrical Engineering
Jamia Millia Islamia
New Delhi
India

Kaushal Kishore
Department of Electrical Engineering
IIT Kharagpur
India

Shahid Malik
Centre for Sensors, Instrumentation, and Cyber Physical System Engineering
IIT Delhi
India

Md Manzar Nezami
Assistant Professor
Department of Electronics and Communication Engineering
GLA University
Mathura-281406
Uttarpradesh
India

Oliver Ozioko
James Watt School of Engineering
University of Glasgow
Glasgow
UK

Ramesh K Pokharel
Faculty of Information Science and Electrical Engineering
Kyushu University
Fukuoka
Japan

Chunkai Qiu
Department of Electrical and Computer Systems Engineering
Monash University
Melbourne
Australia

Dinesh Rano
Moscow Institute of Electronics and Mathematics
National Research University
Moscow
Russia

Siddhartha Sen
Department of Electrical Engineering
IIT Kharagpur
India

Laxmeesha Somappa
Department of Electrical Engineering
IIT Bombay
Mumbai
India

K Sridharan
Department of Electrical Engineering
Indian Institute of Technology Madras
Chennai
India

Mehmet Rasit Yuce
Department of Electrical and Computer Systems Engineering
Monash University
Melbourne
Australia

Chapter 1

A review on flexible sensors for soft robotics

Debasrita Kar, Boby George and K Sridharan

During the last decade, soft robotics has seen an immense amount of growth due to its attractive characteristics, such as environmental adaptability, owing to the flexible structures employed, and operational safety for users, providing a promising human–machine interaction environment. To realize a soft robotic system, smart integration of several subsystems such as sensors, signal conditioning circuits, processors and software, actuators, and control and feedback mechanisms are needed. In this chapter, an overview of the different types of sensors needed for soft robotics is presented. Some of the sensors help to realize haptic exploration, collision avoidance, grasping, object identification, and shape recognition. In addition, this chapter emphasizes the important characteristics of these sensors along with their advantages, disadvantages, applications and different parameters affecting their coherent behavior with the working environment.

1.1 Introduction

Robotic systems of various kinds have been increasingly used in the last few decades to automate activities that are either cumbersome or dangerous for humans. Robot arms have found applications in assembly while automated guided vehicles have been widely used in factories and other zones for the transport of heavy objects.

Conventional robotic systems used in manufacturing and other applications employ rigid parts. With the increasing penetration of robots in domestic and other healthcare environments, there is a need for greater flexibility and safety, especially in the context of robot–human interaction as well as for effective handling of delicate and fragile objects.

Robotic structures made of shape memory alloys, polymers, etc, are slowly replacing metallic parts for a range of applications. This has led to the establishment of a domain popularly known as soft robotics. Sensing, actuation and control of soft robotic structures present various challenges. Techniques hitherto developed for rigid counterparts are generally not applicable as such to soft robots.

doi:10.1088/978-0-7503-3935-3ch1

In this work, we focus on sensing for soft robots. We enumerate the unique features required for sensors in soft robots. We then discuss in detail various types of sensing strategies and highlight recent work in sensing for soft robotics. We also outline issues that remain and indicate promising directions for further exploration.

1.2 Important characteristics required for soft robotics sensors

The desired characteristics of typical sensors are accuracy, precision, sensitivity, resolution, repeatability, negligible loading on the measurand, etc. Soft robotics sensors need additional features as listed below:

1. *Compliance/flexibility*. The sensor should not reduce the compliance of the soft body. This is one of the fundamental characteristics of sensors employed for the coherent working of soft robots. The sensor should have a comparable low modulus (in the range of 10^5–10^9; [33]), matching the underlying structure. Attachment of the sensor to the robotic body should not impart any additional, noticeable impedance change to the whole structure.
2. *Nature of the surface*. It should be soft such that the physical interface between the sensor and soft robotics layers seamlessly work together. If the sensor layer is fitted outside the soft robotic part, then it should not alter the biomimetic capabilities of the robotic element.
3. *Material requirements*. The material should be stretchable as with the other soft robotics parts. The materials most commonly used for fabricating soft sensors are polydimethylsiloxane (PDMS) elastomers and a variety of silicone elastomers. In some cases, to elevate the conductivity in a certain part of the sensor (as part of different sensing schemes), impurities like carbon black particles or blend composites and carbon nanotubes (CNTs) are combined in the mixture of the elastomer. The material should have a sufficient fatigue limit, considering the cyclic loading expected. In general, low-cost materials are preferred to make them more economic.
4. *Multifunctional sensor*. Similar to the features of human skin that can sense multiple parameters (e.g., force, temperature, shape, surface finish, etc), it is beneficial for the soft robotic sensor to have the ability to sense multiple parameters. This will save space and cost.

1.3 Sensing techniques employed to develop soft robotic sensors

In this section, different sensing techniques used to develop soft robotics sensors are presented, along with their corresponding advantages and limitations which make them eligible (or not) to be employed for soft robots. Based on the sensing schemes, parameters to be sensed, size, and the nature of the object in contact with the soft robot, the sensors can be chiefly categorized into the following.

1.3.1 Magnetic sensors

One of the types of sensors used in the field of soft robotics is Hall-effect-based sensors. Most Hall effect sensors equip the soft robot or actuator with a sense of touch and measure the force/pressure imparted on it. They are cost effective and easy to fabricate. Such a sensor mainly consists of a tiny permanent magnet and a Hall sensor embedded in the elastomer. Figure 1.1 depicts the general working principle of the sensor. The Hall sensor senses the magnetic field of the permanent magnet. The permanent magnet has a predefined magnetic field pattern as a function of the distance from it. When the distance between the Hall sensor and the permanent magnet changes, due to the deformation associated with the force or touch being sensed, the intensity of the magnetic field seen by the sensor differs from the initial value. This is used in soft robotics for tactile sensing as, under the influence of touch that exerts some force/pressure on the surface, the magnet is displaced. This positional change alters the value of the magnetic field at the location of the sensor. This type of sensor can be used for the measurements of force, both perpendicular and tangential to the surface. We can also classify the different shapes of the objects being grasped by employing a number of sensors in different orientations on the same surface. As no wired connection is required to the sensor part (a permanent magnet in this case) in the elastomer the reliability and life span of the sensor will be high.

In [19] and [30], some designs of Hall-effect-based soft robotic sensors are presented where the correlation between the applied force and the change in the magnetic field is reported. They also demonstrate the tactile-based obstacle avoidance and haptic classification of objects. Similar work appears in [38], where a prototype, named MagOne, is developed. It provides high-resolution, three-axis force measurement in a form factor that allows it to be easily integrated with other applications. Further, in [4] also, a three-axis soft force sensor was realized by employing a Hall effect sensor attached to a cantilever beam to estimate the displacement of the free end by measuring the magnetic field variation under compression or bending. Similar to Hall effect sensors, one could use giant magnetoresistance (GMR)- or tunnel magnetoresistance (TMR)-based sensing elements [1, 16] as those types of sensors can provide higher sensitivity and lower power consumption.

Though these types of sensors provide good accuracy and sensitivity, their outputs are prone to external magnetic variations. Another shortcoming is that the sensor cannot be used when the robot is used to interact with materials with ferromagnetic properties since they alter the original magnetic field coming from the permanent magnet. This can induce a huge measurement error.

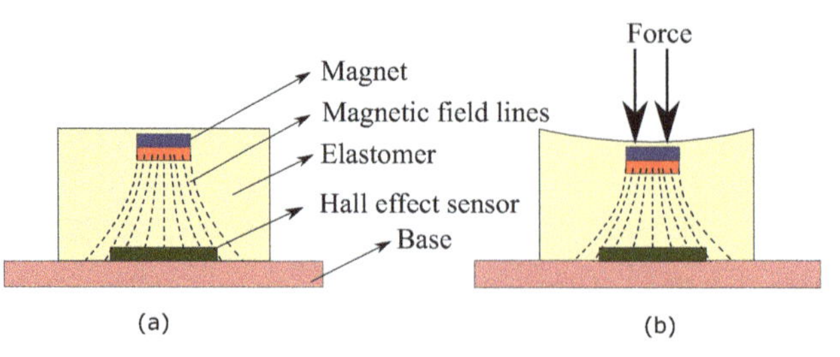

Figure 1.1. Hall-effect-based sensor force sensing. A permanent magnet, integrated with the elastomer, generates a magnetic field which is sensed by the Hall sensor. The magnetic field shown is a pictorial one, and only the field in the region of the Hall sensor is indicated.

1.3.2 Inductive sensors

Inductive sensors are very popular owing to their high reliability and insensitivity to moisture, dust, oil, etc. However, they are not widely employed in the field of soft robotics. Inductive sensors work based on the principle of a change in the self-inductance or mutual inductance of the coil due to dimensional change or change in the relative permeability of the medium. The construction of a sensor using this technique typically results in bulky sensors, less fit for incorporation in soft robotics. With that said, these sensors have a higher dynamic range, provide linear output and can be employed for a higher frequency range. The cost of the sensor is also low. Taking these advantages into account, researchers have found a promising future for this type of sensor in the field of soft robotics.

In [31], the authors developed a sensor comprising a single coil that acts as a sensing coil which is coupled to the elastomer element (actuator). When the elastomer element bends due to external force, the coil also undergoes deformation/modification in dimension. For example, when there is bending or deformation, there is a deflection of the coil from its initial position. As a result, the magnetic flux linked with the coil changes and changes the inductance of the coil.

In addition, in [41], an inductive sensor which is sensitive to larger strain was developed, where the inductor design is similar to a mechanical spring. When it bends, the whole coil stretches and there is a deviation in the length of the coil. After the change, the corresponding change in inductance is measured and correlated to the degree of bending or the extent of stretch. To demonstrate the engineering feasibility of the designed sensor, it was incorporated into pneumatic actuators as well as hand gloves. It is capable of sensing movements in tiny spaces and also provides structural stability. It also shows a better life span and repeatability, though the sensitivity of the sensor is limited.

In [5], microfluidic channels were made within the elastomer layer where a liquid metal alloy was injected. These non-toxic metal alloy channels were made in a spiral design and act as an inductor. When the whole membrane is stretched, the corresponding inductance changes. This change in inductance induces a change in the wirelessly transmitted voltage. This technique shows very negligible hysteresis and can be employed in large stretching or bending measurements. As mentioned before, this area has great potential but realizing reliable coil structures that do not alter the basic characteristic of the soft robotic body is still a challenge.

1.3.3 Capacitive sensors

This type of sensor consists of electrodes and a dielectric medium between them. Capacitive sensors are extensively used in various domains of industry and research areas for rigid sensors [14, 16], and they are a boon for soft robotics as they can include both tactile and proximity sensing in the soft robots together, as shown in figures 1.2(a) and (b). Generally, a parallel plate configuration is used for the tactile sensing scheme and a coplanar configuration is used for proximity sensing. Various capacitive electrode structures [9, 28] and their integration in a soft robotic body have been reported.

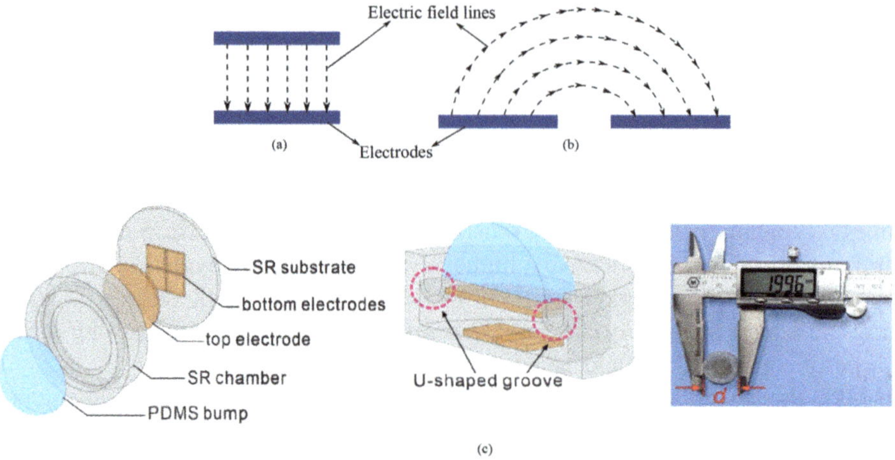

Figure 1.2. Basic capacitive sensor configurations: (a) parallel plates and (b) coplanar. (c) An example of a flexible capacitive 3D-force tactile sensor. Reproduced from [42].

1.3.3.1 Tactile sensing

Tactile sensing gives a proprioceptive quality to soft robots. In a capacitive tactile sensor, when an external force is applied, the distance or the overlapping area between the plates changes or the conductivity of the dielectric changes [28], which in turn changes the capacitance. The authors of [36] reported a capacitive robotic skin developed from conductive elastomers with a dielectric layer between them. When stretched, the overlapping area and the distance between the conductive elastomer plates changes, which in turn changes the capacitance at low frequencies. But at higher frequencies, the changes in electrode resistance under strain, with lumped parameter approximation, can lead to erroneous capacitance measurements. In another example of tactile sensing, the authors developed [6] a capacitive sensor by incorporating a liquid crystal-carbon nanotube (LC-CNT) as a dielectric medium between two gold electrodes. By altering the applied driving frequency and voltage, the dielectric anisotropy changes, which increases the range of measurement. The authors of [44] used conductive fiber electrodes for touch-sensing, which are very flexible and can be easily integrated with soft structures. The authors of [9] also presented a multifunctional and stretchable capacitive sensor made of thin gold films embedded on a soft silicone rubber substrate. Therefore, we can conclude that capacitive sensors may play a very significant role in tactile sensing. The measurement circuits required for capacitive sensors are still relatively complex and the accuracy and resolution that can be achieved with low-cost electronics is limited [3, 21].

1.3.3.2 Proximity sensing

For sensing the proximity of an object using capacitive sensors, two modes of operations are commonly used: (i) using a single capacitive electrode, (ii) using two or more electrodes. In the case of single-electrode sensors, an AC signal is fed to the electrode, which in turn generates an electric field and current flows from it. If any conductive or dielectric object capable of changing the electric field is present in close proximity to that electrode, then it brings a change in the amplitude of the current leaving from the transmitter end in the circuit [8]. Alternatively, when two electrodes are employed, namely transmitting and the receiving electrodes, and an AC voltage is applied to the transmitting electrode and the receiving electrode is kept as close to the ground, an electric field is generated between them. This electric field is responsible for a displacement current flowing in the circuit and it can be measured in the receiving end. When an object, for example a human finger (which is connected to the ground via the human body to ground capacitance), comes into close proximity, it shields some amount of electric field. This again changes the displacement current; this phenomenon is called the shielding effect [17, 29]. These techniques are widely manipulated and electrodes are placed in these configurations in soft actuators or structures to give multidirectional use to capacitive sensors.

1.3.4 Strain gauge/resistance-based sensors

Resistive sensors cover a wide range of sensing parameters, and the soft robotic community has explored these sensors by suitably modifying and fabricating them; most of the existing designs and fabrication steps are meant for making rigid sensors. Soft sensors with microchannels containing conductive liquids embedded in them, or flexible strain or bending sensors made up of flexible and stretchable elastomers with impurities like carbon or other conductive fillers are a few such examples. Under stress or pressure they stretch, which brings about a considerable change in the measured resistance. Figures 1.3 and 1.4 show various types of resistive sensors used in the domain of soft robotics.

Resistive sensors using conductive polymers to measure force find their way back to 1977 [27]. Franklin invented a sensor that consists of a sensing film made of conducting polymers. When it is exposed to any external force/pressure, the resistance of the sensor decreases. This is because the average distance between the particles of the conductive polymer decreases under stress, and experiences a reduction in the effective resistance, which is the resistance between the electrode and the conductive polymer. This work has given a multidimensional work space for researchers in the future. An example of elastomer-based sensors used for pressure measurement is reported in [32]. Here the sensor is manufactured out of both insulating and conductive rubber. The conductive layer acts as the sensing part, made up of a PDMS elastomer with carbon black particles embedded in it. The limitations of this type of sensor are its relatively long recovery time and noticeable hysteresis. The hysteresis is relatively large as during unloading the internal carbon black network also undergoes electromechanical hysteresis. Even after the unloading process, it takes time to recover its initial resistance value. Another example is the use of conductive thermoplastic elastomer sensors for static strain-sensing estimation [37].

Another category of this type of sensor involves piezoresistive polymers or piezoelectric ceramics. They can be proficiently useful for static and dynamic tactile sensing, respectively. When stretched or under tension, the resistance of the

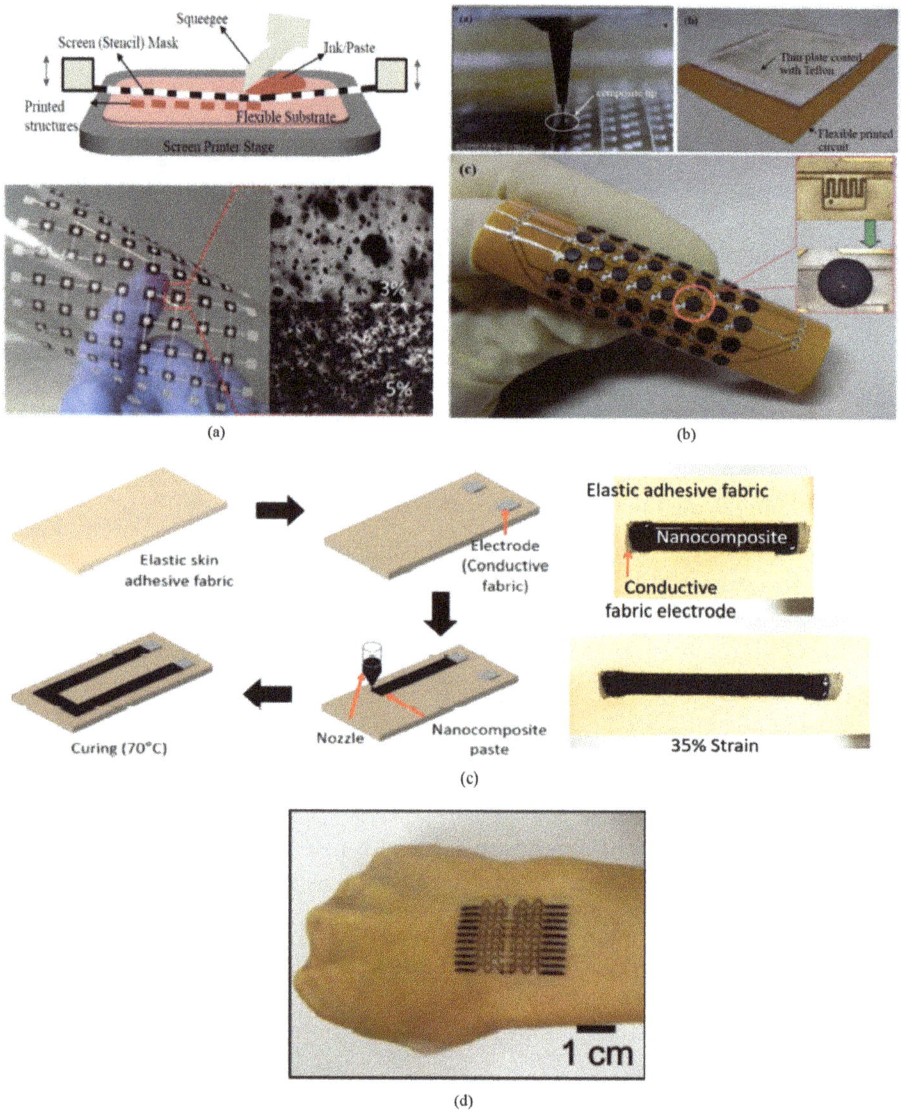

Figure 1.3. Different types of resistive sensors: (a) screen-printed flexible tactile sensor made out of conductive nanocomposites (© 2015 IEEE, reprinted, with permission, from [22]); (b) interdigitated electrodes and graphite/PDMS composites on a flexible temperature sensor array (reproduced from [34], © IOP Publishing Ltd, all rights reserved); (c) piezoresistive nanocomposite printed by a dispenser on woven elastic fabric (reproduced from [24], © IOP Publishing Ltd, all rights reserved); (d) strain gauges based on elastomers mounted on human skin ([26], John Wiley & Sons. Copyright © 2012 Wiley-VCH Verlag GmbH & Co. KGaA, Weinheim).

piezoresistive layer changes, from which the corresponding tension or force it was exposed to can be measured after calibration.

Different types of resistive sensors are available in the market, providing varied functionalities including haptic identification of different objects [18], measuring

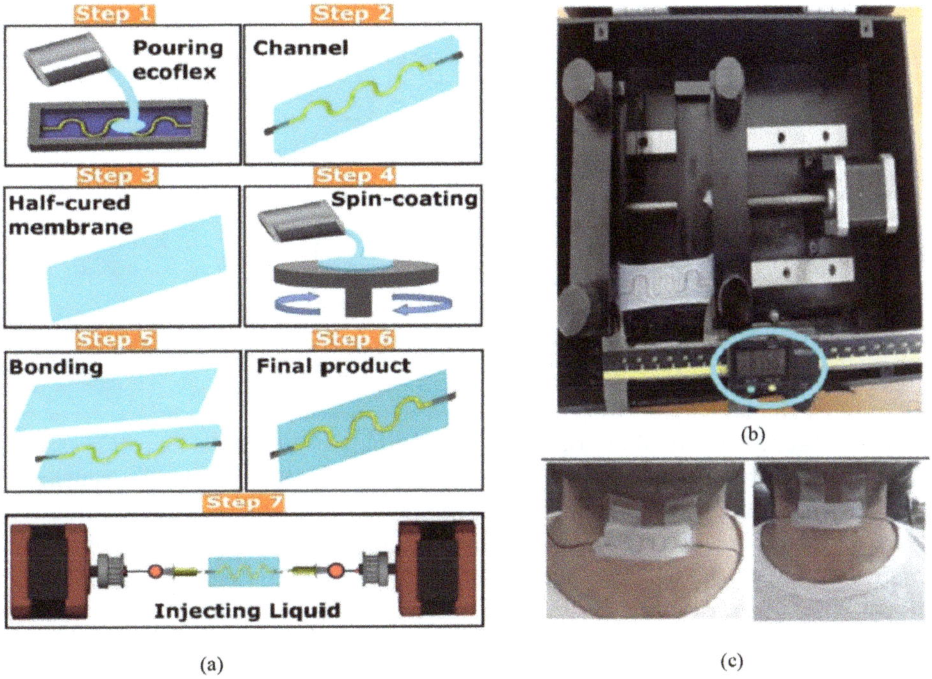

Figure 1.4. (a) Fabrication steps for microchannel-based strain sensor; (b)–(c) sensor prototype. Reproduced from [35].

the grasping force (resistive flex sensors), etc. Some even use screen-printed conductive inks on soft substrates (used for sensing the curvature and strain experienced) [23]. In [45] a coiled conductive polymer fiber was used and a differential sensing scheme employed to measure the angle of a soft system by modeling the resistance change.

Resistive soft sensors have the desirable characteristics that they are free from electromagnetic interference, easy to fabricate, and fairly cost effective. However, most of the existing ones show error due to hysteresis and slow recovery after unloading.

1.3.5 Eddy current sensors

Eddy current sensors are widely used in industry [16]. They are yet to be explored, in detail, in the field of soft robotics. Several qualities which have intrigued researchers to dig further into the use of this technique in soft robotics are as follows:

- Very low hysteresis;
- Good repeatability;
- Robustness to environmental contaminants; and
- Higher sensitivity.

While working with soft robots, this type of sensor can be used for tactile sensing, measuring the external force/pressure applied during contact or bending of the

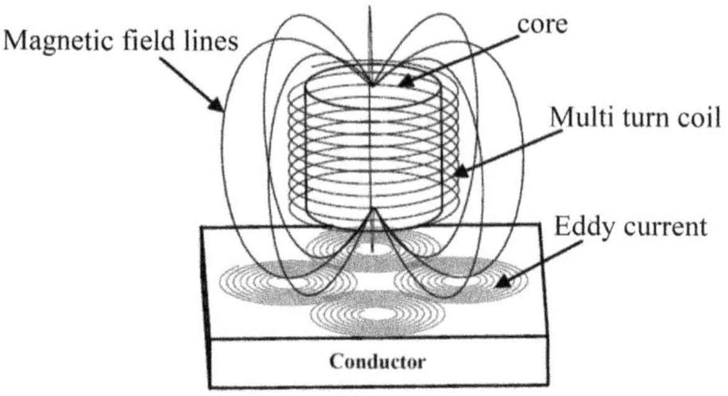

Figure 1.5. Eddy current sensing. Reproduced from [25]. © IOP Publishing Ltd. CC BY 3.0.

actuator attached to it. The operating principle of an eddy current sensor is illustrated in figure 1.5, wherein an AC current is used to excite the sensing coil which in turn generates an alternating magnetic field in its surroundings. This induces an eddy current in the neighboring conductive film (target). This gives rise to a magnetic field that opposes the original/emitted magnetic field from the sensing/primary coil. This coupling between the coil and the conductive film reduces the magnetic flux linked with the coil and the effective inductance, as well. When an external force is applied to the conductive film, it comes closer to the sensing coil. This increases the induced eddy current in the film and the magnitude of the opposing magnetic field. As a result, the inductance of the coil decreases. The basic structure of the eddy current sensor used in soft robots comprises a planar coil (sensing layer), a deformable elastomer layer and a conductive film (sensing target).

In [39], an eddy-current-based soft sensor has been reported. The authors adopted the basic design but incorporated an array of such coils to get an amplified value of the total inductance variation. But this type of sensor will only work if the conductive film acts as a good sensing target. Electromagnetic shielding is required to avoid any interference caused by any ferromagnetic object in the surroundings. The prototype sensor reported showed good resolution with an expanded range over 15 N, low hysteresis, good repeatability and adaptability to the harsh environment. In another example [40], a soft magnetic tactile sensor is developed using the concept of electromagnetic induction. The sensor has a primary coil and a secondary coil with conductive or ferromagnetic sheets embedded in silicone elastomer. The developed prototype shows higher sensitivity with a wider range and lower hysteresis as compared to the existing counterparts of this type of sensor.

1.3.6 Optical sensors

Optical sensors are one the most sophisticated and sought after sensors for efficient, space-limited devices. Recent developments in this area have made them suitable for different applications such as tactile sensing, deflection sensing of any surface, finding bending or stretching in a structure, sensing force or pressure applied to a surface, etc. The advantages of using optical sensors are that there are various techniques which can be manipulated for different application-specific tasks and obtain repeatable data. They are known for their accurate and precise results and mostly provide real-time data, which is very beneficial for feedback and control. In [20] a soft optical waveguide integrated inside a pneumatic actuator is reported, giving proprioceptive information of the actuator. The air chamber inside a pneumatic actuator is coated with a reflective layer of silver. A light-emitting diode (LED) and a photodiode are placed inside the actuator, passing light into the air chamber, and the intensity of this emitted light is detected by the photodiode. When the actuator is bent or stretched, minor micro-cracks are generated on the silver layer, which allows the light to escape the air cavity. This reduces the light intensity inside the chamber. This change in intensity of light can be calibrated to find the bend of the whole structure.

In another example [10], deflection sensing of soft surfaces was realized by using optical sensors. Here also a LED and a photodiode have been employed for finding the deflection angle of a curved surface by correlating the monotonically increasing relation between the emission angle and the light intensity. The shape and bend of the deformed object can be re-created by joining multiple displacement sensors (optical) together. The main advantage of this technique is that the sensor is not in direct contact with the deflection region.

The inclusion of optical fibers has given more liberty to this domain of research. In [7], an angled-tip optical fiber was designed to find out the amount of force/pressure applied to the surface normally. The optical fiber here acts both as an emitting and sensing element. There is also a fluorescence layer that acts as the key element to the sensing scheme. It is kept just below the surface experiencing the stress, with the angled tip of the optical fiber attached to it. The light coming from the optical fiber is absorbed by the fluorescent layer and emitted back with a different wavelength (due to fluorescence emission). When there is a force applied normally on the surface, the fluorescent layer would come closer to the optical fiber and, as a result, higher light intensity will be detected by the camera attached at the end.

A fiber Bragg grating is also an effectively used technique, implemented inside an optical fiber attached to an actuator or any other flexible structure. With this, real-time monitoring is possible and curvature measurement in both directions can be achieved [13, 43].

In addition, in a recent work [12], an eSkin was developed with both self-energy-generating and sensing properties. It is mainly made up of an array of miniaturized solar cells with infrared light-emitting diodes (IREDs) embedded on a soft elastomer substrate. A variety of parameters, for example object location, proximity, edge detection and tactility, can be sensed using this eSkin. The basic working principle of this eSkin is that a solar cell generates less energy upon shadowing. The variation of the generated energy from the solar cell implies the possibility of any contact, proximity, etc. While optical sensing has many advantages, the sensor and associated elements are delicate, need complex measurement systems in general, and the overall cost is relatively high.

1.4 Interfacing circuits and associated measurement systems

Interfacing circuits are key in deriving the best from the sensing element in a sensor unit. When interfaced with an electronic system, each type of sensing element needs a different treatment depending on the different, changing electrical parameters in the sensor owing to the change in the measurand [15, 16]. The importance of interfacing circuit/signal conditioning is listed below:

- the output signal of most sensors is too small to be digitized and needs amplification before feeding to the analogue-to-digital converter in the data-acquisition unit;
- there may be interference/noises which need to be suppressed/filtered in order that the required resolution is achieved by the overall sensor unit;
- the output signal from the sensor may be too sensitive to loading of the measurement circuit or affected by the parasitic elements of the sensing element; hence an appropriate design of the interfacing circuit is needed to obtain a reliable output that is not sensitive to the unwanted parameters; and
- sometimes the output is needed in a precise format that needs a specific conversion.

Depending on the specific needs one can select or develop a suitable signal conditioning/interfacing circuit focusing on the performance parameters such as overall sensitivity, power consumption, resolution, compactness, drift, interference rejection capability, etc [2, 11, 16]. Data acquisition and further signal processing are a large aspect of the latest measurement process. The building blocks of a typical measurement system will have a sensing element that is interfaced to the data acquisition using a suitable interfacing circuit and a processor that processes the digitized data from the data-acquisition system.

1.5 Comparison of different sensing techniques

There are some important parameters that define the performance of the sensors and their efficiency for a required application. These include accuracy, precision, sensitivity, linearity, hysteresis, response time, sensitivity to electromagnetic interference, ease of manufacturing, cost, flexibility, etc. By knowing the ease of manufacturing or fabrication of a sensor, we can get an idea about the complexities involved in the process and thus estimate the cost, reliability, and even estimate its commercial viability. In the domain of soft robotics, flexibility is one of the deciding factors for any sensor to be integrated with an actuator. Another key point is that the sensitivity and range of the sensor are very often conversely related to each other, i.e. a higher range gives lower sensitivity. On the other hand, repeatability gives an assurance that the results provided by the sensor match when tested again and again under the same circumstances. In this regard we can refer to resistive sensors with conductive fillers suspended in an elastomer layer: the sensing element does not show good repeatability though they show good sensitivity. Hence for those applications wherein a shorter range of measurement with higher sensitivity is required, one can opt for these types of sensors. Usually, magnetic sensors have higher sensitivity and

Table 1.1. Comparison table.

Parameters	Magnetic	Inductive	Capacitive	Resistive	Eddy current	Optical
Ease of manufacturing	Moderate	Complex	Good	Complex	Moderate	Complex
Repeatability	Good	Very good	Good	Moderate	Good	Good
Flexible	Moderate	Yes	Moderate	Yes	Yes	Moderate
Hysteresis	Moderate	Good	very Good	Moderate	Good	Good
Sensitivity	High	Moderate	Very mood	Good	Good	High
Cost	Moderate	Low	Low	Moderate	Low	High
Accuracy	Good	Good	Moderate	Moderate	Good	Good
Linearity	Good	Good	Good	Moderate	No	Good
Sensitive to interference	Yes	Yes	Yes	Low	Yes	Moderate

accuracy but, side by side, they are also very sensitive to external electromagnetic interference. In order to ensure reliable output, proper shielding is required. Another relevant parameter is hysteresis, which is attributed to the difference between the loading and unloading characteristics of the sensing element. This is of importance in the case of soft robotic sensors as the actuator is expected to undergo cyclic operations. A brief comparison of the various types of sensors is provided in table 1.1.

1.6 Challenges and future directions

Soft robotics is a developing field motivating researchers to contribute via addressing new challenges. The future of soft robotic sensors stands on the idea of how far they can provide an interactive and compatible environment with humans. With substantial evidence from the literature survey, we see that there is a surge in research on feedback mechanisms, different material classification depending upon the requirement of the sensor, various algorithmic approaches and novel fabrication methods for soft sensors. In addition, they should be cost effective to make them feasible for common and commercial use. We thus require soft sensors with less complex designs that enable them to be used by non-experts, as well. In the future, soft robots should be able to traverse unknown terrain and be able to react according to the environment. To accomplish this we will need advanced sensors synchronized with enhanced feedback mechanisms and control. One of the challenges in this field is the life span of the sensors; sensors with a higher life span are desirable. Soft sensors integrated with soft robots can bring about enormous change in the field of biomedical applications. This is in view of their highly elastic and flexible structure, which enables them to perform well in a biological environment.

References

[1] Anoop C S and George B 2012 Electronic scheme for computing inverse-cosine and its application to a GMR based angle sensor *IEEE Trans. Instrum. Meas.* **61** 1991–9

[2] Areekath L, George B and Reverter F 2019 An extended study on an interference-insensitive switched capacitor CDC *IEEE Sens. J.* **19** 8283–92

[3] Areekath L, George B and Reverter F 2021 Analysis of a direct microcontroller interface for capacitively coupled resistive sensors *IEEE Trans. Instrum. Meas.* **70** 1501010

[4] Chathuranga D S, Wang Z, Noh Y, Nanayakkara T and Hirai S 2016 Magnetic and mechanical modeling of a soft three-axis force sensor *IEEE Sens. J.* **16** 5298–307

[5] Chen H Y and Conn A T 2020 A stretchable inductor with integrated strain sensing and wireless signal transfer *IEEE Sens. J.* **20** 7384–91

[6] Chen L, Liu J, Wang X, Ji B, Chen X and Yang B 2017 Flexible capacitive hydrogel tactile sensor with adjustable measurement range using liquid crystal and carbon nanotubes composites *IEEE Trans. Electron Devices* **64** 1968–72

[7] De Chiara F, Wang S and Liu H 2020 Creating a soft tactile skin employing fluorescence based optical sensing *IEEE Robotics and Automation Letters* **5** 3375–81

[8] Chiurazzi M, Garozzo G G, Dario P and Ciuti G 2020 Novel capacitive-based sensor technology for augmented proximity detection *IEEE Sens. J.* **20** 6624–33

[9] Cotton D P J, Graz I M and Lacour S P 2009 A multifunctional capacitive sensor for stretchable electronic skins *IEEE Sens. J.* **9** 2008–9

[10] Dobrzynski M K, Pericet-Camara R and Floreano D 2011 Contactless deflection sensor for soft robots *IEEE Int. Conf. on Intelligent Robots and Systems* 1913–8

[11] Elangovan K and Sreekantan A C 2021 Evaluation of new digital signal conditioning techniques for resistive sensors in some practically relevant scenarios *IEEE Trans. Instrum. Meas.* **70** 1–9

[12] Escobedo P, Ntagios M, Shakthivel D, Navaraj W T and Dahiya R 2021 Energy generating electronic skin with intrinsic tactile sensing without touch sensors *IEEE Trans. Robot. Autom.* **37** 683–90

[13] Ge J, James A E, Xu L, Chen Y, Kwok K-w and Fok M P 2016 Bidirectional soft silicone curvature sensor based on off-centered embedded fiber Bragg grating *IEEE Photon. Technol. Lett.* **28** 2237–40

[14] George B, Mohan N M and Jagadesh Kumar V 2008 A linear variable differential capacitive transducer for sensing planar angles *IEEE Trans. Instrum. Meas.* **57** 736–42

[15] George B, Roy J K, Jagadeesh Kumar V and Mukhopadhyay S C 2017 *Advanced Interfacing Techniques for Sensors* vol 25 (Cham: Springer)

[16] George B, Tan Z and Nihtianov S 2017 Advances in capacitive, Eddy current, and magnetic displacement sensors and corresponding interfaces *IEEE Trans. Ind. Electron.* **64** 9595–607

[17] George B, Zangl H, Bretterklieber T and Brasseur G 2009 Seat occupancy detection based on capacitive sensing *IEEE Trans. Instrum. Meas.* **58** 1487–94

[18] Homberg B S, Katzschmann R K, Dogar M R and Rus D 2015 Haptic identification of objects using a modular soft robotic gripper *IEEE Int. Conf. on Intelligent Robots and Systems* pp 1698–705

[19] Jamone L, Natale L, Metta G and Sandini G 2015 Highly sensitive soft tactile sensors for an anthropomorphic robotic hand *IEEE Sens. J.* **15** 4226–33

[20] Jung J, Park M, Kim D and Park Y L 2020 Optically sensorized elastomer air chamber for proprioceptive sensing of soft pneumatic actuators *IEEE Robot. Autom. Lett.* **5** 2333–40

[21] Khan A U, Islam T, George B and Rehman M 2019 An efficient interface circuit for lossy capacitive sensors *IEEE Trans. Instrum. Meas.* **68** 829–36

[22] Khan S, Tinku S, Lorenzelli L and Dahiya R S 2015 Flexible tactile sensors using screen-printed P(VDF-TrFE) and MWCNT/PDMS composites *IEEE Sens. J.* **15** 3146–55

[23] Koivikko A, Raei E S, Mosallaei M, Mäntysalo M and Sariola V 2018 Screen-printed curvature sensors for soft robots *IEEE Sens. J.* **18** 223–30

[24] Lee H, Cho H, Kim S J, Kim Y and Kim J 2018 Dispenser printing of piezo-resistive nanocomposite on woven elastic fabric and hysteresis compensation for skin-mountable stretch sensing *Smart Mater. Struct.* **27** 025017

[25] Li C, Zhang W, Zhang Q, Chen R and Ma W 2020 Research on the detection of curved ferromagnetic materials by flexible eddy current array sensor based on ldc1614 *J. Phys.: Conf. Ser.* **1544** 012135

[26] Lu N, Lu C, Yang S and Rogers J 2012 Highly sensitive skin-mountable strain gauges based entirely on elastomers *Adv. Funct. Mater.* **22** 4044–50

[27] Eventoff F N 1992 Electronic pressure sensitive transducer apparatus *US Patent* US4314227A

[28] Narendiran A and George B 2015 Capacitive tactile sensor with slip detection capabilities for robotic applications *Conf. Record—IEEE Instrumentation and Measurement Technology Conf.* pp 464–9

[29] Navarro S E *et al* 2020 A model-based sensor fusion approach for force and shape estimation in soft robotics *IEEE Robot. Autom. Lett.* **5** 5621–8

[30] Paulino T, Ribeiro P, Neto M, Cardoso S, Schmitz A, Santos-Victor J, Bernardino A and Jamone L 2017 Low-cost 3-axis soft tactile sensors for the human-friendly robot Vizzy *Proc. IEEE International Conf. on Robotics and Automation* pp 966–71

[31] Prituja A V, Banerjee H and Ren H 2018 Electromagnetically enhanced soft and flexible bend sensor: a quantitative analysis with different cores *IEEE Sens. J.* **18** 3580–9

[32] Rocha R P, Lopes P A, De Almeida A T, Tavakoli M and Majidi C 2018 Fabrication and characterization of bending and pressure sensors for a soft prosthetic hand *J. Micromech. Microeng.* **28** 034001

[33] Rus D and Tolley M T 2015 Design, fabrication and control of soft robots *Nature* **521** 467–75

[34] Shih W P, Tsao L C, Lee C W, Cheng M Y, Chang C, Yang Y J and Fan K C 2010 Flexible temperature sensor array based on a graphite-polydimethylsiloxane composite *Sensors* **10** 3597–610

[35] Soomro A M, Khalid M A U, Shah I, Kim S W, Kim Y S and Choi K H 2020 Highly stable soft strain sensor based on Gly-KCl filled sinusoidal fluidic channel for wearable and water-proof robotic applications *Smart Mater. Struct.* **29** 025011

[36] Tairych A and Anderson I A 2019 Capacitive stretch sensing for robotic skins *Soft Robot.* **6** 389–98

[37] Thuruthel T G, Hughes J, Georgopoulou A, Clemens F and Iida F 2021 Using redundant and disjoint time-variant soft robotic sensors for accurate static state estimation *IEEE Robot. Autom. Lett.* **6** 2099–105

[38] Wang H, De Boer G, Kow J, Alazmani A, Ghajari M, Hewson R and Culmer P 2016 Design methodology for magnetic field-based soft tri-axis tactile sensors *Sensors (Switzerland)* **16** 1356

[39] Wang H, Kow J, Raske N, de Boer G, Ghajari M, Hewson R, Alazmani A and Culmer P 2018 Robust and high-performance soft inductive tactile sensors based on the Eddy-current effect *Sensors Actuators A* **271** 44–52

[40] Xie S, Zhang Y, Jin M, Li C and Meng Q 2021 High sensitivity and wide range soft magnetic tactile sensor based on electromagnetic induction *IEEE Sens. J.* **21** 2757–66

[41] Xing Z, Lin J, McCoul D, Zhang D and Zhao J 2020 Inductive strain sensor with high repeatability and ultra-low hysteresis based on mechanical spring *IEEE Sens. J.* **20** 14670–5

[42] Yao T *et al* 2020 Highly sensitive capacitive flexible 3D-force tactile sensors for robotic grasping and manipulation *J. Phys. D: Appl. Phys.* **53** 445109

[43] Yin H, Zhang Y, Wang J and Cao J 2021 Position prediction of soft finger driven by SMA based on fiber Bragg grating sensor *IEEE Sens. J.* **21** 2951–62

[44] Yuen M C S, Lear T R, Tonoyan H, Telleria M and Kramer-Bottiglio R 2018 Toward closed-loop control of pneumatic grippers during pack-and-deploy operations *IEEE Robot. Autom. Lett.* **3** 1402–9

[45] Zhou D, Liu Y, Tang X and Zhao J 2021 Differential sensing method for multidimensional soft angle measurement using coiled conductive polymer fiber *IEEE Trans. Ind. Electron.* **68** 401–11

Chapter 2

Recent advances in laser-induced graphene (LIG)-based flexible electronic devices

Khairunnisa Amreen and Sanket Goel

Laser-induced graphene (LIG) is basically a 3D porous carbon material, fabricated through direct laser ablation over carbon-based substrates. Since its discovery, in 2014, it has become an emerging technique for preparation of multilayered and single-layered graphene, and other graphene forms including reduced graphene and graphene oxides. This method of preparation eliminates the process of conventional wet chemical synthesis and simplifies it in a one-step procedure. Through its recent development, different types of lasers and carbon substrate materials have been reported for LIG preparation, the most common being infrared CO_2 lasers and UV lasers. A variety of substrates ranging from chemical-based polymers, like polyimide and polyetherimide, to organic substrates, like cloth, wood, food items, etc, have been demonstrated for LIG formation. Several reports have been presented on the strategic synthesis of LIG to customize its characteristics such as its chemical composition, surface morphology, functional groups, porosity, etc. Further, fine-tuning of the laser parameters, like power and speed, has been proven to enhance its chemical composition, robustness, mechanical strength and electrical conductivity. Due to these tunable properties, LIG has gained significant interest from researchers. Broad field applications utilizing LIG such as flexible electronic devices, microfluidic devices, sensors, electrocatalysts, biosensing, energy devices, electronics, artificial intelligence, electrochemical sensing, etc, have been reported. Owing to its flexible nature and high conductivity, application of LIG in electronic devices has been extensively explored. The present chapter summarizes recent advances in the preparation and application of LIG in diverse electronic devices. It also briefly discusses the future outlook and present research gaps in LIG-based flexible electronics applications.

doi:10.1088/978-0-7503-3935-3ch2

2.1 Introduction

The uncovering of various carbon-based materials, like nanotubes [1], graphene [2], fullerene [3], carbon fiber [4], graphite [5], etc, has paved a route for extensive carbon-nanomaterial-based research and applications [6–9]. Often these materials are associated with outstanding chemical, compositional, physical, thermal, electrical, mechanical and optical properties [10]. These extraordinary properties are morphologically dependent, and can be tuned based on the size and shape of the nanomaterial, the chemical composition of the substrate and the method of preparation. Hence, due to these properties, there exists a broad spectrum of applications of these materials, such as sensors, energy devices, electronic devices, circuits, etc, in both industry and academia. As the applications of carbon nanomaterials are morphologically oriented, different structures, ranging from 1D, 2D, 3D, fibers, ribbons, sheets, etc, has been fabricated. Among these, an exclusive 2D carbon material, graphene, has gained much attention as of late. The arrangement of graphene is a planar hexagonal assembly of sp^2 hybridized carbon atoms. Graphene exhibits exceptional thermal conductivity of ~5000 W m^{-1} K^{-1} [11], electrical conductivity [12] and rigidity due to delocalization of 2p orbitals over the carbon sheets. Further, it is reported to have a large surface area of 2630 M^2 g^{-1} [13]. Figure 2.1 gives a schematic representation of the summarized properties of graphene [14–20].

These excellent properties have increased the commercial market value of graphene to ~$80 million, and the market is continuously growing [21]. Integration of graphene in various commercial products, such as light-emitting diodes (LEDs), supercapacitors, electrodes, sensors, energy generators, etc, to

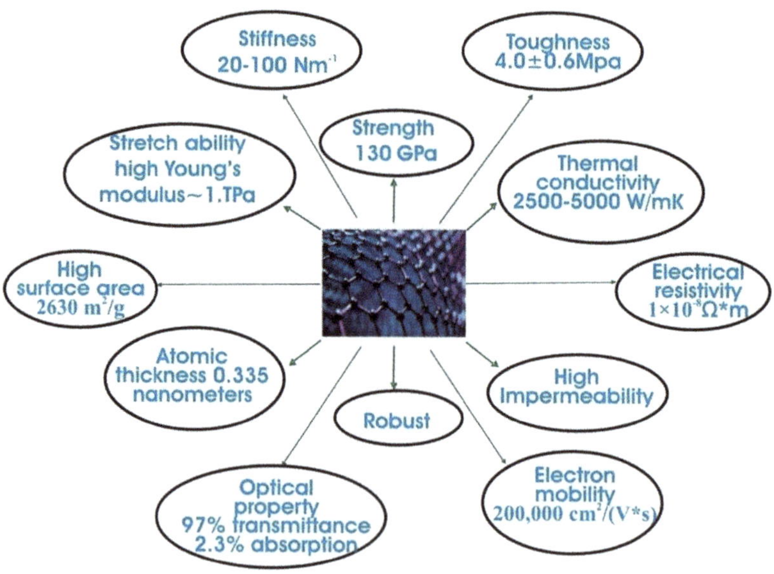

Figure 2.1. Schematic representation of the properties of graphene.

improve their durability and performance, has increased the demand for graphene production. In general, chemical vapor deposition, organic wet chemical synthesis and epitaxial growth are the traditional approaches for its preparation. Lee *et al*, in 2019, summarized the various synthetic approaches of graphene that are employed at the industrial level [22]. Figure 2.2 represents an overview of these methods.

However, these methods are associated with the usage of elevated temperatures, environmental pollutants like hydrazine, high energy, etc, which presents limitations such as uncontrolled morphology, less efficiency and higher cost during industrial-scale preparation. As a result, in the search for methods for fabricating graphene at low temperatures using green precursors, a considerable amount of research has been conducted. Laser direct ablation is one of the outcomes of this search. In this technique, a laser is used to convert carbon precursors into graphene and its derivatives. This is an accurate method, which helps to modify the chemical and physical properties of the material being exposed.

A variety of materials like metals, ceramics and polymers, and carbon-containing substrates like wood, cloth, food items, organic waste, e-waste, etc, can be exposed to laser for graphene generation [23]. This prepared graphene is called laser-induced graphene (LIG).

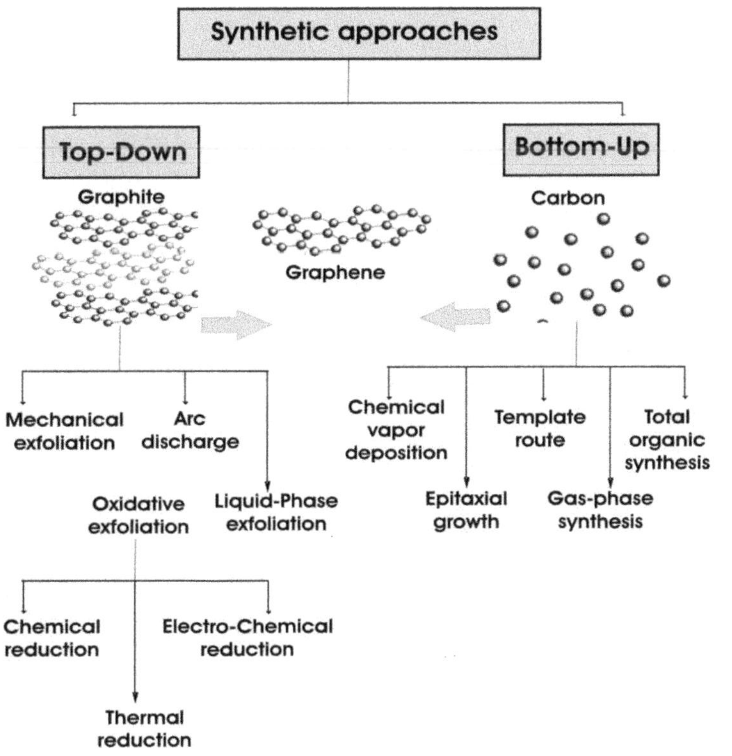

Figure 2.2. Schematic summary of various graphene synthesis approaches.

2.1.1 Characterization techniques

The properties of graphene are dependent on the morphological features and surface composition. Depending upon the preparation method, the chemical composition, morphology and other physical and chemical properties vary. Therefore, prepared graphene has to be physico-chemically characterized prior to use. The following techniques are broadly used for this purpose:

(i) *Scanning electron microscopy (SEM)*, for surface morphology.
(ii) *Atomic force microscopy (AFM)*, for measuring surface thickness and roughness.
(iii) *Transmission electron microscopy (TEM)*, for morphology at the atomic level.
(iv) *X-ray diffraction (XRD) or X-ray crystallography*: crystal size, atomic arrangement, crystallinity.
(v) Raman spectroscopy, for graphitic and distorted nature.
(vi) *Fourier transform infrared spectroscopy (FTIR)*, for chemical composition and functional groups.
vii *X-ray photoelectron spectroscopy (XPS)*, for oxygen, carbon and other elemental analysis
(viii) *Brunauer–Emmett–Teller (BET)*, for measuring pore size.
(ix) *Electrochemical analysis, potentiometry, conductometry*: In order to analyze the electron transfer and conductance behavior.

2.2 LIG fabrication

The laser-based fabrication method has several advantages, such as having no carcinogenic precursors, being non-contact, working in ambient temperatures, one-step synthesis, high speed and uniform morphology [24]. A laser instigates a thermal and photochemical transition over the substrates, which cannot be done using common thermal heaters. The type of laser (CO_2, UV, infrared, pulsed, etc), power and speed determines the formation and characteristics of the LIG. Laser ablation changes sp^3 hybridized carbon to sp^2. Furthermore, to enhance the efficacy and usage, different derivative forms of graphene like oxides, reduced forms, multi-layered, single layered, ribbons, etc, can be prepared via this technique. Our group has successfully attempted preparation of graphene oxide over a polyimide (PI) surface [25]. Figure 2.3 gives a schematic representation of the process showing a PI

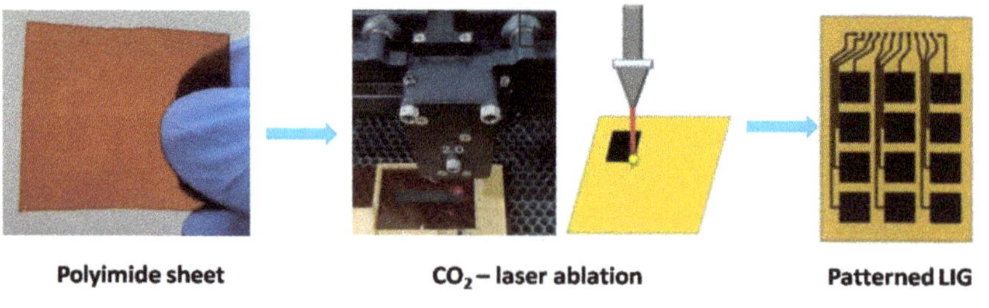

Figure 2.3. Schematic representation of the LIG fabrication process.

sheet of desired dimensions. The design of the pattern was made using AutoCAD Fusion 360 software. This design file was converted to a .dxf file and further transferred to CorelDraw X7. A PI sheet was exposed to a CO_2 laser with varying speed and power. It was discovered that the power of the laser influenced the thickness and elemental composition of the LIG. A 5.4 W power laser gave 91.3% carbon, 7.7% oxygen and 1.0% nitrogen. Further, the laser's power was varied from 1.35 to 1.95 W to obtain various LIGs and a comparative study of their chemical composition was conducted. It was also observed that the laser power affected the conductivity of the LIG.

Likewise, various research groups have reported different lasers for LIG formation for a multitude of applications. For example, El-Kady *et al* reported fabrication of LIG over a DVD disc with a pulsed laser. Further, they applied it in the application of an electrochemical supercapacitor [26]. Similarly, Shi *et al* reported single-step LIG fabrication using a pulsed laser. Herein, two variants of the laser, a 355 nm and a 450 nm variant, were deployed to scribe LIG over the PI. These were applied as a micro-supercapacitor and strain sensor [27]. Kurra *et al* reported a pulsed laser method to get graphene nanoribbons from a pyrolytic graphite substrate. The prepared material had five to six layers of graphene ribbons, as well as graphitic ribbons of 30–40 μM [28]. Zhu *et al* used a 405 nm diode laser to prepare graphene films over polydimethylsiloxane (PDMS) and used it for a pressure-sensing application [29]. Garland *et al* utilized a 405 nm UV laser for direct writing of LIG on PI and further applied it for gas sensing, in their case nitrogen sensing in soil [30]. Wakaya *et al* applied a UV pulsed laser to exfoliate graphene from SiO_2 and Si substrates. The thickness of the graphene was dependent on the power density of the laser [31]. Lin *et al*, in 2017, prepared graphene oxide using a 650 nm laser, which was further utilized for a heating application [32]. Arul *et al* prepared reduced graphene oxide from graphene oxide using three different lasers, an 800 nm fs pulse laser, a 248 nm ns pulse laser and a 788 nm continuous wave laser, and performed a comparative study [33]. Deng *et al* used a 800 nm femtosecond laser with three different power densities, i.e. low, medium and high, and reduced graphene oxide was patterned over different substrates [34]. Teoh *et al* prepared reduced graphene oxide from a thin film of graphene oxide spin-coated over a silicon dioxide substrate using a 532 nm diode laser. Tao *et al* developed LIG over a PI substrate using a 450 nm diode laser; the prepared LIG was used to fabricate an intelligent artificial throat sensor with sound-sensing ability [35].

Similarly, Lin *et al* reported studies on the formation of graphene films from various polymers like PI and polyethermide using a 10.6 m CO_2 pulsed laser, used for a supercapacitor application [36]. Rahimi *et al* utilized a CO_2 laser to fabricate LIG over PI and it was further applied as a strain sensor [37]. Lamberti *et al* developed LIG over a sulfonated poly ether ketone substrate using an Ar–Kr laser of 514.5 nm wavelength, which was also used for a supercapacitor application [38]. Chyan *et al* reported an interesting study of scribing LIG over various carbon-containing substrates like paper, coconut, bread, potato, etc, with a 10.6 m CO_2 pulsed laser [39]. Ye *et al* developed a 9.3 m CO_2 pulsed laser for forming fluorinated graphene over a Teflon substrate [40]. A 532 nm optically pumped laser was employed over methane and single-layer graphene to produce graphene oxide [41]. The same research group also reported the formation of LIG over a wood surface utilizing a 10.6 μM CO_2 pulse laser and used it as an electrocatalyst [42]. Huang *et al* studied the effect of a CO_2 laser with low power density over single-layer graphene to explore the thermal effects on its properties [43]. Further, Kiisk *et al* used a Nd:YAG laser for 5 ns over single-layer graphene to study the effect on its properties [44].

2.2.1 LIG chemical modification

LIG can be utilized as formed (bare) or can be chemically modified as per requirements for a diverse range of applications. The presence of a large number of oxygen groups in the edges and basal plane of graphene oxide eases its interaction with a variety of organic and inorganic chemicals, metals and materials. Furthermore, various other carbon forms can also be either surface-modified or incorporated to form nanocomposites. These chemical modifications often enhance the performance of LIG in terms of chemical reactivity, electron conductivity, mechanical strength, power density, thermal stability, robustness, etc. Lasers with different speed and power can be used for incorporation of hetero atoms like C, N, and O into the graphene matrix [36]. Radiating a PI sheet with a laser in the presence of boric acid resulted in doping of boron [45]. Likewise, the presence of a metal complex solution while ablating PI in ambient environmental conditions resulted in incorporation of nanoparticles into the LIG as nanocomposites of Co_3O_4, Fe_3O_4 and MoO_2 [46]. Exposing the PI during LIG formation to atmosphere containing Ar, O_2, H_2 or SF_6 enhances its hydrophilic and hydrophobic properties [47]. For surface modification alone, several approaches like ablation in the presence of metal salts [48], electrodeposition of metal oxides, polymers, etc, [49] have been developed. All these methods enhance the LIG's properties.

2.3 LIG in electronic devices

2.3.1 LIG-based miniaturized fuel cells

Microfluidics has proven its importance in developing various sensing and monitoring devices in several areas from energy and the environment to health care. However, there is a huge thrust to develop rapidly prototyped, clean-room free and flexible devices to make them inexpensive and integrable with other mechanical and electronic components. Owing to its flexible nature and porous structure, LIG can be used in microfluidic devices as porous microchannels or even electrode material. In order to produce these microchannels, methods such as template approach, curing, optical lithography, self-assembly, etc, have been reported. Several microfluidic devices for energy generation and energy harvesting have been developed. Our group has also reported a significant number of LIG-based microfluidic energy devices. Rewatkar *et al* reported LIG-based bioelectrodes for an enzymatic biofuel cell. LIG was fabricated over a pre-cleaned PI sheet with a CO_2 laser of 10.6 μM and 30 W power. A PI sheet with dimensions of 30 mm × 3 mm was exposed to a laser to obtain LIG for use in bioelectrodes. To improve the roughness and hydrophilic nature of the LIG, it was treated with oxygen plasma for about 1 min. This prepared

LIG was used for enzyme immobilization wherein LIG was immersed for 2 hours in a linker solution followed by drop casting of glucose oxidase enzyme solution and drying. The thus-prepared LIG gave excellent current density and performance [50]. This work was further extended by Jayapiriya *et al*, wherein a microfluidic device with integrated LIG bioelectrodes, prepared over PI through CO_2 laser ablation, was developed. As shown in figure 2.4(a), the surface of such LIG bioelectrodes was chemically modified with multiwalled carbon nanotubes (MWCNTs), which gave improved electrical conductivity and enhanced immobilization of the enzyme. These fabricated electrodes were then incorporated into a PDMS microfluidic device consisting of a Y-shaped microchannel. This device gave 1.37 times enhanced performance than simple LIG and generated 2.2 mW cm^{-2} of power density with just a 200 mL min^{-1} flow rate [51]. Our group itself reported development of a microfluidic device with LIG electrodes (both anode and cathode) enhanced with MWCNTs for a membraneless microfluidic fuel cell. As depicted in figure 2.4(b), LIG was modified with various catalysts like MWCNT, silver nano-ink, single-walled carbon nanotubes (CNTs), MnO_2, TiO_2, etc. A comparative study of the effect of these catalysts on the overall performance of the device was undertaken. Further, these modified electrodes were integrated with a PDMS microfluidic device fabricated via soft lithography. The oxidant used here was atmospheric oxygen; a mixture of formic acid as fuel and sulfuric acid as electrolyte were used. The silver nano-ink increased the performance of the fuel cell and gave a power density of 88.80 μW cm^{-2} at a 24 mL h^{-1} flow rate [52].

Jayapiriya *et al* also reported a microfluidic device with a CO_2 laser ablated microchannel and bioelectrodes (figure 2.5(a)). A PI sheet, with dimensions of 75 mm × 25 mm, was exposed to a laser with 5.1 W power and 0.625 cm s^{-1} speed. Prior to the laser exposure, the substrate was cleaned using isopropyl alcohol (IPA). This sheet was affixed over a glass side. CorelDRAW was used to draw the design of a Y-shaped microchannel and electrodes. Upon laser treatment, 80% of the substrate was ablated and removed using IPA. In this way, a microchannel with 100 μm depth was created. Adjacent to the Y-shaped microchannel, LIG electrodes of 10 mm × 1 mm with a spacing of 1 mm, as well

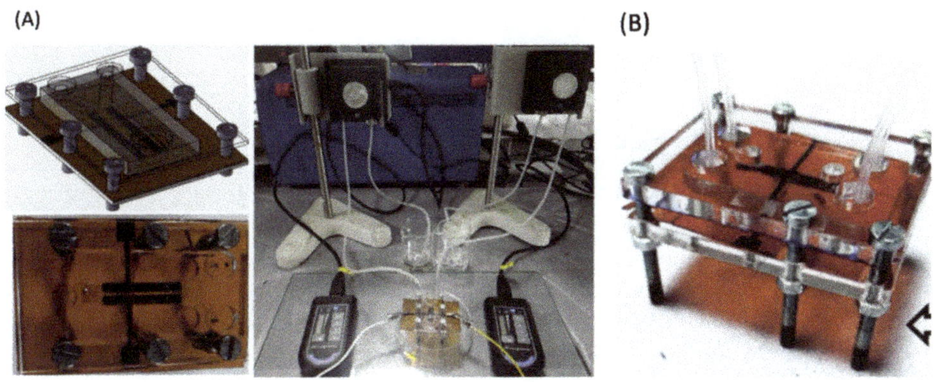

Figure 2.4. (a) Diagrammatic representation. Reprinted from [51]. Copyright (2020), with permission from Elsevier. (b) The device. Reprinted from [52]. Copyright (2020), with permission from Elsevier.

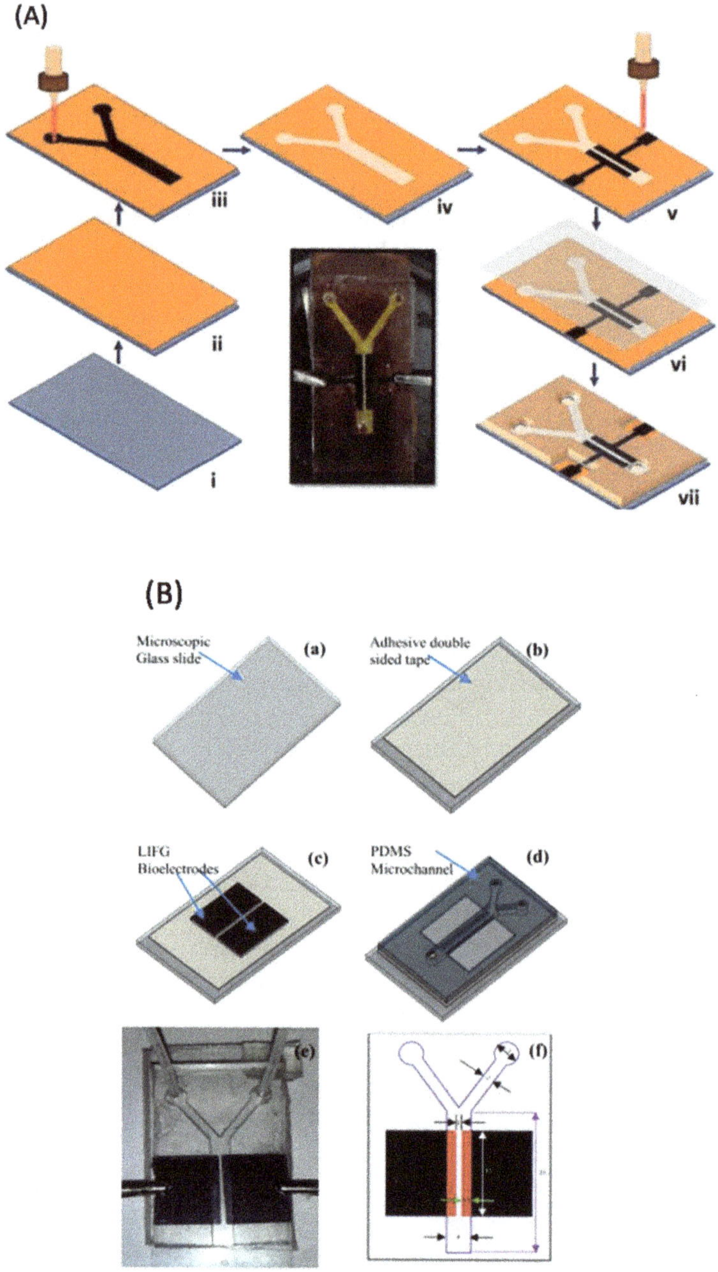

Figure 2.5. (a) Diagrammatic representation of their device fabrication. © 1969 IEEE. Reprinted, with permission, from [53]. (b) Schematic representation of the device fabricated by them. © 2020 IEEE. Reprinted, with permission, from [13].

as an electroactive working surface area of 0.1 cm^2, were created using a CO_2 laser. Further, the LIG electrodes were modified with MWCNTs to enhance enzyme immobilization. These LIG electrodes were converted to bioanode and biocathode by immobilizing glucose oxidase and laccase enzymes, respectively. After fabrication of the microchannel and bioelectrodes, a complete fuel cell was made by sealing these electrodes in a PDMS layer to provide mechanical strength [53]. Rewatkar *et al* also fabricated a PDMS -based microfluidic biofuel cell with glucose and integrated with optimized LIG electrodes. LIG was prepared on a PI sheet using a CO_2 laser of 10.6 μM wavelength and a power of 30 W. Further, the LIG was modified with an enzyme. As shown in figure 2.5(b), the electrodes were then integrated into a PDMS microfluidic device. For this, first a Y-shaped microchannel mode was made using polyethylene terephthalate (PET) with dimensions of 25 mm × 4 mm × 0.2 mm. The inlet and outlet were created by surgical needle for the flow of the anolyte and catholyte [13].

2.3.2 LIG-based miniaturized energy-storage devices

Apart from energy-generation applications, LIG-based devices have also been reported for energy-storage applications, as well. For instance, Song *et al* developed a LIG-based micro-supercapacitor using a PI sheet as substrate, which was exposed to a CO_2 infrared laser of 5 W power to obtain 3D porous LIG. As mentioned in figure 2.8(a), a pair of interdigited electrodes with dimensions of 8.6 mm × 0.86 mm with an intermediate spacing of 0.86 mm was developed and transferred over a silicon rubber support. To enhance the capacitance, LIG was chemically modified by doping nitrogen via coating a conductive layer of poly(3,4-ethylenedioxythiophene). A volumetric capacitance of 7.2 mF cm^{-3} was obtained, which was about 77% higher than the LIG-only-based device [54]. Li *et al* developed interdigital-electrodes-based pseudocapacitive micro-supercapacitors using LIG on a PI sheet using a CO_2 laser. Furthermore, to enhance the capacitance, the LIG was electrodeposited with pseudocapacitive materials like polyaniline, manganese dioxide (MnO_2) and ferric oxyhydroxide. After symmetrical assembly, these devices gave appreciable performance comparable to micro-batteries [49]. Wang *et al* reported using Co_3O_4 nanoparticles and LIG to form micro-supercapacitors on a PI sheet. To incorporate the Co_3O_4 nanoparticles onto the LIG, before exposure to a CO_2 laser a thin film of $CoCl_2$ dissolved in gelatin was coated over the cleaned PI sheet and dried in vacuum. Further assembly of the micro-supercapacitor was done using copper membrane [55].

Likewise, Zhang *et al* presented a methodology to fabricate a stretchable micro-supercapacitor using LIG modified with mesoporous ZnP nanoparticles. ZnP nanosheets were prepared using a fast heating process of an homogenous mixture of cetyl trimethylammonium bromide, ethylene glycol and zinc acetate, followed by phosphorization. PI sheets were cleaned and a PDMS support was provided, followed by CO_2 laser ablation to pattern an interdigited LIG electrode. Finally, the prepared ZnP ink was coated as a film and dried to form chemically modified LIG electrodes as micro-supercapacitors [56]. Similarly, another research group, Li *et al*, reported LIG-based pseudocapacitive electrodes via inkjet printing. A PI sheet was ablated with a CO_2 laser of 10.6 μm wavelength to prepare a bis-substituted cobalt(II) complex of TPy-Co ink, which was printed over LIG using inkjet printing [57]. Wen *et al* reported LIG enhanced with CNTs. A composite of graphene oxide and CNTs was prepared, PET was adhered to a DVD surface, and the prepared composite was drop casted and dried. The prepared disc was subjected to a laser to scribe LIG in an interdigitated pattern. The device gave a volumetric capacitance of 3.10 F cm^3, a current density of 1000 mA cm^3, a power density of 1.0 W cm^3 and a 0.84 mW hr cm^3 energy density [58].

2.3.3 LIG heaters

Heating devices should generally be flexible, mechanically robust and useful in multiple applications ranging from sensing to micro-devices. These heaters are mostly placed below these devices to provide a thermostatic environment for the desired reactions to take place. Therefore, such heaters play a significant role in realizing the fully integrated assembly of various electronic devices. Conventionally, these devices are made from silicon, silver nanoparticles, polymeric films, etc. In some cases, inkjet printing is used over a substrate to deposit a self-heating layer, which leads to limitations in terms of fabrication. However, use of LIG as a heating device in electronics has resolved this issue to a great extent. Bobinger *et al* demonstrated a photothermal method for developing LIG heaters over a PI sheet. A diode laser was used for ablation of a pattern over bare PI. The LIG pattern exhibited a negative temperature coefficient of −0.46 m $°C^{-1}$. The heaters had an operating temperature of 400 °C [59].

Chen *et al* recently developed LIG paper heaters showing multimodal patterned electrothermal behavior, which was further applied for manufacturing low-energy composites [61]. Cho *et al*, in 2021, developed silver-decorated LIG and used it as flexible heaters on a PI sheet. As shown in figure 2.10(b), further, PI was treated with UV ozone, followed by immersion in a potassium hydroxide solution for about 20 min. After thorough washing of these films, they were dipped in a $AgNO_3$ solution for nearly 2 hr to ensure an ion exchange process, after which they were washed thoroughly and exposed to a CO_2 laser of 10.6 μM wavelength. The resulting Ag-LIG exhibited excellent thermal properties and was used as heaters [60]. Srikanth *et al* from our group has also developed a droplet-based lab-on-a-chip platform for synthesizing gold nanoparticles. Herein, to provide the ambient temperatures, LIG heaters were integrated in the device. The prepared nanoparticles were further used for electrochemical and fuel cell applications. Using these heaters, a maximum temperature of 589 °C was achieved with a laser power of 15% and 5.5% scanning speed. A PDMS-based droplet microfluidic device was fabricated as per reported procedures and was bonded over a glass slide to form the microfluidic droplet platform. The PI sheet was exposed to a laser to form LIG in the specified dimensions. Over this, a fabricated PDMS microchannel was bonded via double-sided tape. In order to gain a proper thermal contact and prevent thermal loss, the edges of the heater were appropriately covered with copper tape and sealed using a silver ink. The LIG was given electrical contacts and a voltage was supplied. Owing to the characteristic nature of graphene to heat up easily, the LIG gave a quick

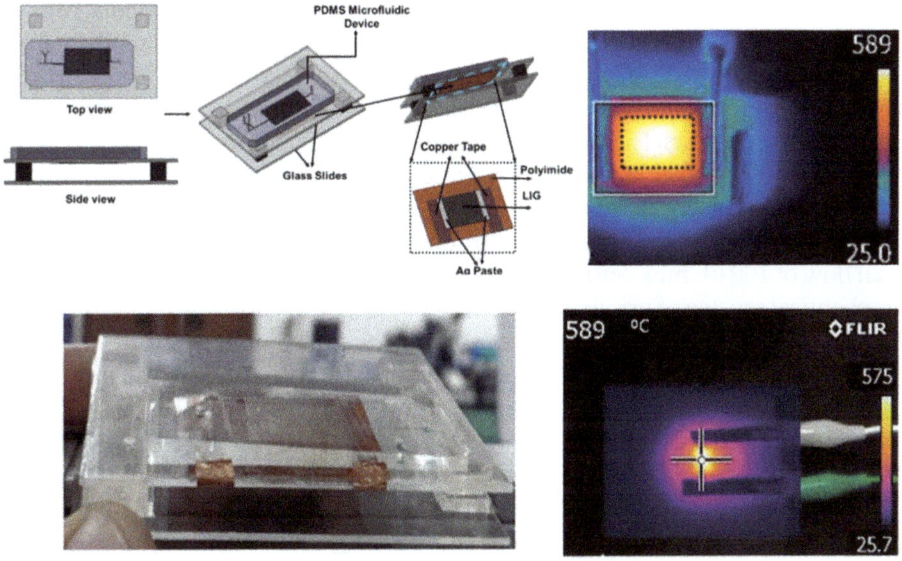

Figure 2.6. Schematic representation of the LIG-heater-based microfluidic device fabrication process, real image of the device and thermal camera images with maximum temperature developed by our group [62]. Copyright © 2021, The Author(s). With permission of Springer.

response and was heated. The temperature was maintained depending upon the desired requirement for gold nanoparticle synthesis. Figure 2.6 provides a pictorial representation of the LIG-heater-based microfluidic platform fabricated by our group, and images using a thermal camera at maximum temperature [62].

2.3.4 LIG sensors

Owing to the ease of preparation, cost effectivity and simplicity, fabrication of sensors using LIG as a base material is an area of continuing development. Different types of sensing devices have been developed using bare and chemically modified LIG. These include electrochemical, electrochemiluminescence, optical, thermal, etc. In this section, different types of LIG-based sensors will be discussed.

2.3.4.1 LIG in electrochemical sensing

Of late, LIG has gathered significant attention for use in the fabrication of electrochemical sensors, biosensors, immunosensors, etc. Bare or chemically modified LIG is applied for sensing of chemicals, biochemicals, biomarkers, pollutants, body fluids, proteins, DNA, RNA, antigens, antibodies, etc. Our group has also developed LIG-based electrochemical sensors for both microfluidic and bulk sensing applications. Some recent reports, including the ones from our research group, are discussed in this section. Kothuru *et al*, from our group, reported bare LIG for electrochemical enzymeless sensing of hydrogen peroxide. A conventional bulk, three-electrode system, comprising a working bare LIG electrode, counter platinum electrode and Ag/AgCl reference electrode, was employed here. The LIG showed excellent electrocatalytic reduction of H_2O_2 in a linear range of

1–10 μM with a limit of detection of 0.3 μM [25]. Our group also reported a LIG-based microfluidic device called an electromicrofluidic device. A PI sheet was used as substrate for both microchannel fabrications as well as for the electrodes. In order to form a microchannel, the LIG formed on the PI sheet after ablation with a CO_2 laser was peeled off from the design, which left an etched pattern behind. To get the LIG electrodes, the PI sheet was ablated again. Figure 2.7 shows a schematic representation of the fabrication and realization of a LIG-based microfluidic platform and electrodes for electrochemical sensing, fuel cell and interdigitated electrodes (IDE). The device was subjected to an optimal flow of 100 $\mu L\ min^{-1}$ using a syringe pump. The fabricated device was applied for electrochemical sensing of uric acid. A linear range of 10 μM–3 mM was obtained with the device and a limit of detection (LoD) of 0.61 μM was also obtained. Furthermore, the same device was also subjected for a fuel cell application. A power density of 3.027 nW cm^{-2} was achieved. An array of interdigitated microelectrodes was also fabricated and used for nitrite detection [63].

Likewise, bare LIG was used by Sarda *et al* for electrocatalytic detection of hydrazine using a bulk three-electrode system [64]. Chao *et al* utilized bare LIG for detection of trans-resveratrol in a real sample of grape wine. [65]. Dearaujo *et al* developed a LIG-based three-electrode miniature sensor for sensing caffeic acid, ascorbic acid and picric acid in clinical, food and forensic samples [66]. Apart from utilizing bare LIG as a working electrode in electrochemical sensor devices, the LIG surface was chemically modified with various polymers, metals, enzymes, carbons, etc, to increase the electron transfer behavior. The surface modification led to high selectivity, sensitivity and stability of the electrode. For instance, LIG modified with copper nanoparticles was reported by Lin *et al* for enzymeless glucose sensing [67].

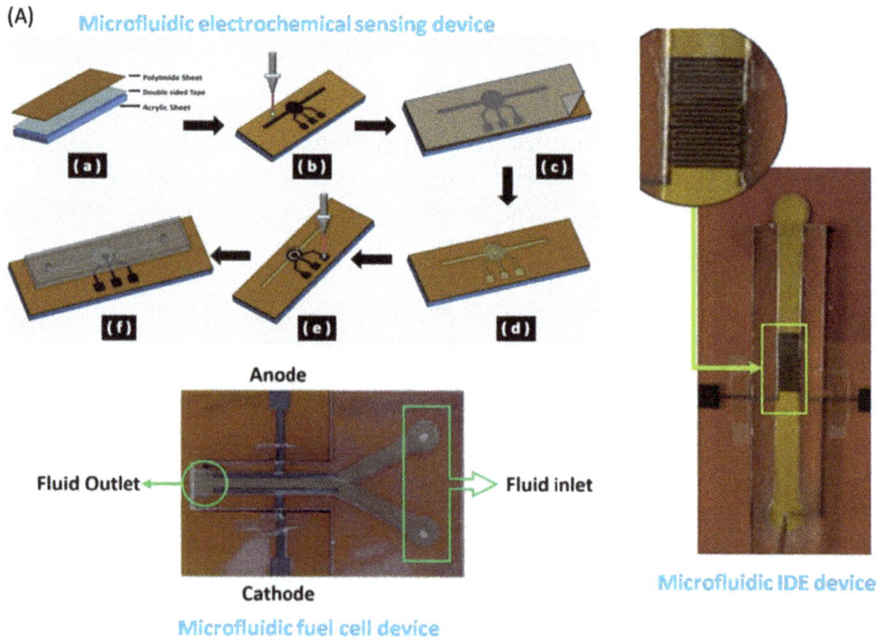

Figure 2.7. Schematic representation of the LIG-based sensors developed by [63]. © 2020 IEEE. Reprinted with permission.

Similarly, Tehrani *et al* developed a glucose0sensing strip using LIG chemically modified with copper nanocubes [68]. Phenolic resin was used for LIG preparation by Zhang *et al* for glucose biosensing [69]. Martinez *et al* electrodeposited silver nanoparticles over bare LIG and prepared an enzymeless sensor for hydrogen peroxide in milk samples [70]. Yuhan *et al* developed platinum nanoparticles over bare LIG for sensitive non-enzymatic hydrogen peroxide detection [24]. Likewise, a LIG electrode over a gold plate was demonstrated for an electrochemical-sensing application by Hyosang *et al* [71]. The same research group also reported LIG chemically modified with platinum nanoparticles and glucose oxidase. This was further used for detection of glucose in a real sweat sample. Chang *et al* developed a UV-laser-based LIG over a glass substrate and modified with glucose oxidase for glucose sensing [72]. Nitrogen-doped LIG was developed by Zhengfen *et al* for detection of micro-RNA in a three-electrode-based system [73]. LIG, chemically modified with copper microparticles and diamine oxidase enzyme, was prepared for biosensing and detection of histamine by Vanegas *et al* [74]. In a similar work, Mamleyev *et al* fabricated a LIG chemically modified electrode with chitosan and urease enzyme for detection of urea [21]. Nayak *et al* reported a very simple approach of modifying platinum nanoparticles over LIG for multiplexed detection of ascorbic acid, dopamine and uric acid [75]. Lei *et al* reported a TiO_2-modified LIG over an indium tin oxide electrode to detect acetylcholinesterase inhibitor enzyme [76]. Furthermore, various composites of more than one metal nanoparticle with LIG have also been reported for electrochemical sensing. Xue *et al*, for example, reported a mixture of platinum and gold nanoparticles over LIG and incorporated in a PDMS substrate and used it for detection of dopamine [77]. Apart from using metal nanoparticle, various carbon nanomaterials and their composites, a mixture of carbon and metal nanoparticles, etc, have been reported for surface modification of LIG. For example, Salam *et al* developed an electrode with a combination of gold nanoparticles and MWCNT over LIG for nitrite sensing [78].

Furthermore, chemical modification of LIG with chemicals such as ionic liquids, polymers and synthetic resins, etc, have also been reported. For example, Zhiwei *et al* developed a combination of ionic liquid and poly-L-cysteine-modified LIG and used it for detecting lead [79]. In another report, LIG was chemically modified with polyaniline and used for pH sensing [80]. Similarly, Xu *et al* reported poly(3,4-ethylenedioxythiophene)-modified LIG for selective sensing of dopamine in the presence of ascorbic and uric acid [81]. A biosensor for thrombin was made by Fenzl *et al* with 1-pyrenebutyric acid and amino-acid-modified LIG [82]. LIG with a modified ion-selective membrane was made by Garland *et al* for sensing of ammonia and nitrate ions in a soil sample [30]. LIG has been reported as a potential candidate for use as an immunosensor. A noble metal gold, silver and platinum decorated LIG was developed by Zhiheng *et al* and further modified with antibodies for detection of *Escherichia coli* [83]. Likewise, Soares *et al* developed a metal-catalyst-modified LIG and antibody-modified electrode for detection of the pathogen *Salmonella* [84]. Overall, the use of bare and modified LIG for electrochemical-sensing devices has shown significant growth in the last few years.

2.3.4.2 LIG in electrochemiluminescence sensing

In addition to electrochemical-sensing devices, LIG has also shown potential to be used in electrochemiluminescence (ECL) sensing. Our group has also reported a couple of LIG-based, ECL-based microfluidic devices for multiple analyte sensing. Bhaiyya *et al*, from our group developed a miniaturized platform with integrated LIG bipolar electrodes for multiplexed sensing of hydrogen peroxide and glucose. As shown in figure 2.8(a), the reaction was carried out based on luminal chemistry, and an Android smartphone was used to capture and process the ECL response [85]. Bhaiyya *et al* also developed a single-LIG-electrode-based miniaturized device for interference-mitigated detection of hydrogen peroxide, dopamine, xanthine and glucose. As depicted in figure 2.8(b), the developed device could be used as a point-

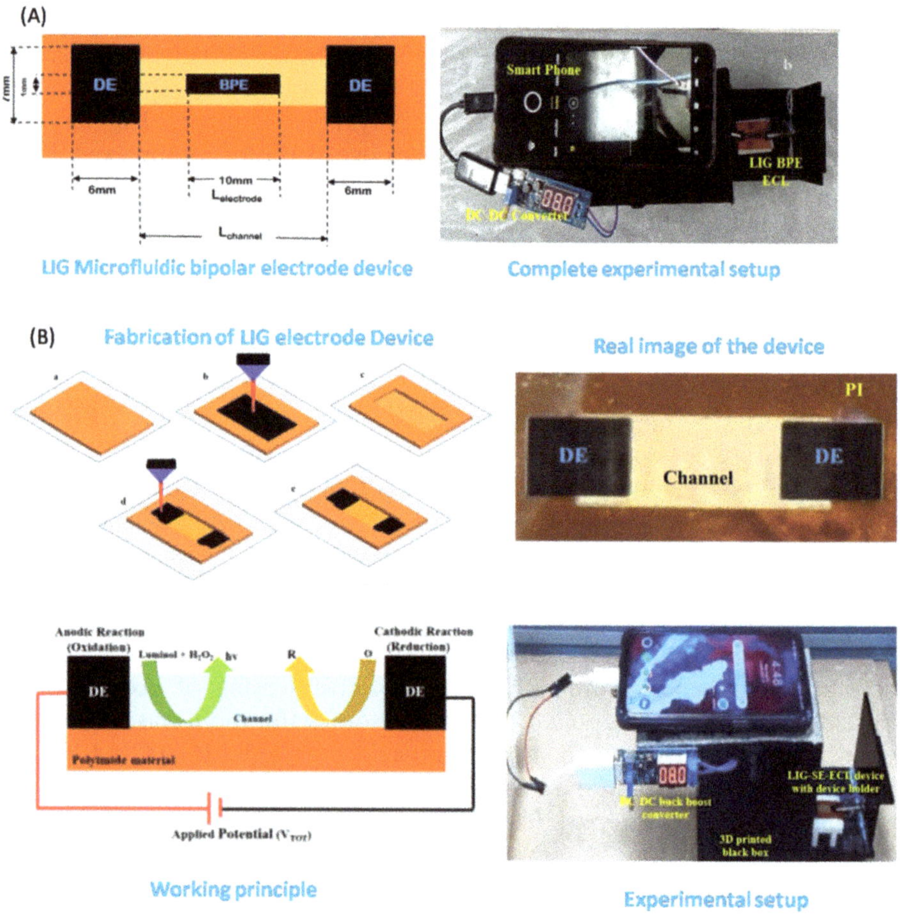

Figure 2.8. (a)–(d) Schematics and real images of the miniaturized ECL devices fabricated with LIG [85–88]. (a) © 2021 IEEE. Reprinted, with permission, from [85]. (b) © 2021 IEEE. Reprinted, with permission, from [86]. (c) [87] (2021) Copyright © 2021, The Author(s), under exclusive licence to Springer-Verlag GmbH, Germany, part of Springer Nature. With permission of Springer. (d) © 2021 IEEE. Reprinted, with permission, from [88].

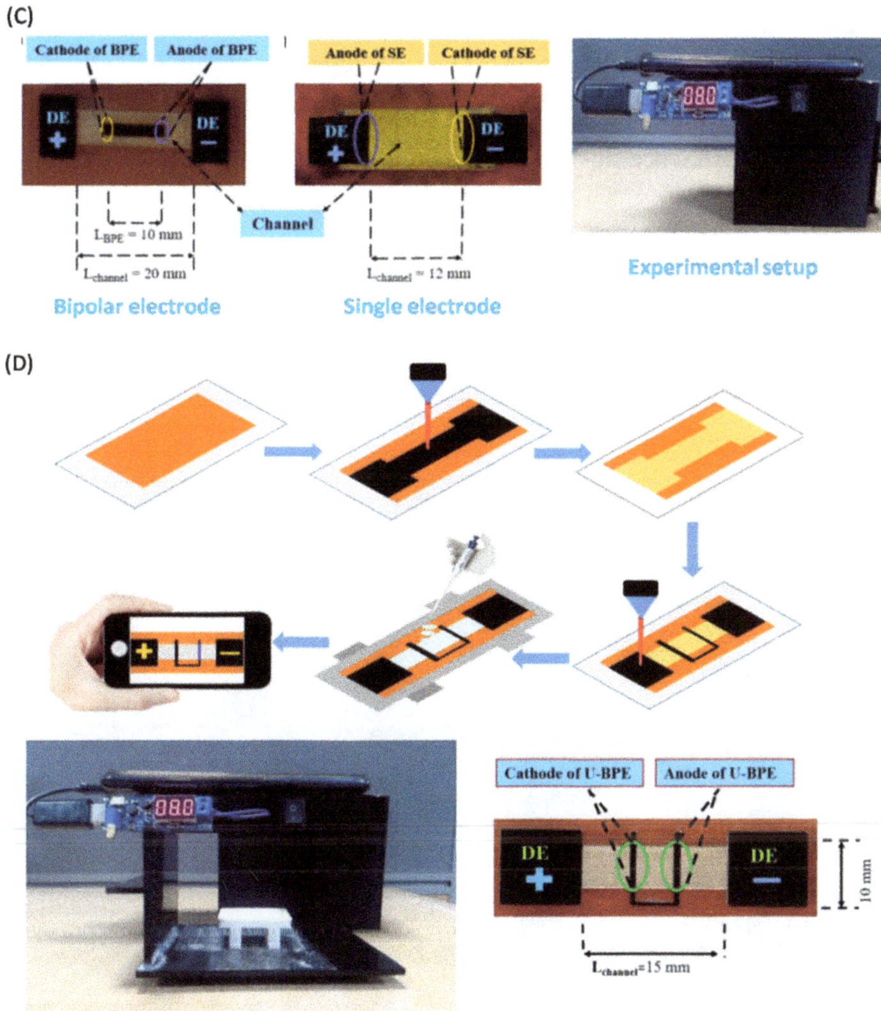

Figure 2.8. (Continued.)

of-care testing device with an Android smartphone. Linear ranges of 0.1–70 μM for H_2O_2 and glucose, 0.1 to 100 μM for dopamine and xanthine were obtained with a LoD of 1.71 μM for H_2O_2, 3.76 μM for glucose, 1.25 μM for xanthine and 3.40 μM for dopamine [86]. Bhaiyya *et al* also reported ECL sensing of vitamin B12 with LIG in a miniature 3D-printed device. Two types of electrodes were prepared, in which single and bipolar electrodes were developed over PI using a CO_2 laser. A 3D-printed miniature system was also fabricated for monitoring the ECL signal, and an Android smartphone was used to power the ECL platform and capture the images. A linear range of 0.5–700 nM for the bipolar electrode and a LoD of 0.107 nM was obtained, as well as a range of 0.5–1000 nM and a LoD of 0.094 nM for the single electrode was obtained. Figure 2.8(c) shows real images of both electrodes and the complete 3D-printed ECL device [87]. Bhaiyya *et al* reported a U-shaped LIG bipolar electrode for multiplexed analyte sensing using a similar platform. As shown in figure 2.8(d), analytes like hydrogen peroxide, glucose, lactate and choline were tested in the device within linear ranges of 0.1–50, 0.1–70, 1–100 and 1–100 μM with a LoD of 4.36, 2.51, 5.32 and 4.01 μM, respectively. A less than 5% error was obtained from interference due to other chemicals.

2.3.4.3 LIG-based physical sensors

Gas sensors. Gas sensors are essential for air-quality assessment and leakage analysis for industrial use. LIG has proven to be a useful candidate for the selective detection of various gases. Recently, our group has reported a 3D-printed nasal wearable device, named an electronic nasal pod (ENP), as a filter and electrochemical sensor. As illustrated in figure 2.9(a), the ENP has a high-efficiency particulate air filter for

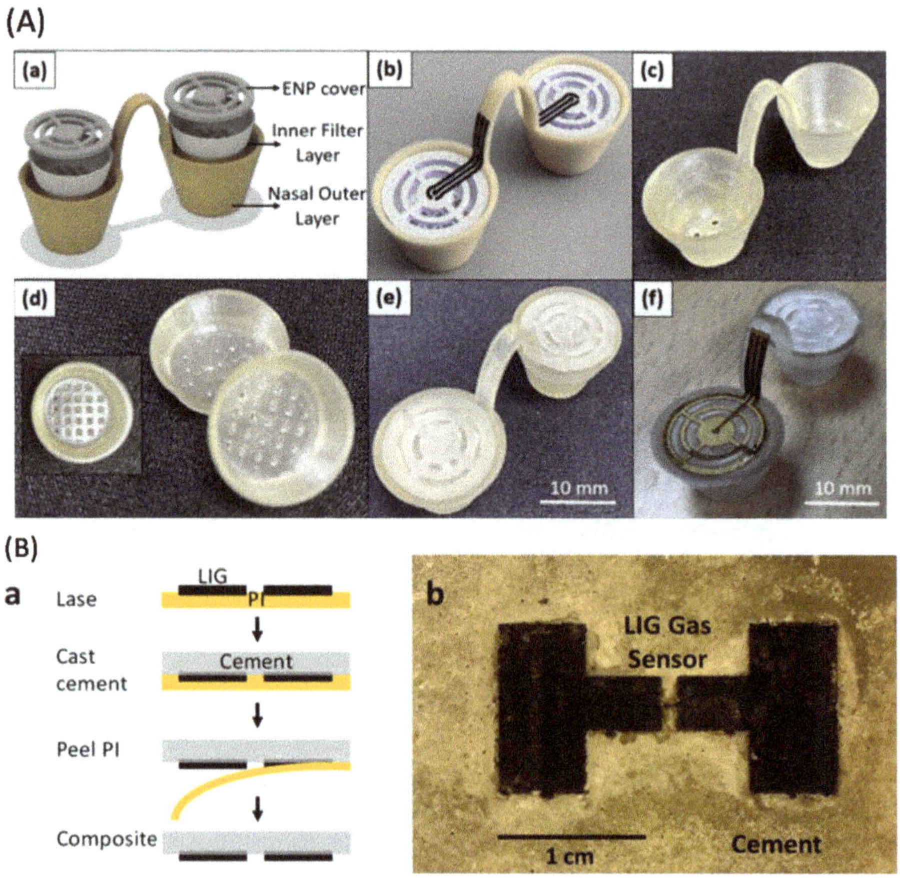

Figure 2.9. (a)–(d) Schematic and real images of the devices. (a) © 2020 IEEE. Reprinted, with permission, from [89]. (b) Reprinted with permission from [84]. Copyright (2019) American Chemical Society. (c) Reprinted with permission from [90]. Copyright (2019) American Chemical Society. (d) © 2020 IEEE. Reprinted, with permission, from [25].

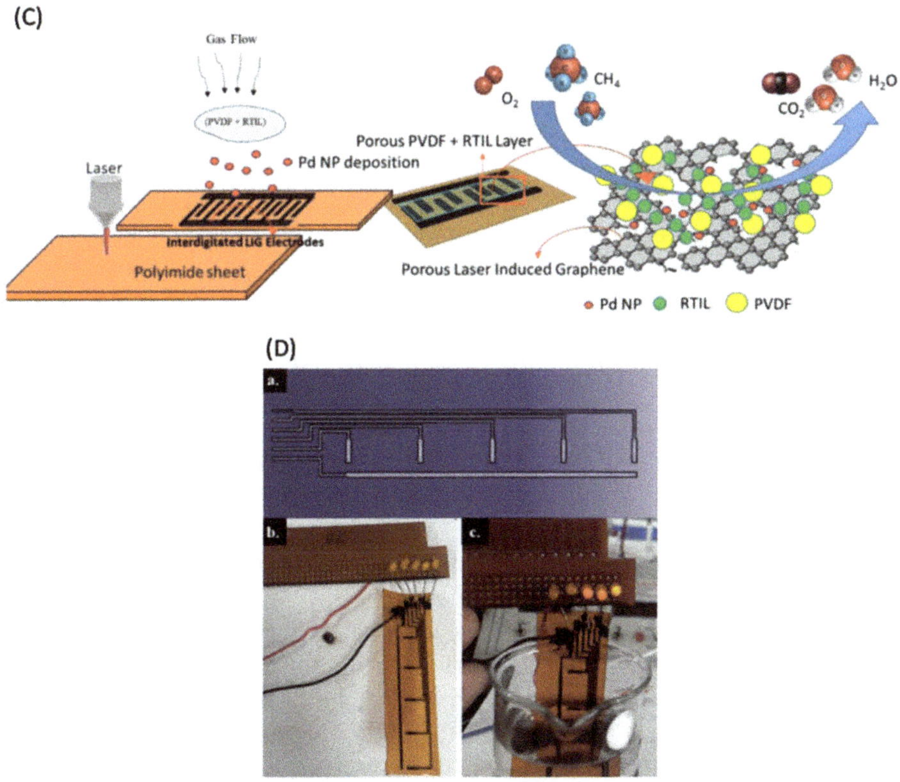

Figure 2.9. (Continued.)

filtering and three LIG electrodes for electrochemical sensing. One of the electrodes was coated with Ag/AgCl ink, which was used as a reference electrode; bare LIG electrodes were used as working and counter electrodes. The fabricated device was used for electrochemical detection of oxygen. The experiments were done using a potentiostat, cyclic voltammetry, in a potential window of −1 V to +1 V, with a scan rate of 10 mV s^{-1} in a phosphate buffer solution of 0.1 M concentration purged with various oxygen concentrations. A total 100 μL sample volume was used for testing. A sharp reduction peak corresponding to electrocatalytic reduction of oxygen was observed, which was not present in the absence of oxygen [89]. Likewise, LIG-based gas sensors were developed by Michael *et al* for determination of different concentrations of CO_2 and N_2 in the presence of noble gases helium and argon. As shown in figure 2.9(b), across the LIG channel a potential was applied, which applied heating to the LIG and, thus, thermal conductivity of an array of gases. This sensor was embedded in a cement substrate to give robust structure to the device for real-time application [84]. Further, a methane sensor based upon interdigitated LIG electrodes decorated with palladium nanoparticles was reported by Manan *et al.* An ionic liquid and polyvinylidene fluoride electrolyte was used and amperometric studies were performed by varying methane concentrations. Figure 2.9(c) shows a schematic diagram of their device [90]. Yang *et al* fabricated a unique gas-sensing platform using a stretchable LIG platform with a self-heating property. Further, the device was used for NO_2 sensing [91]. Yen *et al* fabricated a MoS_2 and LIG electrodes combination for NO_2 sensing. The effect of the morphology of the structure was also studied [92].

Liquid sensors LIG has also shown potential in application as a liquid sensor. Our group has also reported a LIG-based liquid-level sensor wherein a positive terminal is placed in a straight length and negative terminals are joined together with five intervals on a horizontal plain. Further, LED lights were attached as indicators along with a 9 V battery. Figure 2.9(d) shows the device fabricated by Kothuru *et al* [25]. Likewise, Yanan *et al* prepared a paper-based liquid sensor using LIG and changed its shape and structure, and analyzed the performance [93].

2.3.4.4 Humidity sensors

Lan *et al* recently reported a single-step, LIG-based flexible and wearable humidity sensor and further applied it for plant transpiration monitoring in real time. A semiconductor diode laser of 450 nm wavelength, 3085 W power and 2 cm s^{-1} speed was used to scribe LIG IDE over a PI sheet. The IDE was 500 µm wide, with electrode gaps set to 100, 300 and 500 µm. For real-time application, *Epipremnumaureum* was used. The sensor was attached at the backside of a leaf with tape, and the plant was maintained and allowed to grow at a constant incubator temperature [94]. Standford *et al* reported a small-size LIG humidity sensor which was not visible to the naked eye. These were fabricated over a PI sheet using a 405 nm laser. The sensor was further used to detect humidity in real human breath. A response time of 250 ms was observed by the sensor [95]. Zhu *et al* developed a respiration flexible electronic detector device based on LIG porous electrodes. This humidity sensor showed fast response and high stability, and exhibited excellent response within a relative humidity range of 11%–97%. In addition, this sensor was applied for real-time analysis of breathing and non-contact fingertip humidity monitoring. Thus, this prepared sensor had the capability of being used as a wearable electronic device in health-care management [96]. Recently, Lee *et al* developed LIG over a PI sheet with a 355 nm UV pulse laser. The LIG obtained had high surface area and was porous in nature, and hence could be utilized as a humidity sensor. This sensor was further attached to the inner side of a mask filter. Changes in the capacitance and resistance were used as indicators for humidity measurements [97]. Li *et al* fabricated a LIG-based humidity sensor by direct laser scribing over a low-power DVD drive coated with a graphite oxide layer. The resulting sensor was tested for sensing relative humidity in the 19%–97.4% range with rapid response and excellent recovery value [98]. Lee *et al* prepared hair-like LIG patterns over a PI sheet using a 355 nm pulsed laser. This method gave a porous multilayered pattern with defects that were rich in oxygen groups. Over this designed pattern, silver paste was applied on both sides, which were used as electrodes. Further, the prepared sensor was checked for real-time application in a water vapor box. Water vapor was let inside the chamber through a humidifier and the response was recorded. The electrode demonstrated excellent sensitivity [99]. Thamaraiselvan *et al* reported a review article wherein they gave recent updates about a composite of LIG and CNTs as conductive carbon material for environmental monitoring sensors for temperature and humidity as well as various environmental pollutants, nutrients, metals, dew-points, biochemicals, etc [100].

Pressure, strain and capacitive sensors. LIG is also being used to develop flexible pressure and strain sensors. Tian *et al* developed an excellent pressure sensor inspired by the shape of bean pod's structure. Two layers of LIGs were used and polystyrene beads were sandwiched between them. The developed sensor was extremely sensitive and demonstrated the property of self-healing. It showed sensing up to 100 kPa. For practical application, the sensor was subjected to arterial monitoring and gave good results [101]. Chhetry *et al* prepared a strain sensor using a piezoresistive method by chemically modifying LIG with MoS_2. A PI sheet was coated with a MoS_2 layer and then subjected to a laser to obtain a MoS_2-embedded LIG. A remarkably sensitive sensor was obtained [102]. Jeong *et al* reported a LIG-based strain sensor over PDMS using a 355 nm pulsed laser. The sensor was flexible and was prepared by a single-step process and gave a high gauge value of ~160. It was also useful for human finger motion monitoring [103]. Our group has reported LIG-based strain and capacitive sensors. Kothuru *et al*, from our group, developed a LIG-based capacitive touch pad. As shown in figure 2.10(a), a touch pad sensor was designed and printed over a PI sheet using a CO_2 laser, and the capacitance and resistance were measured using an LCR meter. The sensor was mounted over MPR121, which was used for charge transfer reading with a single-wire electrode. When the electrode was touched, the capacitance increased and the MPR121 microcontroller results were showcased on the monitor [25]. The same author reported the fabrication of LIG over a phenolic resin and alcohol composite sheet using a CO_2 laser. As illustrated in figure 2.10(b), a mixture of phenolic resin powder, polyvinyl alcohol and deionized water in appropriate proportions was made and stirred for 24 hr at 60 °C. The mixture was poured over a petri dish and left to form a sheet after drying. After this, the sheet was ablated with a laser to get LIG. This prepared LIG was used for fabricating the capacitive touch pad sensor. Here the sensor showed significant change upon being touched by the user. The change in capacitance was obtained as a readout on a computer monitor. Furthermore, as shown in figure 2.10(c), the same LIG was used for fabricating a pressure sensor, in which a change in internal resistance was measured as a function of pressure application [104].

Sindhu *et al*, from our group, reported a LIG-based wearable antenna, operated at 5.8 GHz, which was further utilized for monitoring human motion wirelessly. It also showed increased sensitivity of 14.08 for compressive and 11.34 for tensile strength. As shown in figure 2.10(d), human motion in real time was monitored using this sensor by adhering it to a human hand. Further, it was Internet-of-Things (IoT) enabled for wireless monitoring. Owing to the porous nature of LIG, upon applying strain the pore dimension changed, leading to a change in frequency. While applying compressive strain, the pores became more compact and hence changes in conductivity took place leading to a change in frequency. When applying tensile strain, the pores were enlarged, leading to lesser conductivity and hence lowering frequency [105]. Kaidarova *et al* reported a LIG-based flexible pressure sensor. The performance of this sensor

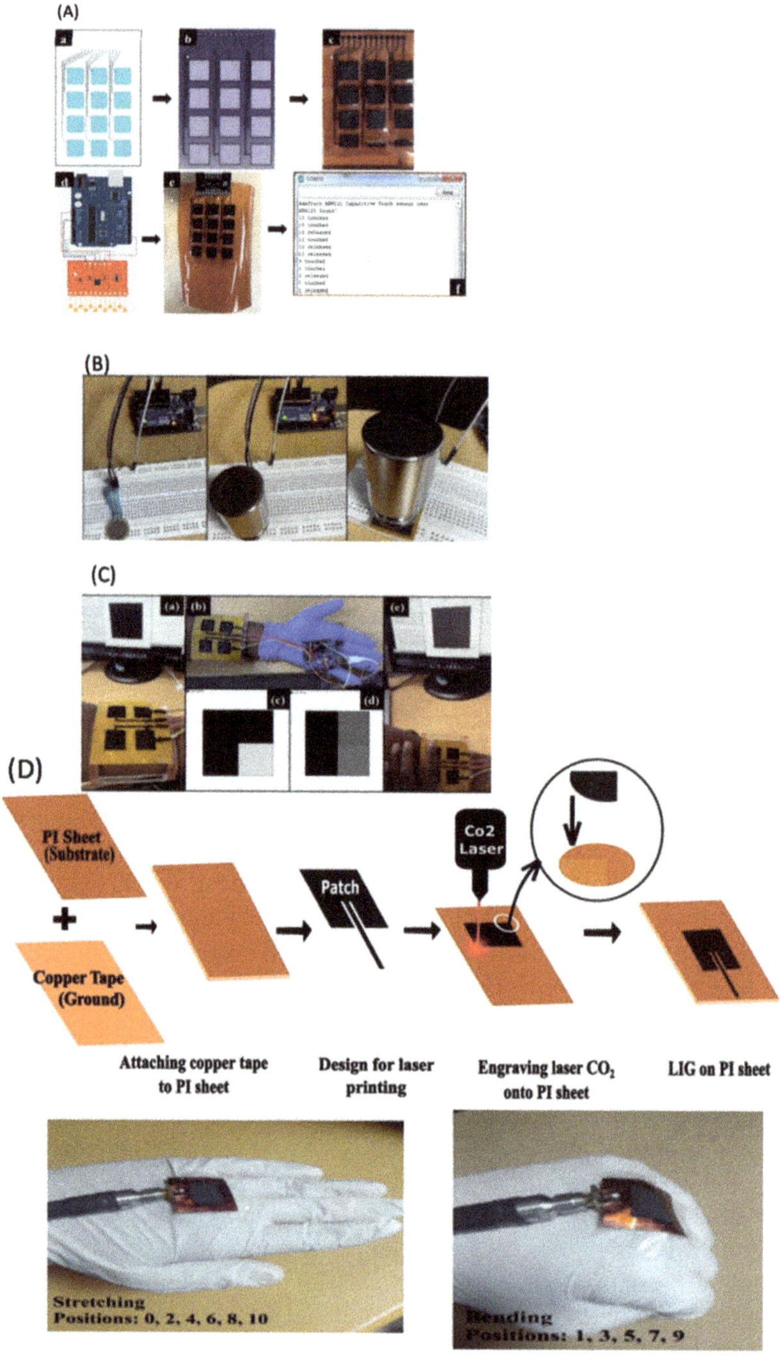

Figure 2.10. (a) Real image of the capacitive touch sensor. © 2020 IEEE. Reprinted, with permission, from [25]. (b) and (c) Real image of the pressure sensor and capacitive sensor. Reproduced from [104] © IOP Publishing Ltd. All rights reserved. (d) Schematic and real image of the strain antenna sensor. © 2021 IEEE. Reprinted, with permission, from [105].

depended upon the geometrical parameters of the sensor. It showed sensitivity in the range of 1.23×10^{-3} kPa with a LoD of 10 Pa. They also checked underwater and for the biocompatibility of the sensor with cytotoxicity [106].

Huang *et al* reported a wearable, flexible LIG-based strain sensor with a 3D silicone rubber and LIG. LIG was fabricated over a PI sheet. The sensor could sense stain up to a 37.8 gauge factor in a range from 0% to 31.8%. The sensor was robust and stable, and could be further used for virtual or augmented electronic or real skin wearable devices [107]. Groo *et al* reported a LIG-based strain sensor to be used in fiberglass composites for detection of damage due to strain. LIG was embedded with piezoresisitivity inside fiberglass composites and such a piezoresistive interlayer monitored the strain [108].

2.4 LIG in wearable and smart devices

Owing to its flexible nature and one-step fabrication process, LIG has been widely adapted for use in smart and wearable electronic devices. Many research groups have also reported devices such as an artificial nose, electronic skin, intelligent throat, etc, using LIG. Generally, this is accomplished by measuring changes in piezoresistance. For example, Tao *et al* reported an excellent LIG wearable device as an intelligent artificial throat. The device could not only sense but also generate sound. Simple sounds like humming, screaming and coughing with various intensities could be detected and further tuned into controlled noises. A 450 nm laser was used for formation of LIG over a PI sheet. The sound testing platform used was a microphone and signal analyzer. Different LIG samples were prepared for detecting different frequency sounds. When the sensor was attached to the throat, the vibrations of the throat cords changed the resistance. The signals were detected in combination with machine learning. This could be useful for physically challenged persons. Figure 2.11(a) shows their device [35]. Sun *et al* developed on-skin electronic devices based on LIG. The prepared LIG sensors were attached to the chest, forehead and finger of a human to measure the electrophysiological processes such as electrocardiogram (ECG), electroencephalogram and electromayogram (EMG). The fabricated LIGs were mixed with an elastomer to improve their mechanical stretchability. The sensors mounted at the forehead gave an alpha rhythm of 10 Hz frequency, indicating that waves from the brain were recorded. ECG waves, T and P, and QRS complex were also recorded. Furthermore, the EMG signals changed significantly with movement of the finger. Thus, this smart sensor acted as a human–machine interface device [109]. Zhang *et al* developed a sound device with flexible LIG over a PI sheet with laser of a 10.6 μM. The thermoacoustic measurements were done using a microphone, generator and signal analyzer. The sensor was mounted on a cylinder. The pressure of the sound was up to 49 dB and was stable for up to 60 min [110]. Xu *et al* recently reported a wearable LIG-based sensor for wirelessly monitoring body conditions like respiration rate, sleeping posture and moisture in the diapers of infants, elderly and challenged people. The sensor could give feedback notifications in the form of alarms. The device has a tilted sensor with metal liquid droplets in a cavity which can

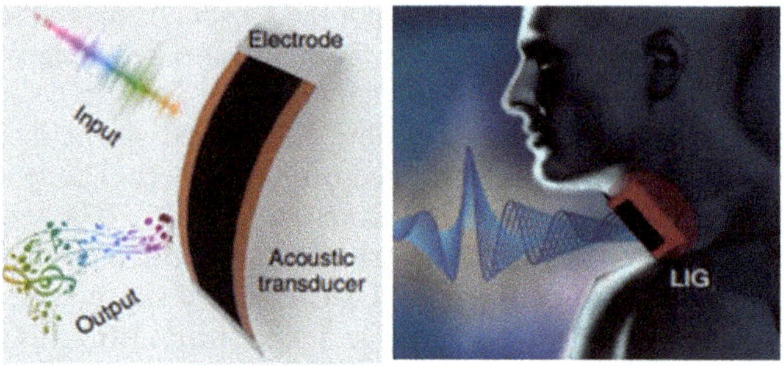

Figure 2.11. Reprint of the images of the devices from [35]. Copyright © 2017, The Author(s). With permission of Springer.

detect about 18 slant positions. Furthermore, strain sensing and humidity sensing can also be done simultaneously; hence this device acted as a multiplex wearable sensor [111]. Huang *et al* presented a LIG-based health-monitoring device that gave notifications. A single-step preparation is reported wherein the electrothermal effect of the LIG is monitored. This device was used to monitor respiration rate. Breathing led to a change in temperature periodically, which also gave change in the thermal energy. Thus, abnormal breathing gave sudden rise in temperatures and the device could give alarm notifications. This could be highly useful for critical patients and those with apnea [112].

2.5 Conclusion and future outlook

It is noteworthy that since its discovery, in 2014, LIG has been extensively explored in a diverse range of applications. Owing to its remarkable chemical and physical properties and its flexible nature, usage of LIG in wearable electronic devices, sensors, point-of-care devices, lab-on-chip, etc, has been significantly reported upon. Such versatility in the applications of LIG is due to its customizable and modifiable properties. A single-step preparation also eases the fabrication process. Furthermore, LIG can be chemically modified and integrated with other metallic, non-metallic, organic and inorganic nanoparticles, molecules, biomolecules, bio-chemicals, etc, to enhance their electrical conductivity, mechanical strength, stability and longtime durability. In addition, LIG can also be incorporated with implantable materials for biomimicking studies. The current emerging trend of LIG in wearable devices is due to the rapid increase in the demand for wearable electronics for consumer products in health monitoring, environmental protection and robotics. Its success is due to the unique bendable nature and fracture resistance of LIG. Physical sensors like temperature, sound, humidity, strain, pressure and bodily signal sensors based on LIG have been developed. Due to the light weight of LIG, these sensors are compact and portable. In addition, the cost effectivity, especially in large-scale industrial-level fabrication of LIG-based sensors, has promoted the development of LIG electronic devices. Bare and pristine LIG developed over polymer substrates are flexible and can also easily be transferred to other substrates with simple adhesive and lamination, hence it is highly useful for smart sensing devices.

Studies have shown that the speed, power and type of laser used play a central role in the properties of the LIG. From sensors to electronic circuits and other display devices, LIG has been adapted to their requirements. Wireless detection and incorporation and enabling of the IoT is the future of LIG-based sensors. For example, Yang *et al* reported a wearable sensor for detection of biomarkers in sweat. They integrated a microcontroller and printed circuit board with their LIG sensor. This helped them to record the signals wirelessly, further coverting the digital signal into an analog result. Thus, this could lead to a non-invasive method of detection [113]. Similarly, Alahi *et al* developed a microcontroller-based, wireless LIG nitrate sensor for practical application [114].

This chapter highlighted the basic properties of LIG, its fabrication methods using different lasers and recent advances reported in their applications. These include temperature controllers, environment quality sensors such as for air quality and humidity, physical sensors like strain, pressure, liquid-level and sound detection, and electrochemical sensors for various analytes, biosensors based on LIG and immunosensors. Further, other applications like energy harvesting in miniaturized LIG devices such as fuel cells, energy-storage devices like micro-supercapacitors and miniaturized, ECL-based sensor micro-devices for multiplexed sensing have been discussed. Upcoming wearable, non-invasive, physiological sensors have also been described.

Despite the fact that these developed LIG-based electronic devices have shown their remarkable applicability and feasibility in real-time applications, there are certain research gaps and limitations that have to be addressed in future to improve the scope of LIG applications for practical usage. For instance, bonding of LIG with substrates for its incorporation in devices is quite challenging. For this, ways to transfer LIG into the substrate are presently being explored. LIG–human interfaces, especially in implantable devices, presently has been little explored. Study of cytotoxicity and biocompatibility should be tested. For wireless monitoring and accurate measurement, smart sensing, IoT incorporation and data management are crucial but challenging steps. Furthermore, adverse and complex environmental conditions need to checked for better real-time monitoring in the case of IoT. Another challenge for real-time application is selective sensing. The reported pristine LIG-based sensors are found to be sensitive toward humidity, sound, temperature and signals. Therefore, this could give cross-reactivity or physical sensors might give misleading outputs in non-ambient environments.

Nevertheless, despite these limitations, LIG is still feasible for production at large scale. The emerging trends in the fabrication and transferring of LIG, its incorporation into the IoT as well as chemical modifications of LIG have proven worthwhile in rewards for overcoming these challenges. The future scope of LIG-based electronic devices is anticipated to be quite significant both economically as well as in terms of the products rolled-out to market.

References

[1] Rao C N R, Satishkumar B C, Govindaraj A and Nath M 2001 Nanotubes *ChemPhysChem.* **2** 78–105

[2] Geim A K 2009 Graphene: status and prospects *Science* **324** 1530–5

[3] Liu J *et al* 1998 Fullerene pipes *Science* **280** 1253–6

[4] Suzuki M 1994 Activated carbon fiber: fundamentals and applications *Carbon* **32** 577–86

[5] Chung D D L 2002 Review: graphite *J. Mater. Sci.* **37** 1475–89

[6] Wen J, Xu Y, Li H, Lu A and Sun S 2015 Recent applications of carbon nanomaterials in fluorescence biosensing and bioimaging *Chem. Commun.* **51** 11346–358

[7] Loh K P, Ho D, Chiu G N C, Leong D T, Pastorin G and Chow E K H 2018 Clinical applications of carbon nanomaterials in diagnostics and therapy *Adv. Mater.* **30** 1–21

[8] Notarianni M, Liu J, Vernon K and Motta N 2016 Synthesis and applications of carbon nanomaterials for energy generation and storage *Beilstein J. Nanotechnol.* **7** 149–96

[9] Verma S K, Das A K, Gantait S, Kumar V and Gurel E 2019 Applications of carbon nanomaterials in the plant system: a perspective view on the pros and cons *Sci. Total Environ.* **667** 485–99

[10] Hayashi T, Kim Y A, Natsuki T and Endo M 2007 Mechanical properties of carbon nanomaterials *ChemPhysChem.* **8** 999–1004

[11] Papageorgiou D G, Kinloch I A and Young R J 2017 Mechanical properties of graphene and graphene-based nanocomposites *Prog. Mater Sci.* **90** 75–127

[12] Wang F, Wang K, Zheng B, Dong X, Mei X, Lv J, Duan W and Wang W 2018 Laser-induced graphene: preparation, functionalization and applications *Mater. Technol.* **33** 340–56

[13] Rewatkar P, Kothuru A and Goel S PDMS-based microfluidic glucose biofuel cell integrated with optimized laser-induced flexible graphene bioelectrodes *IEEE Trans. Electron Dev.* **67** 1832–8

[14] Nicholl R J T, Conley H J, Lavrik N V, Vlassiouk I, Puzyrev Y S, Sreenivas V P, Pantelides S T and Bolotin K I 2015 The effect of intrinsic crumpling on the mechanics of free-standing graphene *Nat. Commun.* **6** 1–7

[15] Lee C, Wei X, Kysar J W and Hone J 2008 Measurement of the elastic properties and intrinsic strength of monolayer graphene *Science* **321** 385–8

[16] Zandiatashbar A, Lee G H, An S J, Lee S, Mathew N, Terrones M, Hayashi T, Picu C R, Hone J and Koratkar N 2014 Effect of defects on the intrinsic strength and stiffness of graphene *Nat. Commun.* **5** 1–9

[17] Zhang P *et al* 2014 Fracture toughness of graphene *Nat. Commun.* **5** 1–7

[18] Chen J H, Jang C, Xiao S, Ishigami M and Fuhrer M S 2008 Intrinsic and extrinsic performance limits of graphene devices on SiO_2 *Nat. Nanotechnol.* **3** 206–9

[19] Berry V 2013 Impermeability of graphene and its applications *Carbon* **62** 1–10

[20] Sheehy D E and Schmalian J 2009 Optical transparency of graphene as determined by the fine-structure constant *Phys. Rev. B: Condens. Matter. Mater. Phys* **80** 2–5

[21] Mamleyev E R, Heissler S, Nefedov A, Weidler P G, Nordin N, Kudryashov V V, Länge K, MacKinnon N and Sharma S 2019 Laser-induced hierarchical carbon patterns on polyimide substrates for flexible urea sensors *NPJ Flex. Electron.* **3** 2

[22] Lee X J, Hiew B Y Z, Lai K C, Lee L Y, Gan S, Thangalazhy-Gopakumar S and Rigby S 2019 Review on graphene and its derivatives: synthesis methods and potential industrial implementation *J. Taiwan Inst. Chem. Eng.* **98** 163–80

[23] Caballero-Lucas F, Florian C, Fernández-Pradas J M, Morenza J L and Serra P 2015 Precise surface modification of polymethyl-methacrylate with near-infrared femtosecond laser *Appl. Surf. Sci.* **336** 170–5

[24] Wang L, Wang Z, Bakhtiyari A N and Zheng H 2020 A comparative study of laser-induced graphene by CO_2 infrared laser and 355 nm ultraviolet (UV) laser *Micromachines* **11** 1–9

[25] Kothuru A, Rao C H, Puneeth S B, Salve M, Amreen K and Goel S 2020 Laser-induced flexible electronics (LIFE) for sensing applications *IEEE Sensors J.* **20** 7392–9

[26] El-Kady M F, Strong V, Dubin S and Kaner R B 2012 Laser scribing of high-performance and flexible graphene-based electrochemical capacitors *Science* **335** 1326–30

[27] Shi X, Zhou F, Peng J, Wu R, Wu Z S and Bao X 2019 One-step scalable fabrication of graphene-integrated micro-supercapacitors with remarkable flexibility and exceptional performance uniformity *Adv. Funct. Mater.* **29** 1–9

[28] Kurra N, Sagade A A and Kulkarni G U 2011 Ultrafast direct ablative patterning of HOPG by single laser pulses to produce graphene ribbons *Adv. Funct. Mater.* **21** 3836–42

[29] Zhu Y, Cai H, Ding H, Pan N and Wang X 2019 Fabrication of low-cost and highly sensitive graphene-based pressure sensors by direct laser scribing polydimethylsiloxane *ACS Appl. Mater. Interfaces* **11** 6195–200

[30] Garland N T, McLamore E S, Cavallaro N D, Mendivelso-Perez D, Smith E A, Jing D and Claussen J C 2018 Flexible laser-induced graphene for nitrogen sensing in soil *ACS Appl. Mater. Interf.* **10** 39124–33

[31] Wakaya F, Kurihara T, Yurugi N, Abo S, Abe M and Takai M 2015 Maskless laser processing of graphene *Microelectron. Eng.* **141** 203–6

[32] Lin S Y, Zhang T Y, Lu Q, Wang D Y, Yang Y, Wu X M and Ren T L 2017 High-performance graphene-based flexible heater for wearable applications *RSC Adv.* **7** 27001–6

[33] Arul R, Oosterbeek R N, Robertson J, Xu G, Jin J and Simpson M C 2016 The mechanism of direct laser writing of graphene features into graphene oxide films involves photo-reduction and thermally assisted structural rearrangement *Carbon* **99** 423–31

[34] Mohammad M A *et al* 2016 Tunable graphene oxide reduction and graphene patterning at room temperature on arbitrary substrates *Carbon* **109** 173–81

[35] Tao L Q *et al* 2017 An intelligent artificial throat with sound-sensing ability based on laser-induced graphene *Nat. Commun.* **8** 1–8

[36] Lin J, Peng Z, Liu Y, Ruiz-Zepeda F, Ye R, Samuel E L G, Yacaman M J, Yakobson B I and Tour J M 2014 Laser-induced porous graphene films from commercial polymers *Nat. Commun.* **5** 1–8

[37] Rahimi R, Ochoa M, Yu W and Ziaie B 2015 Highly stretchable and sensitive unidirectional strain sensor via laser carbonization *ACS Appl. Mater. Interfaces* **7** 4463–70

[38] Lamberti A, Serrapede M, Ferraro G, Fontana M, Perrucci F, Bianco S, Chiolerio A and Bocchini S 2017 All-SPEEK flexible supercapacitor exploiting laser-induced graphenization *2D Mater.* **4** 035012

[39] Chyan Y, Ye R, Li Y, Singh S P, Arnusch C J and Tour J M 2018 Laser-induced graphene by multiple lasing: toward electronics on cloth, paper, and food *ACS Nano.* **12** 2176–83

[40] Ye R, Han X, Kosynkin D V, Li Y, Zhang C, Jiang B, Martí A A and Tour J M 2018 Laser-induced conversion of teflon into fluorinated nanodiamonds or fluorinated graphene *ACS Nano.* **12** 1083–8

[41] Park J B, Xiong W, Gao Y, Qian M, Xie Z Q, Mitchell M, Zhou Y S, Han G H, Jiang L and Lu Y F 2011 Fast growth of graphene patterns by laser direct writing *Appl. Phys. Lett.* **98** 16–9

[42] Ye R, Chyan Y, Zhang J, Li Y, Han X, Kittrell C and Tour J M 2017 Laser-induced graphene formation on wood *Adv. Mater.* **29** 1–7

[43] Huang T, Long J, Zhong M, Jiang J, Ye X, Lin Z and Li L 2013 The effects of low power density CO_2 laser irradiation on graphene properties *Appl. Surf. Sci.* **273** 502–6

[44] Kiisk V, Kahro T, Kozlova J, Matisen L and Alles H 2013 Nanosecond laser treatment of graphene *Appl. Surf. Sci.* **276** 133–7

[45] Peng Z, Ye R, Mann J A, Zakhidov D, Li Y, Smalley P R, Lin J and Tour J M 2015 Flexible boron-doped laser-induced graphene microsupercapacitors *ACS Nano.* **9** 5868–75

[46] Ye R, Peng Z, Wang T, Xu Y, Zhang J, Li Y, Nilewski L G, Lin J and Tour J M 2015 *In situ* formation of metal oxide nanocrystals embedded in laser-induced graphene *ACS Nano.* **9** 9244–51

[47] Li Y, Luong D X, Zhang J, Tarkunde Y R, Kittrell C, Sargunaraj F, Ji Y, Arnusch C J and Tour J M 2017 Laser-induced graphene in controlled atmospheres: from superhydrophilic to superhydrophobic surfaces *Adv. Mater.* **29** 1–8

[48] Zhang J, Ren M, Li Y and Tour J M 2018 *In situ* synthesis of efficient water oxidation catalysts in laser-induced graphene *ACS Energy Lett.* **3** 677–83

[49] Li L *et al* 2016 High-performance pseudocapacitive microsupercapacitors from laser-induced graphene *Adv. Mater.* **28** 838–45

[50] Rewatkar S G P and Kothuru A , 1967 Laser induced flexible graphene bioelectrodes for ezymatic fuel cells *Angew. Chem. Int. Ed.* **6** 951–52

[51] Rewatkar J U S P and Goel S 2021 Miniaturized polymeric enzymatic biofuel cell with integrated microfluidic device and enhanced laser ablated bioelectrodes *Int. J. Hydrogen Energy* **46** 3183–92

[52] Rao L T, Dubey S K, Javed A and Goel S 2021 Laser-induced graphene electrodes enhanced with carbon nanotubes for membraneless microfluidic fuel cell *Sustain. Energy Technol. Assess.* **45** 101176

[53] Jayapiriya U S, Rewatkar P and Goel S Direct electron transfer based microfluidic glucose biofuel cell with CO_2 laser ablated bioelectrodes and microchannel *IEEE Trans. Nanobiosci.* **1241** 1–1

[54] Song W, Zhu J, Gan B, Zhao S, Wang H, Li C and Wang J 2018 Flexible, stretchable, and transparent planar microsupercapacitors based on 3D porous laser-induced graphene *Small* **14** 1–7

[55] Wang W, Lu L, Xie Y, Mei X, Tang Y, Wu W and Liang R 2020 Tailoring the surface morphology and nanoparticle distribution of laser-induced graphene/Co_3O_4 for high-performance flexible microsupercapacitors *Appl. Surf. Sci.* **504** 144487

[56] Zhang C, Peng Z, Huang C, Zhang B, Xing C, Chen H, Cheng H, Wang J and Tang S 2021 High-energy all-in-one stretchable micro-supercapacitor arrays based on 3D laser-induced graphene foams decorated with mesoporous ZnP nanosheets for self-powered stretchable systems *Nano Energy* **81** 105609

[57] Li G, Meng Z, Qian J, Ho C L, Lau S P, Wong W Y and Yan F 2019 Inkjet printed pseudocapacitive electrodes on laser-induced graphene for electrochemical energy storage *Mater. Today Energy* **12** 155–60

[58] Wen F, Hao C, Xiang J, Wang L, Hou H, Su Z, Hu W and Liu Z 2014 Enhanced laser scribed flexible graphene-based micro-supercapacitor performance with reduction of carbon nanotubes diameter *Carbon* **75** 236–43

[59] Bobinger M R, Romero F J, Salinas-Castillo A, Becherer M, Lugli P, Morales D P, Rodríguez N and Rivadeneyra A 2019 Flexible and robust laser-induced graphene heaters photothermally scribed on bare polyimide substrates *Carbon* **144** 116–26
[60] Cho E C, Chang-Jian C W, Huang J H, Wu R T, Lu C Z, Lee K C, Weng H C and Hsu S C 2021 Laser scribing of Ag-decorated graphene for high-performance and flexible heaters *J. Taiwan Inst. Chem. Eng.* **119** 224–31
[61] Chen J, Wang Y, Liu F and Luo S 2020 Laser-induced graphene paper heaters with multimodally patternable electrothermal performance for low-energy manufacturing of composites *ACS Appl. Mater. Interfaces* **12** 23284–97
[62] Srikanth S, Dudala S, Jayapiriya U S, Mohan J M, Raut S, Dubey S K, Ishii I, Javed A and Goel S 2021 Droplet-based lab-on-chip platform integrated with laser ablated graphene heaters to synthesize gold nanoparticles for electrochemical sensing and fuel cell applications *Sci. Rep.* **11** 1–12
[63] Kothuru A, Amreen K and Goel S 2020 Electromicrofluidic device on multilayered laser-induced polyamide substrate for diverse electrochemical applications *IEEE Trans. Electron Dev.* **67** 5097–103
[64] Sharma S, Ganeshan S K, Pattnaik P K, Kanungo S and Chappanda K N 2020 Laser induced flexible graphene electrodes for electrochemical sensing of hydrazine *Mater. Lett.* **262** 127150
[65] Zhang C, Ping J and Ying Y 2020 Evaluation of trans-resveratrol level in grape wine using laser-induced porous graphene-based electrochemical sensor *Sci. Total Environ.* **714** 136687
[66] de Araujo W R, Frasson C M R, Ameku W A, Silva J R, Angnes L and Paixão T R L C 2017 Single-step reagentless laser scribing fabrication of electrochemical paper-based analytical devices *Angew. Chem.—Int. Ed.* **56** 15113–7
[67] Lin S, Feng W, Miao X, Zhang X, Chen S, Chen Y, Wang W and Zhang Y 2018 A flexible and highly sensitive nonenzymatic glucose sensor based on DVD-laser scribed graphene substrate *Biosens. Bioelectron.* **110** 89–96
[68] Tehrani F and Bavarian B 2016 Facile and scalable disposable sensor based on laser engraved graphene for electrochemical detection of glucose *Sci. Rep.* **6** 1–10
[69] Zhang Z, Song M, Hao J, Wu K, Li C and Hu C 2018 Visible light laser-induced graphene from phenolic resin: a new approach for directly writing graphene-based electrochemical devices on various substrates *Carbon* **127** 287–96
[70] Aparicio-Martínez E, Ibarra A, Estrada-Moreno I A, Osuna V and Dominguez R B 2019 Flexible electrochemical sensor based on laser scribed Graphene/Ag nanoparticles for non-enzymatic hydrogen peroxide detection *Sens, Actuators B: Chem.* **301** 127101
[71] Yoon H, Nah J, Kim J, Xuan X and Park J 2019 Laser-induced graphene stamp for high performance electrochemical sensing applications *Proc. of the IEEE Int. Conf. on Micro Electro Mechanical Systems (MEMS)* 537–40
[72] Chang T L and Chen Z C 2015 Surface patterning of multilayer graphene by ultraviolet laser irradiation in biomolecule sensing devices *Appl. Surf. Sci.* **359** 543–9
[73] Wan Z, Nguyen N T, Gao Y and Li Q 2020 Laser-induced graphene for biosensors *Sustain. Mater. Technol.* **25** e00205
[74] Vanegas D C, Patiño L, Mendez C, de Oliveira D A, Torres A M, Gomes C L and McLamore E S 2018 Laser scribed graphene biosensor for detection of biogenic amines in food samples using locally sourced materials *Biosensors* **8** 42

[75] Nayak P, Kurra N, Xia C and Alshareef H N 2016 Highly efficient laser scribed graphene electrodes for on-chip electrochemical sensing applications *Adv. Electron. Mater.* **2** 1600185
[76] Ge L, Hong Q, Li H and Li F 2019 A laser-induced TiO_2-decorated graphene photoelectrode for sensitive photoelectrochemical biosensing *Chem. Commun.* **55** 4945–8
[77] Hui X, Xuan X, Kim J and Park J Y 2019 A highly flexible and selective dopamine sensor based on Pt–Au nanoparticle-modified laser-induced graphene *Electrochim. Acta* **328** 135066
[78] Nasraoui S, Al-Hamry A, Anurag A, Teixeira P R, Ameur S, Paterno L G, Ben Ali M and Kanoun O 2019 Investigation of laser-induced graphene electrodes modified by MWNT/AuNPs for detection of nitrite *16th Int. Multi-conference on Systems, Signals and Devices, SSD 2019* 615–20
[79] Lu Z, Lin X, Zhang J, Dai W, Liu B, Mo G, Ye J and Ye J 2019 Ionic liquid/poly-L-cysteine composite deposited on flexible and hierarchical porous laser-engraved graphene electrode for high-performance electrochemical analysis of lead ion *Electrochim. Acta* **295** 514–23
[80] Rahimi R, Ochoa M, Tamayol A, Khalili S, Khademhosseini A and Ziaie B 2017 Highly stretchable potentiometric pH sensor fabricated via laser carbonization and machining of carbon–polyaniline composite *ACS Appl. Mater. Interfaces* **9** 9015–23
[81] Xu G, Jarjes Z A, Desprez V, Kilmartin P A and Travas-Sejdic J 2018 Sensitive, selective, disposable electrochemical dopamine sensor based on PEDOT-modified laser scribed graphene *Biosens. Bioelectron.* **107** 184–91
[82] Fenzl C, Nayak P, Hirsch T, Wolfbeis O S, Alshareef H N and Baeumner A J 2017 Laser-scribed graphene electrodes for aptamer-based biosensing *ACS Sens.* **2** 616–20
[83] You Z, Qiu Q, Chen H, Feng Y, Wang X and Wang Y 2019 Biosensors and bioelectronics laser-induced noble metal nanoparticle-graphene composites enabled flexible biosensor for pathogen detection *Biosens. Bioelectron.* 111896
[84] Stanford M G, Yang K, Chyan Y, Kittrell C and Tour J M 2019 Laser-induced graphene for flexible and embeddable gas sensors *ACS Nano.* **13** 3474–82
[85] Bhaiyya M, Rewatkar P, Salve M, Pattnaik P K and Goel S 2021 Miniaturized electrochemiluminescence platform with laser-induced graphene electrodes for multiple biosensing *IEEE Trans. Nanobiosci.* **20** 79–85
[86] Bhaiyya M L, Pattnaik P K and Goel S 2021 Miniaturized electrochemiluminescence platform with laser-induced graphene-based single electrode for interference-free sensing of dopamine, xanthine, and glucose *IEEE Trans. Instrum. Meas.* **70** 1–8
[87] Bhaiyya M, Pattnaik P K and Goel S 2021 Electrochemiluminescence sensing of vitamin B12 using laser-induced graphene based bipolar and single electrodes in a 3D-printed portable system *Microfluid. Nanofluid.* **25** 1–8
[88] Bhaiyya M, Pattnaik P K and Goel S 2021 Portable electrochemiluminescence platform with laser-induced graphene-based u-shaped bipolar electrode for selective sensing of various analytes *IEEE Trans. Electron Dev.* **68** 2447–54
[89] Kothuru A and Goel S 2020 Electronic nasal pod: a 3D printed device to filter and electrochemically detect pollutants *Proc. of IEEE Sensors* 17–20
[90] Dosi M, Lau I, Zhuang Y, Simakov D S A, Fowler M W and Pope M A 2019 Ultrasensitive electrochemical methane sensors based on solid polymer electrolyte-infused laser-induced graphene *ACS Appl. Mater. Interfaces* **11** 6166–73

[91] Yang L, Yi N, Zhu J, Cheng Z, Yin X, Zhang X, Zhu H and Cheng H 2020 Novel gas sensing platform based on a stretchable laser-induced graphene pattern with self-heating capabilities *J. Mater. Chem.* A **8** 6487–500

[92] Yan W, Yan W, Chen T, Xu J, Tian Q and Ho D 2020 Size-tunable flowerlike MoS_2 nanospheres combined with laser-induced graphene electrodes for NO_2 sensing *ACS Appl. Nano Mater.* **3** 2545–53

[93] Wang Y, Niu Z, Chen J, Zhai Y, Xu Y and Luo S 2019 Freestanding laser-induced graphene paper based liquid sensors *Carbon* **153** 472–80

[94] Lan L, Le X, Dong H, Xie J, Ying Y and Ping J 2020 One-step and large-scale fabrication of flexible and wearable humidity sensor based on laser-induced graphene for real-time tracking of plant transpiration at bio-interface *Biosens. Bioelectron.* **165** 112360

[95] Stanford M G, Zhang C, Fowlkes J D, Hoffman A, Ivanov I N, Rack P D and Tour J M 2020 High-resolution laser-induced graphene. flexible electronics beyond the visible limit *ACS Appl. Mater. Interfaces* **12** 10902–7

[96] Zhu C, Tao L Q, Wang Y, Zheng K, Yu J, Xiandong L, Chen X and Huang Y 2020 Graphene oxide humidity sensor with laser-induced graphene porous electrodes *Sens. Actuators B: Chem.* **325** 128790

[97] Lee J, Shin Y, Yang H and Shin B 2020 A study on the fabrication of laser-induced graphene humidity sensor for mounting on a disposable mask **23** 693–9

[98] Li X, Feng W, Zhang X, Lin S, Chen Y, Chen C, Chen S, Wang W and Zhang Y 2020 Facile fabrication of laser-scribed-graphene humidity sensors by a commercial DVD drive *Sens. Actuators B: Chem.* **321** 128483

[99] Lee J U, Ma Y W, Jeong S Y and Shin B S 2020 Direct fabrication of ultra-sensitive humidity sensor based on hair-like laser-induced graphene patterns *Micromachines* **11** 476

[100] Thamaraiselvan C, Wang J, James D K, Narkhede P, Singh S P, Jassby D, Tour J M and Arnusch C J 2020 Laser-induced graphene and carbon nanotubes as conductive carbon-based materials in environmental technology *Mater. Today* **34** 115–31

[101] Tian Q, Yan W, Li Y and Ho D 2020 Bean pod-inspired ultrasensitive and self-healing pressure sensor based on laser-induced graphene and polystyrene microsphere sandwiched structure *ACS Appl. Mater. Interfaces* **12** 9710–7

[102] Chhetry A, Sharifuzzaman M, Yoon H, Sharma S, Xuan X and Park J Y 2019 MoS_2-decorated laser-induced graphene for a highly sensitive, hysteresis-free, and reliable piezoresistive strain sensor *ACS Appl. Mater. Interfaces* **11** 22531–42

[103] Jeong S Y, Ma Y W, Lee J U, Je G J and Shin B S 2019 Flexible and highly sensitive strain sensor based on laser-induced graphene pattern fabricated by 355 nm pulsed laser *Sensors* **19** 1–12

[104] Kothuru A and Goel S 2020 Laser-induced graphene on phenolic resin and alcohol composite sheet for flexible electronics applications *Flex. Print. Electron.* **5** 042001

[105] Sindhu B, Kothuru A, Sahatiya P, Goel S and Nandi S 2021 Laser-induced graphene printed wearable flexible antenna-based strain sensor for wireless human motion monitoring *IEEE Trans. Electron Dev.* **68** 1–6

[106] Kaidarova A, Alsharif N, Oliveira B N M, Marengo M, Geraldi N R, Duarte C M and Kosel J 2020 Laser-printed, flexible graphene pressure sensors *Glob. Chall.* **4** 2000001

[107] Huang L, Wang H, Wu P, Huang W, Gao W, Fang F, Cai N, Chen R and Zhu Z 2020 Wearable flexible strain sensor based on three-dimensional wavy laser-induced graphene and silicone rubber *Sensors* **20** 1–14

[108] Groo L A, Nasser J, Zhang L, Steinke K, Inman D and Sodano H 2020 Laser-induced graphene in fiberglass-reinforced composites for strain and damage sensing *Compos. Sci. Technol.* **199** 108367

[109] Sun B, McCay R N, Goswami S, Xu Y, Zhang C, Ling Y, Lin J and Yan Z 2018 Gas-permeable, multifunctional on-skin electronics based on laser-induced porous graphene and sugar-templated elastomer sponges *Adv. Mater.* **30** 1–8

[110] Zhang P, Tang X, Pang Y, Bi M, Li X, Yu J, Zhang J, Yuan M and Luo F 2020 Flexible laser-induced-graphene omnidirectional sound device *Chem. Phys. Lett.* **745** 137275

[111] Xu K, Fujita Y, Lu Y, Honda S, Shiomi M, Arie T, Akita S and Takei K 2021 A wearable body condition sensor system with wireless feedback alarm functions *Adv. Mater.* **33**

[112] Huang Y, Tao L Q, Yu J, Li X, Wang Z, Zheng K and Chen X 2020 Health monitoring and automatic notification device based on laser-induced graphene *IEEE Trans. Electron Dev.* **67** 4488–92

[113] Yang Y *et al* 2020 A laser-engraved wearable sensor for sensitive detection of uric acid and tyrosine in sweat *Nat. Biotechnol.* **38** 217–24

[114] Alahi M E E, Nag A, Mukhopadhyay S C and Burkitt L 2018 A temperature-compensated graphene sensor for nitrate monitoring in real-time application *Sens. Actuators A: Phys.* **269** 79–90

Chapter 3

Printable flexible sensors for hydration monitoring and moisture measurement in concrete structures

Tarikul Islam and Kazi Jabed Akram

This chapter reviews some electrical techniques for the health assessment of concrete structures. These electrical sensing systems are used to monitor the hydration and determination of initial and final setting times, and moisture measurement in concrete samples. Different sensors with proper electrical connections working on electrical properties and their associated electronic interfaces can be utilized to determine important concrete health parameters. The most common electrical method is the impedance measurement technique. Desirable features of such electrical techniques are that they be non-invasive or minimally invasive with the possibility of on-line and remote measurement, at the quickest time as far as possible. Destructive methods cause damage to the structure. In this chapter, we first discuss some commonly used techniques for hydration monitoring and moisture measurement. We further highlight some merits and demerits of the conventional techniques and some new techniques proposed in the literature. The chapter then explains uses of a fringing field impedance method for hydration monitoring and moisture measurement in concrete using flexible electronics. The performance of fringing field sensors is compared with the commonly used techniques. The impedance sensors are simple in terms of operation, fabrication and are cost effective for on-line measurement purposes.

3.1 Introduction

Health monitoring of a civil structure is essential to protect loss of human life and to avoid loss of money invested in the construction of the structures. Two types of techniques, known as non-destructive and destructive techniques, are used for assessing the quality of structures. Generally, for structural health monitoring (SHM), non-destructive techniques which cause minimum damage to the building are preferred for damage detection and performance evaluation of a structure.

doi:10.1088/978-0-7503-3935-3ch3

Destructive techniques which cause damage to the building are less preferred. The sensing-technology-based assessment process involves selection of a suitable sensor and its fabrication, installation, data acquisition, data transmission, data storage and data analysis to evaluate the conditions of a structure. Life span prediction and timely maintenance of the structure are essential to avoid potentially fatal disasters. The advancement of technology and invention of different building materials inspire people to construct tall, robust and earthquake-immune buildings to cope up with increasing land scarcity and rapidly growing population. In addition, aesthetic advancements in architecture demand ever more complex applications of concrete and building materials, which sometimes causes failure to the structures due to quality compromise. As such, modern structures require more precaution during masonry work and cost-effective maintenance from time to time. Newly built structures such as business hubs, offices, market complexes, flyovers and bridges are important for leveling up the living standard of citizens. In addition, for healthy living attention must be paid to sanitary units and waste-management systems. In another respect, the rich historical and cultural background of a nation is reflected in its historical monuments and structures, therefore evaluation of such historical structures health conditions is important to ensure that appropriate action is taken before any damage or failure. SHM is focused on providing fruitful and on-time information about the condition and performance of a structure, which includes the following:

- damage detection during construction by confirming any unexpected changes in design properties;
- quantitative and continuous data monitoring for quality assurance;
- understanding the performance of a new structure when constructed using novel materials or technologies; and
- early-stage detection of deterioration to ensure the safety of the surrounding environment, nature and people.

Concrete is composed of graded and fine coarse aggregates, sand, cement, water and a large number of voids. When water is added with fine cement particles, different chemical and physical bonds are formed with the cement powder. Cement is one of the most important ingredients for construction purposes [1, 2]. It contains various inorganic oxides like tricalcium silicate, dicalcium silicate, tricalcium aluminate, tetracalcium aluminoferrite, etc. These oxide materials react with water during hydration and lead to the formation of some new oxides, prominent among them calcium hydroxide, calcium silicate, ettringite, and monosulfate [3, 4]. During physical bonding the water molecules do not react with cement but give rise to porosity in the concrete. These pores become the source of moisture traps. The trapped water molecules in the pores can evaporate when the concrete is exposed to the environment or heated at some time to a temperature around 105 °C. However, some water molecules form chemical bonds with the cement and some are physically bonded strongly, which requires a very high temperature to evaporate [5].

The basic principle of SHM is automated continuous monitoring, inspection and damage detection for structures with minimum effort and cost. A typical SHM system requires three basic operations, such as a sensor system for the detection and quantification of various crucial parameters of the structures, a data-acquisition system for continuous relay of data on-line to a digital system, and a diagnostic tool for analysis and inspection of data to determine any abnormalities in the structure. There are many factors associated with damage to a structure. These factors come under two main categories: damage due to physical effects and damage due to chemical effects.

Some examples of damage due to physical effects are vibration, shock, erosion due to wind, cavitation due to flowing water, freezing and thawing of water in cracks, high temperatures, etc. Among chemical damage effects, the most notable are aggression by carbon dioxide, sulfates, chlorides, and alkali-aggregates reactions, etc. The presence of moisture hastens the chemical reaction and results in faster deterioration of the structure. All the aforementioned problems belong to the case of in-build structures. Some important parameters to be measured for the SHM application are strain, stress, crack, corrosion in steel reinforcing bars, temperature, and moisture content in the concrete, etc. Temperature and moisture play an important role in concrete's durability, shrinkage and strength. These parameters greatly influence the hydration process at the early stage [5, 6]. Not only do these two parameters influence the strength in the beginning but also at later stages for the deterioration of the concrete. Water molecules from various sources like from sea, rain, soil, snow and floods are responsible for the transfer of high-chloride-content compounds, which lead to corrosion in steel rebars. Moisture transfer also promotes harmful chemical processes such as carbonation and alkali-aggregate reactions [7, 8].

According to some recent investigations, moisture and temperature are the two major causes of extensive failures in concrete structures. Failures start when the humidity level in the concrete reaches in the range of 50% to 70% relative humidity (RH) [9, 10]. Therefore monitoring hydration, and determining the temperature and humidity of concrete are essential for condition monitoring at the early stage of construction as well as at later stages during its utilization. Therefore sensors to measure these parameters, when embedded in concrete, provide more useful information about the progress of curing as well as strength gain by the concrete.

In recent times, printed electronics integrated with sensing elements on flexible substrates is a rapidly growing development in diverse fields, including SHM, because of its many attractive features such as inexpensive fabrication, rapid prototyping and the possibility of applications in both planar and non-planar surfaces. It is now possible to merge age-old printing technology with established microelectronics technology to rapidly develop electronics systems integrated with electronic circuits, multi-sensing elements and wireless data communication modules at low cost, but with superior performance parameters. Printable flexible electronics can be used to develop different sensors including capacitive, piezoresistive/resistive, field effect transistors and ultrasonic sensors for measuring the various parameters for SHM. These parameters may include hydration monitoring, measurement of hydration temperature, moisture/humidity in concrete, strain, stress, crack detection and corrosion in reinforced bars, etc. A variety of substrates can be used to fabricate the printable electronics like polyethylene terephthalate (PET), polymethyl methacrylate (PACr), polyethylene naphthalate (PEN), polycarbonate (PC), polystyrene

(PS) and polyimide (PI). Prominent piezoelectric substrates like polyvinylidene fluoride (PVDF) and its copolymer, trifluoroethylene (TrFE), are widely used for their desirable mechanical flexibility, and stable piezo, pyro and ferroelectric properties. The attractive features of P(VDF-TrFE) are its high piezoelectric coefficient, wide frequency response, flexibility, cost effectiveness and ease of fabrication [11, 12]. Copper-clad PTFE substrate is another very useful flexible film, which is highly hydrophobic and has exceptional insulation properties; it can work at up to 175 °C for long times and up to 225 °C for short durations without damage to the film and copper cladding. The substrate also works well at very low temperatures. Most of the plastic substrates are available with or without a metal cladding printed circuit board (PCB). The metal cladding to the substrate may be made of copper, silver, gold or some other metal films like indium tin oxide, fluorine-doped tin oxide, etc. These metal-cladded substrates can directly be used to form different electrode structures of the sensors including interdigital, concentric and parallel plate, for mounting the sensors on different planar and non-planar surfaces by a simple etching process. However, conducting paste can also be patterned on the bare flexible substrate using different patterning methods. For printed flexible electronics, silver (Ag) pastes and its solutions are extensively used because of its better adhesion and better electrical performance on the plastic substrate. Even conducting polymers, closely matching the properties of inorganic metals with better adhesion, are now easily available for the fabrication of sensors suitable for SHM applications. For example, PEDOT:PSS transparent conducting polymer, with a composition of 3,4-polyethylenedioxythiopene-polystyrene sulfonic acid, has been widely investigated because of its high conductivity. The conductivity of pristine PEDOT:PSS films is only $\sim$0.2 S cm^{-1}, and it can be increased to higher than 2100 S cm^{-1} [13]. A nanocomposite conducting polymer can be made by mixing metal nanoparticles in an organic elastomer like (poly)dimthylsiloxane (PDMS). A bare PDMS film can also be used as a flexible substrate for fabricating the sensors. A variety of sensing films can be deposited on the substrate by applying different pattering techniques such as spin coating, dip coating, micromachined mold transfer, single or multiplayer inkjet printers and screen printing. Spin coating and inkjet printing pose some limitations for depositing multiplayer films but the screen printing method performs well in terms of the uniformity of film over large areas in a cost-effective manner. Some important properties of the plastic substrates are mentioned in table 3.1.

Various contact and non-contact methods are reported in the literature. However, some of them suffer from electrode polarization, and others suffer from misalignment of electrodes due to contraction of concrete samples; thus, the measurement results are inaccurate. The non-contact method like time domain reflectometry, requires a high-frequency signal source and analysis of the reflectogram with a complex and relatively costly signal-processing unit. Also, the performance parameters are affected by large gaps between the electrodes, surface area and the geometrical curvature of the electrode arrangement. Moreover, the stated measuring schemes so far are not viable for a portable prototype due to the large size of the primary-sensing element. These are mainly laboratory-based techniques. Therefore, real-time measurement is quite difficult. Hence, an improvement is necessary in the design and development of hydration-monitoring systems.

Table 3.1. Some useful properties of polymer substrates.

Substrate properties	Polyimide (PI)	Polystyrene (PS)	Polytetra-fluroethylene (PTFE) (Cuflow)	Polymethyl methacrylate (PAcr)	Polycarbonate (PC)	Polyethylene napthalate (PEN)	Polyethylene terephthalate (PET)
Transparency (%)	35–60	89	40–60	>90	92	88	90
Water adsorption (%)	2–3	1.4	<0.01%	0.2	0.25	0.4	0.6
Solvent resistance	Good	Poor	Good	Good	Poor	Good	Good
Surface roughness	Good	Good	Fair	Fair	Good	Poor	Poor
Stability	Fair	Fair	Good	Good	Fair	Poor	Good
Dielectric constant	3.5 (1 kHz)	2.50	2.1 (18 GHz)	3.5	2.9 (1 MHz)	3.2 (10 kHz)	3.58 (1 kHz)
Mechanical strength (Y) $10^9 N/m^2 GPa$	2.5		0.024 (GPa)	2.4–3.4	2.6	0.1–0.5	2–2.7
Glass trans. temp. (°C)	270	203	119	105	145	120	70
Coeff. of thermal expansion (CTE) ppm $°C^{-1}$	8–20	54	129	70	75	20	33

The main motivation of this chapter is to discuss the design, fabrication and investigation of flexible sensors for hydration monitoring, and measurement of moisture and hydration temperature at an early stage of construction. These flexible sensors can easily be fabricated as a system with integration of associated electronics suitable for non-invasive or minimally invasive SHM applications.

3.2 Hydration and its monitoring using flexible sensors

If prevention can be taken at the early stages of masonry work then, to a large extent, catastrophic events can be avoided. When water is added to cement an exothermic reaction takes place, which is called hydration. Hydration of cement involves the chemical reaction of various kinds and composition of cement with water. Dissolving cement increases the level of calcium (Ca) and silicon (Si) in the solution. When the level of Ca and Si reaches a maximum level then a precipitation reaction happens, which leads to the formation of a solid. Improper hydration leads to poorer durability and strength. Concrete setting time is a crucial parameter for monitoring the quality of construction progress as well as concrete quality control. With advanced knowledge of setting time, the construction of concrete can be preplanned. The time of mixing of mortar with cement, transporting of the mixture to the actual site, casting and finishing can be properly controlled and various time set controlling admixtures can be decided. There are two critical parameters to be determined during hydration. These parameters are the initial setting time (IST) and the final setting time (FST). According to ASTM 403, the IST is defined as the time taken by the cement when it starts hardening by forming different oxides. At this stage, the concrete loses plasticity but it can be molded to any shape without losing its strength. In contrast, at the FST, the concrete loses its plasticity and becomes hard, though still lots of reactions are left. The FST refers to the time taken by the cement paste/concrete to completely lose its plasticity and become hard. By continuous monitoring of hydration of fresh mixtures, we can determine the time to proceed with subsequent treatments on the structure and to estimate the readiness of the structures for its service.

Conventionally, the initial and final setting times are measured by penetration resistance methods like a Vicat apparatus (VA) [14]. The penetration method was standardized according to ASTMC 403. In the VA, different types of standard plungers are used and allowed to free fall in the cement sample. The time for penetration of a Vicat plunger to a distance of 5–7 mm in the sample is considered as the IST. The FST corresponds to the time when a mark of a tip of another plunger on the surface of the sample as per the Indian standard is obtained. At the initial and final setting times, the penetration resistance of a fixed dimensional needle reaches 3.5 and 27.7 MPa, respectively. However, the penetration method suffers from the following aspects:

- requirement of skilled manpower, thus accuracy varies from person to person;
- requirement of a sieved mortar to acquire the proper size of mortar, which is hard to achieve;
- long test time, which becomes longer when a retarder is added into the mortar;

- it is a manual method and test results may vary according to the fall position of the needle;
- difficulty of continuous measurement;
- penetration resistance depends only on the top layer of the concrete instead of the bulk sample; and
- a large quantity sample is required, thus there is a wastage of concrete materials.

To overcome these demerits, some alternative techniques have been developed by researchers and engineers. Prominent methods among them are (i) the impact-echo method, (ii) ultrasonic measurement, (iii) electrical resistivity. In the impact-echo method, the setting time is determined by the time when the echo-wave velocity begins to increase. However, this is difficult to apply to determine the IST because of the plasticity of concrete (transition of phases from plasticity to solid) [15]. The ultrasonic method is based on the generation, transmission and reflection/receiving of high-frequency mechanical waves through the concrete sample under testing. The ultrasonic wave can be generated by a piezoelectric device, wherein an electrical signal can be converted into a mechanical wave. In the presence of a test sample, the received wave is modified according to the conditions of the concrete sample. Thus, there is a change in phase or amplitude or the frequency of the transmitted signal [16]. The frequency mode of measurement is more suitable as the frequency of an output signal can be easily measured with high accuracy. Electrical resistivity measurement is another popular non-destructive method for the determination of the setting time and hardness of the concrete. Electrode arrangement and the technique of resistance measurement are simple yet accurate. Measurement can be taken directly on the concrete instead of using a sieved mortar. The sample requirement amount is much smaller (0.0016 m^3) than the penetration method. In addition, continuous fast on-line remote measurement of the concrete is possible [16]. However, there is no quantitative relationship among these techniques and the conventional penetration resistance method. Some techniques for hydration monitoring using optical fiber sensor have also been reported in the literature [17].

3.2.1 Ultrasonic technique

The ultrasonic technique is one of the useful methods used in structural health assessment. It is a useful tool to analyze the early-age hydration of building material as well as for damage detection of existing structures. Early-age hydration of cement-based materials can be characterized by using high-frequency ultrasonic waves. The ultrasonic wave reflection factor (WRF) has been used to monitor the setting of cement [18]. When an ultrasonic wave transmitted through a material enters a second material having different mechanical properties, a part of the wave is reflected back into the first material and the remainder passes through the second material; then, the WRF is the ratio of the amplitude of the reflected wave to the transmitted wave incident to the interface. It depends on the relative mechanical properties of the two materials and can be used to determine the early setting and hardening of concrete. The WRF can be given by

$$WRF = \frac{\rho_2 V_2 - \rho_1 V_1}{\rho_2 V_2 + \rho_1 V_1}, \tag{3.1}$$

where ρ_1 and ρ_2 are the mass density of the first and second materials, and V_1 and V_2 are the velocity of the incident and reflected mechanical ultrasonic waves, respectively. The product of density and velocity is called acoustic impedance. If the fresh concrete sample is placed inside a cylindrical steel mold then the steel mold forms the first material and concrete acts as the second material. Since during hydration only the properties of the concrete change but not of the test chamber, the reflected wave amplitude will be changed according to the status of the concrete. There may be some other factors which may cause a reduction of the amplitude of both incident and reflected waves, but the concrete's status is the dominating factor in the amplitude change. Ultrasonic mechanical waves may be longitudinal or shear. The velocity of a longitudinal wave through a sample can be given by

$$V_l = \sqrt{\frac{E(1 - \upsilon)}{\rho(1 + \upsilon)(1-2\upsilon)}}, \tag{3.2}$$

where E is Young's modulus, υ is the Poissons's ratio and ρ is the mass density. For a shear wave, the velocity can be given by

$$V = \sqrt{\frac{G}{\rho}}, \tag{3.3}$$

where G is the shear modulus.

The WRF between a steel mold and a cement-based material is measured for normally incident longitudinal and shear elastic waves. WRF data is collected soon after the mixing of concrete and placing in the mold, as shown in figure 3.1. The data is acquired up to the final setting of the mixtures, and the initial and final setting times from the response plot are determined. The setup requires a steel chamber of

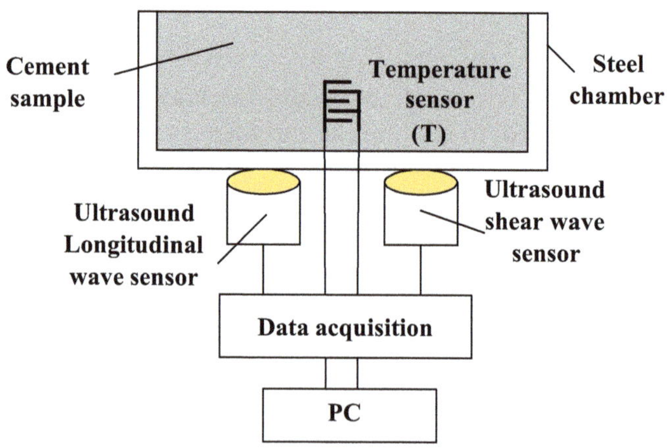

Figure 3.1. Experimental setup for ultrasound-wave-reflection-based hydration monitoring.

suitable dimensions to hold a concrete sample of sufficient quantity. Two piezoelectric transducers are placed on the wall of the chamber to generate the longitudinal and shear waves of frequencies in the range of 4–8 MHz and 2–4 MHz, respectively. These waves are transmitted through the steel chamber to the concrete and reflected back from the steel–concrete interface. The received reflected waves are reconverted into an electrical signal by the same transducers and then acquired by a PC through a data-acquisition system. To measure the temperature of the concrete, a platinum resistance temperature detector (RTD) temperature sensor is embedded in the sample and its data are also acquired. To improve accuracy and reduce other loss factors, another identical dummy test chamber without a concrete sample is employed. The output signals of the active transducers with a cement sample and without sample are compared to get the signal for the sample only. However, in practice the amplitude ratio of the first and third reflected waves are more useful and hence these two waves are compared in frequency domain, as given by

$$WRF = \sqrt{\frac{KR_{13}{}^{S}}{KR_{13}{}^{W}}}, \tag{3.4}$$

where $KR_{13}{}^{S}$ and $KR_{13}{}^{W}$ are the frequency domain amplitude ratio of the first and third reflected waves with a concrete sample and without a concrete sample, respectively. Finally, the IST and FST values are compared with the penetration resistance method using a VA and thermal hydration-monitoring method. The advantages of the ultrasound technique are (i) its non-destructive nature, (ii) more penetration depth than other non-destructive techniques, (iii) single-side test possibilities, (iv) and an image of the sample can be obtained. But the main limitations of this method are (i) that only early-age mechanical stiffness can be determined but not compressive strength, (ii) it provides localized properties of the concrete specimen near the interface, (iii) the surface should be regular in shape and size, that is, the concrete should not be too thin, i.e. it needs a sufficient depth, and (iv) that it should be able to transmit the sonic wave.

In [19], a multi-sensing platform integrating three devices such as an interdigital electrode (IDT) for lamp wave generation and reception, a circular piezoelectric device for detection and localization of low-velocity impacts, and a RTD is reported. The temperature sensor is used for the measurement of the ambient temperature of the structure where the impact sensor is placed and to compensate for the temperature error for impact measurement. Three devices are fabricated on a flexible metal-clad piezoelectric PVDF substrate using a laser-etching process. The substrate has a gold chromium alloy metal film deposited on either side of an 110 µm-thick poled PVDF film with a nominal piezoelectric strain coefficient d_{33} = – 30 pC/N, in an expensive manner for SHM application. A schematic of the structure is shown in figure 3.2. The RTD sensor was fabricated with a wire length and width of 93 mm and 250 µm, respectively, for a nominal resistance of 350 Ω at 25 °C. The IDT sensor was made with a gap wavelength of a finger, 8 mm (gap between two same-side fingers) and a finger pitch of 4 mm (gap between two opposite fingers) and finder

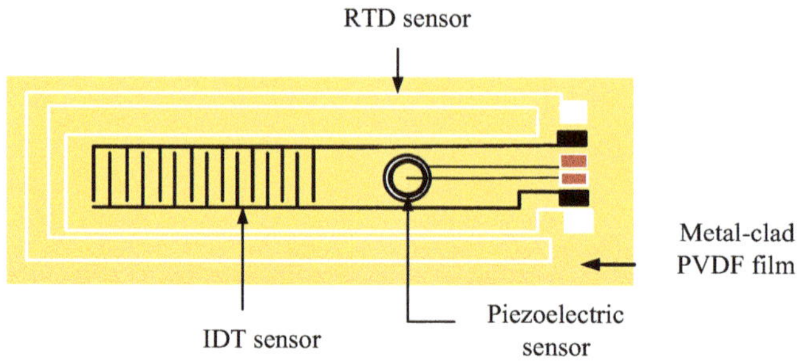

Figure 3.2. Multiparameters sensing unit on a plastic substrate for SHM application.

width 1.6 mm. The actual size of the substrate is 40 mm × 24 mm. Such a type of structure may also be used for hydration monitoring and stress measurement.

3.2.2 Hydration monitoring by temperature measurement

When water is mixed with ordinary Portland cement, two types of reactions occur during hydration: solution-state hydration and solid-state hydration. The first reaction involves dissolution of anhydrous compounds to their ionic constituents, formation of hydrates in the solution and eventual precipitation of hydrates. The second reaction occurs directly at the surface of the anhydrous cement compounds without the compounds going into solution. An ordinary Portland cement contains four types of minerals, namely alite (C3S, 55% by weight), belite (C2S, 18% wt), aluminate (C3A, 10% wt) and aluminoferrite (C4AF, 8% wt). The composition of cement is shown in table 3.2 [20]. When these products are mixed with water, different hydration products are formed. The main hydration products are calcium silicate (C–S–H) and calcium hydroxide. The C–S–H gel plays the role of a binder for the durability and strength of the concrete. C2S and C3S minerals compose the C–S–H gel of about 82% and 61%, respectively. The ultimate strength and durability of high-C2S-content cement will be higher than high-C3S-content cement. C3A reacts immediately after addition of water but its reactions can be slowed down by adding gypsum. The products of the C3A reaction are commonly ettringite in the first stage and monosulfoaluminate at later stages. The precipitation of ettringite is the main contributor of stiffening, setting and early strength development. The hydration of C4AF is similar to C3A with addition of gypsum. However, hydration reactions with these minerals cause a unique heat-rate evolution pattern that can be utilized to determine the setting and hardening behavior of cement and predicting temperature rise. It may be possible to determine the setting times and hydration level by measuring the temperature of the concrete sample. A typical rate of heat evolution curve with the progress of time is shown in figure 3.3. There are five stages of hydration. The first stage is the dissolution stage, in which ions dissolved in water react with C3A and gypsum. Formation of ettringite reduces the reaction at the end of Stage I. This stage has little contribution to the strength of the concrete. Stage II is the dormant stage, in which the concentration of ions gradually increases. The cement concrete remains in a plastic state at this stage. No strength is gained but it is

Table 3.2. Composition of cement.

Components	CaO	SiO_2	SO_3	Al_2O_3	Fe_2O_3	MgO	LOI	Na_2O	K_2O
%Age	69.66	20.01	3.78	3.27	1.39	1.12	0.72	0.03	0.02

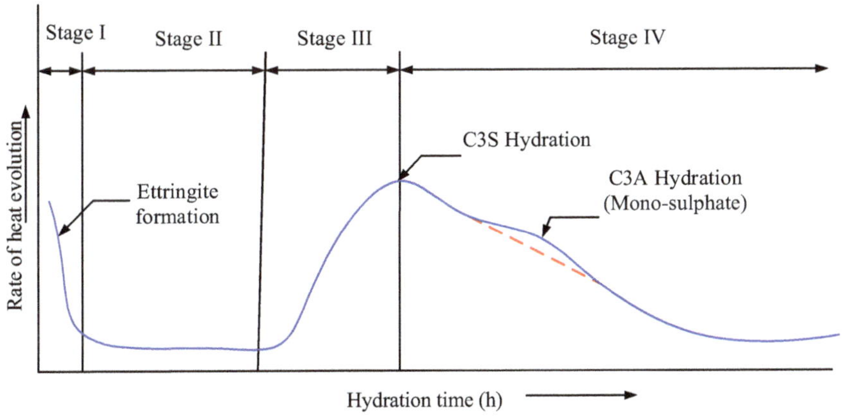

Figure 3.3. Rate of heat evolution with the passage of time [21].

important for transporting the concrete to construction sites. Stage III is the acceleration stage, wherein the alite (C3S) and belite (C2S) in the cement start to hydrate and release heat. In this stage concrete setting begins and heat generation is rapidly accelerated, as shown in the curve. The rate of heat evolution reaches its maximum, the hydration reaches a maximum level, the concrete's strength is developed, the FST is finished and the process of hardening starts. Thus, this is a very important stage. Stage IV is called the deceleration stage, in which the rate of heat generation decreases again and shifts to a diffusion-controlled process. The thickness of hydrated particles increases and the surface area of the unhydrated parts decreases and ettringite is converted into monosulphate, which contributes to a minor temperature rise as shown in the graph.

Stage V is a steady state, in which the formation of thick layers of hydrates around the cement particles hinder the hydration reaction to a great extent. Hydration now completely depends on the diffusion of water molecules to the cement particles and it is a slow process. It leads to the formation of hydration products slowly. The amount of heat generated is governed by cement composition, water-to-cement ratio and supplementary materials like fly ash, chemical admixers and curing process.

Therefore, during hydration an initial exothermic reaction period is followed by an induction period, then a nucleation period and then a growth period at which the temperature of the concrete reaches its peak. The material reaction rate is then reduced and the temperature starts to fall until the temperature reaches ambient temperature. At this point, the hydration reaches a steady state. During the initial exothermic reaction the temperature of the sample increases; during the induction period, it decreases. The end of the induction period indicates the beginning of the stiffening of the concrete. Thus continuous temperature measurement plays an important role. During hydration the temperature of concrete may rise by 25 °C from its initial value when water is added to the cement. Although the temperature rise is not much, it is significant and it provides useful information about the progress of the hydration reaction and the nature of its strengthening and durability. This may be another useful method for determining the setting time and hydration monitoring. Distributed temperature sensors at different locations may give more useful information about the condition of the concrete. However, there are several challenges for the application of sensors for the measurement of SHM parameters. Destructive methods as conventionally practiced are discouraged due to the damage caused to the structure, limitation of on-line monitoring and the damage also caused to the sensors. Thus any sensors embedded into the structure should be minimally invasive and should cause minimum disturbance to the structure, but will very much be useful to gain actual information of hydration, moisture ingress, temporary changes, cracks, corrosion in bars, etc. Another issue with embedded sensors is the possibility of non-contact with the concrete mass or the appearance of air gaps due to the shrinkage of the concrete with the progress of formation of hydrated products. Once a sensor is embedded into concrete for condition monitoring, it is difficult to reuse it, so it should be installed permanently. So, the durability of the sensor is an important consideration of the sensor design. Because of the limited use of the SHM sensors, the cost of the sensor should be inexpensive, though it should provide the necessary information accurately and reliably.

3.2.2.1 A flexible temperature sensor suitable for hydration monitoring

A linear capacitive temperature sensor is fabricated using a PDMS substrate suitable for hydrothermal temperature measurement in [22]. The sensor is a parallel plate capacitor with PDMS that works as a temperature-sensitive dielectric film. It works with change in relative permittivity due to the change in density of the dielectric with variation of temperature. The sensor has a temperature sensitivity of around 150 fF $°C^{-1}$ with a resolution of 0.08 °C and maximum non-linearity of ~1%. The fabrication of the sensor is easy, inexpensive and identical sensors can be fabricated in batch. The materials cost for the fabrication is low. The sensor can be embedded in the concrete sample for continuous on-line temperature monitoring of the hydration thermal reaction.

3.2.2.2 Fabrication of the sensor

A diagram of the sensor is shown in figure 3.4. The structure has a double-sided copper-clad PI substrate (UBE, Japan). The substrate was properly cleaned in distilled water and in acetone for some time to remove impurities. A PDMS film of nearly 50 μm thickness was deposited on a copper plate of size 40 mm × 10 mm × 0.125 mm by a dip-coating method using PDMS solution. The PDMS solution was prepared by mixing silicone elastomer with cross linker (Sylgard 184) at a ratio of 10:1.

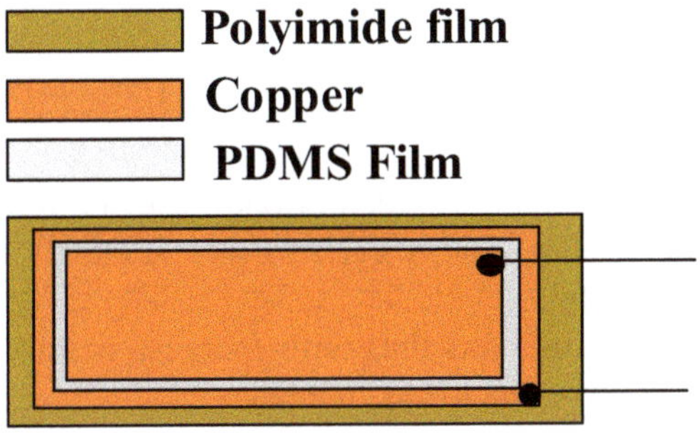

Figure 3.4. Schematic of the capacitive sensor suitable for hydrothermal temperature measurement.

Precautions were taken to maintain uniformity and to avoid trapped air bubbles in the film. Since the sensor is a parallel plate capacitor, non-uniformity will cause an unequal capacitance value across its dimensions. The top electrode of the capacitor, of size 35 mm × 5 mm × 0.125 mm, was then laminated on the PDMS film. The PDMS film was then cured gradually at 200 °C with a heating slope of 40 °C/30 m.

The capacitance value of the sensor decreases due to a decrease in the density of the PDMS film according to the Clausius–Mossotti equation [23], and due to thermal expansion of the film. The temperature-dependent capacitance of the sensor, $C(T)$, can be represented by [22]

$$C(T) = \varepsilon_0 \varepsilon_{pd}(T) \frac{A}{t_{PD}(T)} + C_{fr}, \tag{3.5}$$

where $\varepsilon_{PD}(T)$ is the permittivity of the sensing film (PDMS) that depends on the temperature, $t_{PD}(T)$ is the thickness of the sensing film, which also depends on the temperature due to the positive coefficient of temperature expansion (CTE), A is the cross-section area of the electrodes and C_{fr} is the fringing field capacitance. The fringing capacitance is approximately given by

$$C_{fr} = \frac{\varepsilon_0}{\pi} l \left(1 + \ln\left(\frac{2\pi w}{t_{PD}} \right) \right). \tag{3.6}$$

For the dimensions used for the capacitor, the C_{fr} is found to be approximately 0.7 pF. This value is taken to be negligible. The change in $\varepsilon_{PD}(T)$ due to change in density ($\rho_{PD}(T)$) can be given by

$$\frac{\Delta \varepsilon_{PD}(T)}{\Delta \rho_{PD}(T)} = k \frac{(\varepsilon_{PD}(T) - 1)(\varepsilon_{PD}(T) + 2)}{3\rho_{PD}(T)}, \tag{3.7}$$

where k is a fitting parameter. The density of PDMS can be related to the temperature by the second-order equation [24]

$$\rho_{PD}(T) = 0.992 – 8.925 \times 10^{-4} T + 2.650 \times 10^{-7} T^2. \tag{3.8}$$

The change in thickness can be given by

$$\Delta t_{PD}(T) = t_{PD0}\left(1 + \frac{\beta_{PD}}{3}T\right), \tag{3.9}$$

where β_{PD} is the volume expansion, which is related to temperature by $\gamma_{PDMS}(\text{deg}^{-1}) = 0.900 \times 10^{-3} + 2.76 \times 10^{-7}T + 1 \times 10^{-10}T^2$, t_{PD0} is the initial thickness of the dielectric film.

In the expressions, it is observed that with increase in temperature, the relative permittivity decreases and the thickness increases.

3.2.2.3 Determination of the response characteristics of the sensor

Initial experiments were conducted by placing the sensor in a hot medium and varying its temperature from 20 °C to 100 °C with increments of 5 °C. The temperature of the hot medium was measured by a calibrated temperature sensor. The terminals were connected to a Keysight handheld LCR meter to measure its capacitance value at different temperatures. The sensor was excited by an AC voltage of 500 mV (rms) and frequency 10 kHz. Necessary precautions were taken to ensure the repeatability of the reading. The variation of the capacitance value with increase in temperature is shown in figure 3.5. The capacitance value decreases with rise in temperature and the decrease in capacitance value follows almost a straight line with a linear correlation coefficient of 0.99. The mean sensitivity and the standard deviation of the readings are 150 fF/°C and ±30 fF, respectively.

The impedance response shows that the maximum deviation of the phase angle for the full-scale temperature range from its 90° phase angle value is only 0.1°. The sensor's equivalent series resistance (ESR) at 60 °C is ~100 Ω, but this ESR value changes negligibly compared to the capacitive reactance. This study ensures that the sensor is capacitive in nature.

Experiments were conducted to determine various response parameters such as hysteresis error, response time, repeatability, resolution, etc. The obtained parameters are shown in table 3.3.

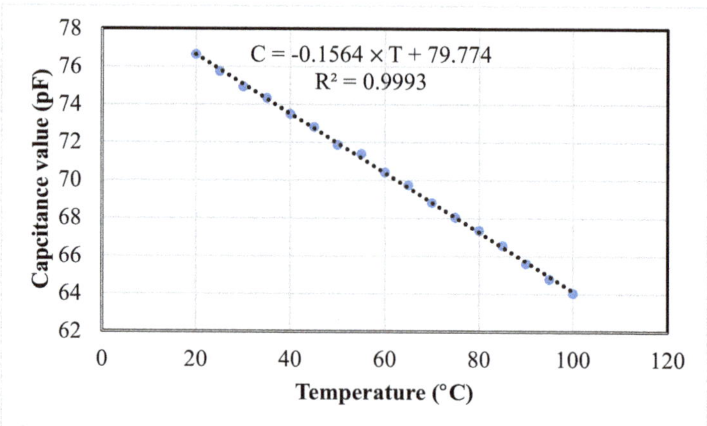

Figure 3.5. Capacitance variation of the sensor with increase in temperature (f = 10 kHz).

Table 3.3. Response parameters of the temperature sensor (10 kHz) [22].

Response parameters	Obtained value
Sensitivity	150 fF °C^{-1}
Resolution	0.08 °C
Linearity	1.1%
Repeatability (50 °C)	0.5%
Response time ($\tau_c(s)$)	0.6 s/°C
Hysteresis	±1.3%

This is a simple, inexpensive and reliable capacitive sensor fabricated using a flexible substrate and flexible film for the measurement of temperature for different applications. The sensing film is hydrophobic and has a resolution of 0.08 °C. The response parameters shown in table 3.3 are very much suitable for the measurement of hydration temperature in concrete. This is because, as shown in figure 3.2, the most crucial Stage III has a maximum temperature rise of about 25 °C, which can be easily measured by this capacitive sensor. Being flexible, thin and lightweight, it can be embedded in concrete for continuous monitoring of hydration temperature without causing any damage to the structure. As such, the sensor is minimally invasive. Further, the cross-sensitivity due to humidity/moisture is kept to a minimum as PDMS is a highly hydrophobic material. In addition, multiple sensors can be integrated on the PI substrate for multiplex measurement of parameters such as humidity, temperature, hydration, strain, etc.

3.3 Hydration of concrete using transformer principle

Conventionally, the resistivity of concrete is measured by placing two electrodes across opposite faces of a sample, then measuring the current passing through it for a fixed excitation voltage. The resistance of the test sample is the ratio between the applied voltage and the current. In such a method, the contact between the electrodes and the test sample is important; if there is a loose contact then the reading will be inaccurate and not reliable. Thus, this method is reliable in the beginning of measurement as hydration of the concrete will cause shrinkage, leading to cracks and fissures and the possibility of the metal contact with the sample being loosened. In addition, cement-based material is alkaline, which corrodes the electrode and thus requires frequent electrode replacement. In such situations, a non-contact resistivity measurement technique for measuring the setting and hydration conditions is preferable. One such technique, based on Faraday's law of induction (transformer principle), was proposed in [25]. In this method, the primary of a transformer is excited by an input voltage of 1 kHz signal frequency and the secondary coil of the transformer is the concrete specimen under investigation in the form of a toroid or rectangular shape. Because of primary excitation, a voltage is induced according to Faraday's law of electromagnetic induction in the secondary coil. The concrete specimen behaves as a lossy dielectric, which can be represented by a parallel combination of resistance (R)

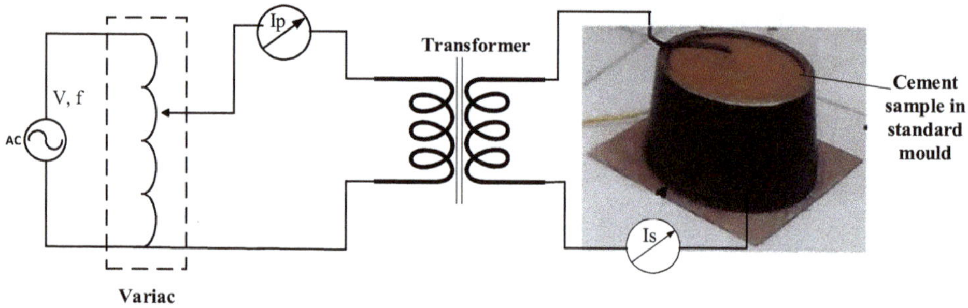

Figure 3.6. Experimental setup to measure the setting time of concrete using the transformer principle. © 2018 IEEE. Reprinted, with permission, from [2].

and capacitance (C). The components values of the parallel combination depend on the state of the concrete sample. Thus, the current produced in the secondary coil will vary according to the condition of the sample.

A simple experimental setup for hydration monitoring using the transformer method is shown in figure 3.6. Here, the concrete sample is connected as a lossy capacitive load across the transformer's secondary winding. The primary winding of the transformer is excited by a certain power frequency AC voltage (V,f). Let the primary current be I_p, the secondary current I_s and the no-load current when the secondary is open I_0. The amount of primary current can then be written as

$$\vec{I_p} = \vec{I_0} + \vec{I_s}. \tag{3.10}$$

The no-load current can be given by $I_0 = V_1(G_0 + jB_0)$, where $(G_0 + jB_0)$ is the admittance of the core, and the secondary current referring to the primary side can be given by

$I_s = \dfrac{(V_1 - V_2)}{Z_{sp}}$, where V_2 is the voltage across the load and Z_{sp} $(=\dfrac{1}{Y_{sp}})$ is the secondary impedance referred to the primary. So, the primary current can be given by

$$I_p = V_1(G_0 + jB_0) + V_1 Y_{sp} - (I_s \times Z_c) Y_{sp}, \tag{3.11}$$

where Z_c is the impedance of the cement sample. The I_0 current is a fixed current required to support an eddy current and hysteresis loss and magnetization of the core. This current varies by 2%–4% of the rated current depending on the transformer rating. So, the change in primary current mainly depends on the change in the impedance of the cement sample. This method was successfully utilized to monitor the hydration of concrete and to determine the critical points of the response curve, so as to measure the initial and final setting times.

3.4 Determination of the response parameters of concrete

Experiments were conducted to determine the characteristic curve of hydration for two cement samples with water-to-cement ratios of 0.27 and 0.30, respectively. The sample cell is a standard Vicat mold, shown in figure 3.6. The cell is a tapered cylinder made with Teflon-coated brass. It has 90 mm of outer diameter and 80 mm of inner diameter at the bottom side and 80 mm of outer diameter and 70 mm of inner diameter at the top side. The height of the mold is 40 mm. The thickness is around 5. For electrical contacts, two copper-clad FR4 substrates of diameter 80 mm and 70 mm, respectively, are placed on the top and bottom of the sample. Necessary arrangements were made to avoid shorting problems. The terminals of load were connected to the secondary of a transformer of rating 220 V/18 V, 3.6 A. The primary rms input voltage was 130 V, 50 Hz. The variation of primary current was measured with a high-resolution digital multimeter. Initially, the primary current was measured when the secondary was open; this is a no-load current and its value was ~12 mA. The variation of primary current was noted for nearly 900 mins for both the samples and the response curves are shown in figure 3.7. It is observed that for each sample, initially the primary current increases for some time, then becomes almost constant for some time, and then the current value decreases with the progression of time. For better visualization, the primary current values from 0 to 180 mins are plotted separately. The current value initially increases from 0 to 65 mins for the 0.27 cement sample and from 0 to 80 mins for the 0.3 sample, then it decreases and then increases and, finally, the current decreases continuously from 140 mins onward for the 0.27 sample and from 110 mins onward for the 0.30 sample respectively. When water is added in cement, different ions such as calcium, iron, silica ions, etc, are released in the sample. Thus the impedance of the cement is reduced, which causes an increase in the primary current (point 'a' for 0.27 and point 'b' for 0.30). Then formation of calcium silicate hydrate ($3CaO{\cdot}2SiO_2{\cdot}4H_2O$, C–S–H gel), around the cement particle prevents reaction with water for some time and the solution becomes supersaturated. At this point, the sample has maximum ionic concentration and then the formation of oxides starts. Then, breaking of the C–S–H gel walls results in an increase in primary current, as shown by the 'a' and 'b' points of the curves. After this phase, due to the formation of rapid hydration products, the impedance of the sample continuously increases, which leads to a continuous decrease in primary current. From the response curve initial and final setting times

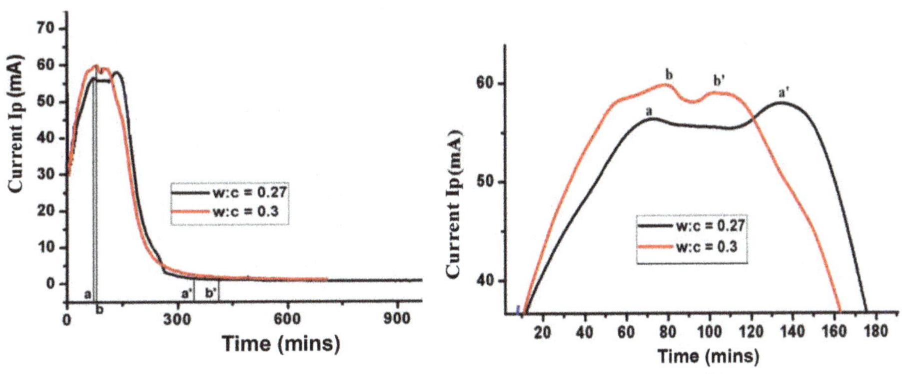

Figure 3.7. (a) Variation of primary current with the time-dependent progression of concrete. (b) Zoom-in view of the primary current variation from 0 to 180 mins. © 2020 IEEE. Reprinted, with permission, from [32].

are determined and results are compared with Vicat apparatus. The IST and FST of the 0.27 sample were found to be 166 mins and 218.5 mins, respectively. These values differ from the IST and FST values of the Vicat apparatus by ± 5 mins. So, the results closely match the conventional method. Experimental results with several cement samples having different water-to-cement ratio (w-c), well confirm the Vicat results. So the method can effectively be used for hydration monitoring and determination of setting times. It is simple, requires minimum hardware components and also on-line data acquisition is possible.

3.4.1 Fringing field flexible capacitive sensors for hydration monitoring

3.4.1.1 Impedance spectroscopy

Impedance spectroscopy refers to the study of the electrical response parameters of certain materials under the application of small AC signals with input frequency varying over a wide range. Useful physicochemical properties of the materials under test are obtained by frequency domain analysis of the response. Time domain measurements are also carried out sometimes, and then converted in the frequency domain using a Fourier transform. Impedance spectroscopy is not limited to only impedance measurement and data analysis, but can be successfully applied to any of the four basic immittance measurements. Various contact and non-contact methods have been reported to monitor the progress of hydration and to measure the setting times using impedance spectroscopy. For example, contact methods such as direct current method, electrical resistivity method, AC impedance spectroscopy method and the Wenner probe method [26–29] have been reported in the monitoring of hydration and determination of setting times.

The fringing field is the non-parallel electric field generated at the edges of a parallel plate capacitor when it is excited by a voltage source. It results in undesired fringing capacitance, which modifies the actual capacitance value of the capacitor. However, this undesired field can be used to develop a non-contact impedance sensor or minimally invasive impedance sensor to investigate the formation of hydration products and to accurately measure the initial and final setting times. One attractive feature of the fringing field sensor is single-side testing of the sample. In the recent past, fringing field capacitive impedances have been extensively used for the measurement of physical and chemical parameters [30] including liquid level, moisture in soil, grain, paper pulp and transformer oil, thickness and flow measurement, measurement of position of moving objects, hydration monitoring and moisture measurement in cement [31]. In this section, a fringing field impedance sensor fabricated using a flexible PI substrate has been investigated to monitor hydration in a cost-effective manner and to determine the setting times. Results are also compared with the conventional VA results.

3.4.1.2 Working of the sensor

A schematic of the sensor is shown in figure 3.8(a). It has a copper-clad PI substrate of dimensions of approximately 60 mm × 20 mm × 0.075 mm. The thickness of the PI film is 50 μm and the thickness of the copper cladding on each side of the substrate is

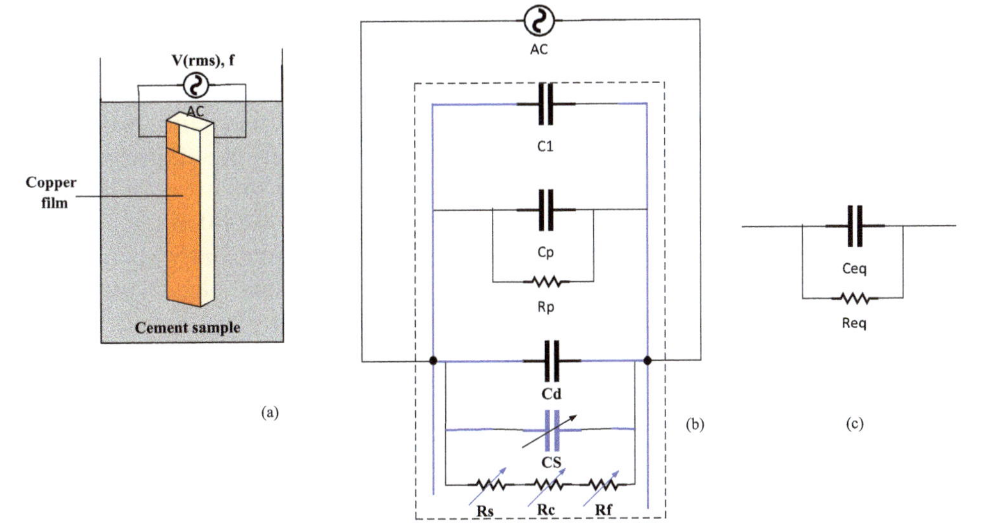

Figure 3.8. Fringing field impedance sensor. (a) The sensor in cement sample; (b) equivalent representation of the sensor; (c) simplified equivalent circuit. © 2020 IEEE. Reprinted, with permission, from [32].

12.5 μm. The PCB substrate with copper film is immersed in cement samples with a fixed amount of Portland land cement in water. The water-to-cement ratio by weight is 0.30. The cement sample can be treated as a lossy dielectric (complex), so the resistance and capacitance values will change due to hydration of the cement products. The hydrating cement consists of ions, which contribute to the resistive part of the impedance, and the hydration products of different oxides, which contribute to the capacitive part of the impedance. If the device is excited by an AC signal, the impedance value of the sensor will change. The device can be represented by an equivalent electrical circuit as shown in figure 3.8(b). The PI substrate sandwiched between copper layers can be represented by a parallel form of capacitance, Cp, and resistance, Rp. The fringing field capacitance can be represented by Cs. The capacitance that is formed at the electrode–electrolyte interface (metal–cement sample) can be represented by Cd. The resistance due to the movement of charges across the metal film and solution is represented by Rs, the resistance due to the movement of ions in the bulk cement sample by Rc. The resistance Rf indicates the diffusion control reaction after FST. The impedance value due to hydration of the cement sample can be represented by a parallel combination of Cd, C_f and ($R_s + R_c +$ Rf). The other components in the circuit contribute to the offset fixed value of the impedance, which remains almost constant provided the substrate material and parasitic capacitance are properly taken care of. The PI substrate as shown in table 3.1 has a water adsorption of 2%–3% and its dielectric constant is 3.5. The greatest contributing factors to the change in impedance value of the sensor are Cf and Rc. The double-layer capacitor also has a significant contribution but its effect is more pronounced at low frequency. If the excitation frequency of the input signal is high (few tens of kHz), then Cd has a small contribution to the total impedance value. The device is finally approximately represented by figure 3.8(c).

3.4.1.3 Determination of the response in the cement sample

The experiment was conducted by dipping the sensor in a 0.3 w-c ratio cement sample at a dipping length of 4 cm. The impedance value of the sensor was measured by a handheld Keysight LCR meter at 1 kHz frequency. Impedance values were noted for nearly 24 hr. Figure 3.9(a) shows the variation of the impedance value with time. The impedance value initially decreases up to 300 mins then starts increasing beyond this time. A better view of the impedance change for 0–300 mins is shown in figure 3.9(b). In figure 3.8(b) different phases of hydration are clearly visible. In the beginning, due to release of ions such as calcium, iron, silica, etc, in the sample, the impedance value decreases, as shown from 'a' to 'b' of the response curve. Then, the impedance value increases due to formation of C–S–H gel around the cement molecules, preventing water from reacting with the cement powder. This phenomenon is observed by the response curve from 'b' to 'c'. After this situation, calcium hydroxide is formed, which breaks the thin wall of C–S–H gel, thus allowing water molecules to react with the cement; thus the impedance value decreases from the 'c' to 'd' point. Then ettringite, a calcium aluminum sulfate compound floating in the mixture of cement and water, suddenly attacks the C–S–H gel and absorbs water. This phenomenon causes a sharp increase in the amount of hydration products. This is shown by the 'd' to 'e' portion of the curve. After this phase, the hydrate crystals start sticking to one another, which is shown by the 'e' to 'f' portion of the curve. Beyond the point 'f', the samples start to gain mechanical strength and IST starts. This initial set time is shown by the point t_i in figure 3.8(b). Using the impedance values, the initial and final setting times of the samples were determined; the values are shown in table 3.4. The experimental results of setting times measured by the sensor closely match the values measured by the VA method. Experiments were conducted to determine the impedance value at 100 kHz frequency. The change in impedance value and the setting time values closely match the variation at 1 kHz, but the value of impedance at the 100 kHz frequency is low. Further, the standard deviation of repeated readings for the same cement sample at 100 kHz is lower than

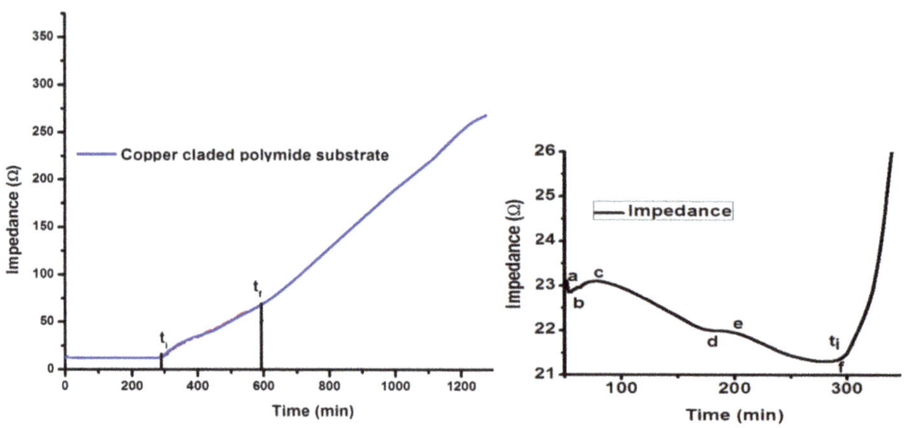

Figure 3.9. (a) Variation of impedance of the device with time-dependent hydration; (b) zoom-in view of the response from 0 to 350 mins. © 2020 IEEE. Reprinted, with permission, from [32].

Table 3.4. Setting time of the cement sample (W:C = 0.30, 1 kHz).

Initial setting time (IST)		Final setting time (FST)	
Measured by flexible impedance probe (min)	Measured by standard Vicat method (VA) (min)	Measured by flexible impedance probe (min)	Measured by standard Vicat method (VA) (min)
290	280 ± 5 mins	580	590 ± 10 mins

Table 3.5. Initial and final setting times of different cement samples (1 kHz).

	Initial setting time (min)			Final setting time (min)		
Cement samples (w:c)	Impedance sensor	VA	Relative error	Impedance sensor	VA	Relative error
0.204	180	175	0.028	310	305	0.016
0.300	287	280	0.025	590	590	0
0.450	866	870	0.04	1090	1100	0.009

the deviation at 1 kHz. This is due to a smaller contribution of the double-capacitance Cd at high frequency, so a high frequency is more suitable to avoid the interface behavior and to maintain precision of the readings.

Experiments were also conducted with a rigid PCB substrate of dimensions 50 mm × 5 mm × 1.5 mm. Three different cement samples were prepared. The first sample had a water-to-cement ratio of 0.30, the second one had a ratio of 0.204 and the third one had a ratio of 0.450. The sample with the lowest ratio has more water than the sample with the highest ratio. Hence the amount of ions in the 0.204 ratio is much higher than in the 0.450 ratio. The impedance value of the lowest ratio sample is minimum and the highest ratio sample is maximum. But the variation of the impedances of all three samples are almost identical, though the impedance value varies widely for different cement samples. From the impedance response curves, initial and final setting times of all the samples were determined. The results are shown in table 3.5. Monitoring the hydration of concrete is important for analyzing the quality of concrete and its strength. This is based on impedance measurement using a flexible copper-clad PCB substrate. The device is very inexpensive, minimally invasive and embedded in cement samples but very much effective. The hydration response curve closely matches the hydrothermal stages of concrete obtained by other reported methods and the setting time values closely match the values obtained by a VA. The device can be easily interfaced to PC through a basic data-acquisition system for on-line hydration monitoring purpose and determination of setting times.

3.5 Flexible sensors for moisture measurement in concrete

Measurement of moisture ingress is another important task to take appropriate measures for, in order to protect from premature corrosion of reinforcement bars [32]. As mentioned earlier, the presence of moisture accelerates the damage of buildings; moisture with more ionic salts content, particularly in coastal areas, is more detrimental. Micro-electromechanical systems sensors have also been used to measure moisture and internal humidity, yet more attention needs to be paid to the longevity of the sensors. Radio frequency identification (RFID) sensors using planner LC circuit have also been used to investigate the local moisture content. Some non-destructive methods like ground-penetrating radar, nuclear magnetic resonance and evanescent field dielectrometry are used in the special case of heritage buildings. A time domain reflectometry (TDR)-based sensor has been proposed recently to measure the hydration process and the moisture ingress of concrete, but the proposed technique requires complex signal conditioning and costly equipment.

3.5.1 Fringing field capacitive sensors for concrete moisture measurement

The moisture in concrete accelerates different chemical reactions, which cause faster deterioration of the structure. It is one of the primary factors in the fast degradation of the physical and mechanical properties of a structure. So, its measurement and continuous monitoring is a healthy practice for SHM. In this section, an effort is made to introduce a fringing field sensor for efficient and cost-effective moisture measurement in concrete.

3.5.2 Theory of the proposed method

A coplanar structure, shown in figure 3.10, can be used for the measurement of moisture in concrete samples. The structure has two identical rectangular electrodes separated by a small gap. Let the dimension of each electrode be $b \times l$ and the gap between the electrodes be g. All the dimensions are labeled in figure 3.10. The coplanar electrodes with suitable gap and size are made on double-sided copper-clad flexible PI substrate by the chemical etching method. On the backside there is a copper film on the PI film, which can be used for shielding purposes (active high or grounded). When the sensor is excited by an AC source of suitable amplitude and frequency, the electric field passes through the positive electrode to the negative electrode. A fringing electric field exists at the edges of the coplanar electrodes and at these edges the field lines bulge out through a long path to reach the negative electrode. This structure can conveniently be used for the single-sided investigation of changes of the dielectric property in a building's structure. If a concrete sample is placed on the sensor, some of the electric fields will pass through the concrete sample. Any variation in the concrete such as caused by moisture content, density, composition, cracks, etc, will change the field strength of the sensor. So, the capacitive impedance of the device will vary according to the condition of the sample [33] (figure 3.10).

The capacitance of the coplanar structure can approximately be given by [34]

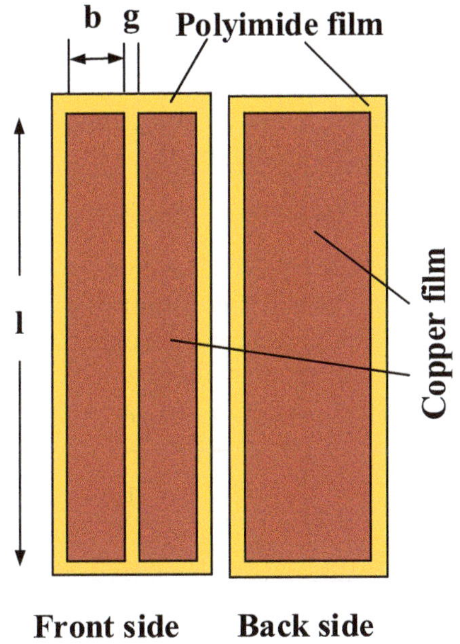

Figure 3.10. Coplanar structure of the sensor.

$$C = \frac{\varepsilon_r l \ln\left(-\frac{2}{\left(\sqrt[4]{1-\frac{g^2}{(g+2b)^2}}\right)-1}\left(\sqrt[4]{1-\frac{g^2}{(g+2b)^2}}\right)+1\right)}{377\pi v_0} \text{for } 0 < \frac{g}{(g+2b)} \leqslant \frac{1}{\sqrt{2}}. \tag{3.12}$$

For $\frac{1}{\sqrt{2}} < \frac{g}{(g+2b)} \leqslant 1$, the capacitance value can be given by

$$C = \frac{\varepsilon_r l}{120 v_0 \ln\left(\frac{2}{\sqrt{\left(\frac{g}{(g+2b)}\right)}-1}\left(\sqrt{\left(\frac{g}{(g+2b)}\right)^{b}}+1\right)\right)}, \tag{3.13}$$

where ε_r is the relative permittivity of the sample and v_0 is the speed of light in vacuum. The expression shows that for a fixed configuration, the value of capacitance is directly proportional to the relative permittivity of the sample. For a typical structure with $l = 40$ mm, $b = 8$ mm and $g = 4$ mm,

$$\frac{g}{g+2b} = \frac{4}{22} = 0.18,$$

which follows the condition of equation (3.12) like $0 < \frac{4}{20} \leqslant \frac{1}{\sqrt{2}}$.

The typical capacitance value for a concrete sample of relative permittivity, say 5, theoretically calculated using expression (3.12) is 3.36 pF. A concrete sample is a somewhat porous material and it has a tendency to absorb water vapor; the relative permittivity change with moisture can approximately be given by Loyanga's empirical equation given below in (3.14) [35, 36]:

$$\varepsilon_r^{1/3} = \gamma((\varepsilon_w^{1/3} - \varepsilon_c^{1/3}) + \varepsilon_c^{1/3}), \tag{3.14}$$

where γ is the fraction of water absorbed by the concrete sample, $\mathcal{E}_w$ is the dielectric constant of water and $\mathcal{E}_c$ is the dielectric constant of the concrete sample in a dry condition. Therefore, it can be observed that the capacitance value of the structure will change according to the variation in moisture concentration in the concrete sample. The electric field penetration depth (T) can be given by

$$F_d = \frac{g}{2}\sqrt{\left(1 + \frac{2b}{g}\right)^2 - 1}\,. \tag{3.15}$$

The penetration depth plays an important role in measuring the moisture content in the sample. It increases with an increase in the b/g ratio of the sensor. On the other hand, with an increase in penetration depth, the value of the capacitance C decreases. So, an optimum penetration depth should be maintained depending on its application. For this typical configuration, the field penetration depth is ~9.8 mm.

3.5.3 Modeling of the sensor

Initially, the proposed structure was simulated using a finite element software to determine the field penetration. The same structure which was used to measure moisture experimentally was designed using this software. The electrodes were excited by an electric potential of 750 mV with a backplane at ground potential. The electric field distribution is shown in figure 3.11. The field lines originate from the positive potential electrode and go to the negative potential electrode. The field penetration depth is 10.8 mm for an electrode width of 9 mm and a gap of 4 mm.

3.5.4 Experimental moisture measurement results in cement slab

A cement slab of adequate size with different compositions of concrete was prepared. The ratio of water to cement in the slab was 0.4. It was dried thoroughly in a laboratory environment via a natural mechanism for a few months. The sensor was placed on a digital balance and then the concrete slab was placed on the sensor head. The terminals of the sensor were then connected to a FDC1004 capacitance to digital conversion evaluation board (Texas Instruments). The high-potential electrode, the sensing electrode (low potential) and the guard electrode in the backplane of the sensor were connected to the board. The board can measure the capacitance value digitally with a maximum measurement range of 0–15 pF. The resolution of the board is 0.5 fF. The high-potential driving electrode was excited with a square AC voltage of 3.3 V and frequency 25 kHz. The evaluation board was interfaced to a laptop for data acquisition and analysis of the data acquired with the help of

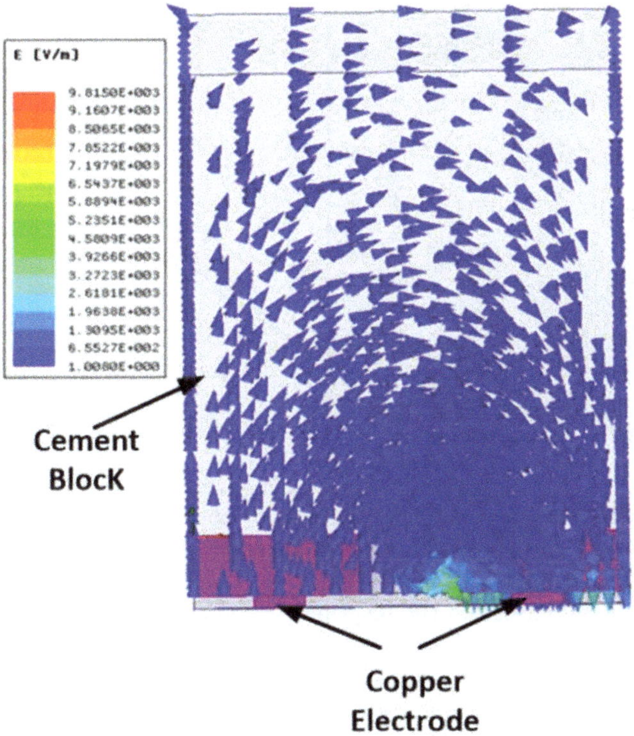

Figure 3.11. Electric field distribution of the coplanar structure.

software provided by the manufacturer. The moisture in the concrete was measured gravimetrically by placing the sensor with a concrete slab on the balance. The moisture level (m_c) was measured by weighing the slab with (w_{sm}) and without moisture (w_{sd}). The sample moisture was varied from 15% to 22% (absolute). The capacitance change of the sensor was noted using the FDC board for different values of moisture concentration. Figure 3.12(a) shows a photograph of the experimental setup. The variation of the capacitance with the weight percentage of moisture in the slab is shown in figure 3.12(b). It shows that the capacitance value increases with an increase in moisture concentration and that the variation as shown in the graph is almost linear. The sensitivity of the capacitance change is almost 50 fF per 1% increase in moisture. The dynamic response of the sensor was also determined with decrease in moisture concentration for nearly three days. Initially the slab was dried in an oven and then it was submerged in water for one day. The slab was then removed from the water and then excess water from the surface of the slab was cleaned by tissue paper. It was then placed on the sensor and the capacitance variation with decrease in moisture was noted. Due to the evaporation of water, the capacitance value gradually decreases.

An experiment was also conducted by depositing a thin layer of hydrophobic insulating film of PDMS on the electrodes and then placing the slab with different moisture levels. The sensitivity of the device with respect to the device without

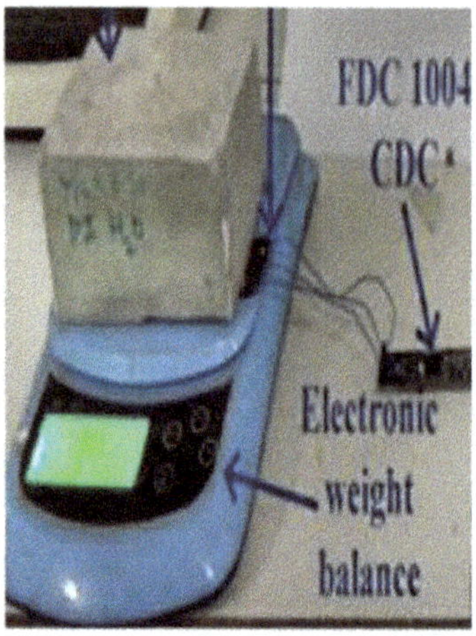

(a)

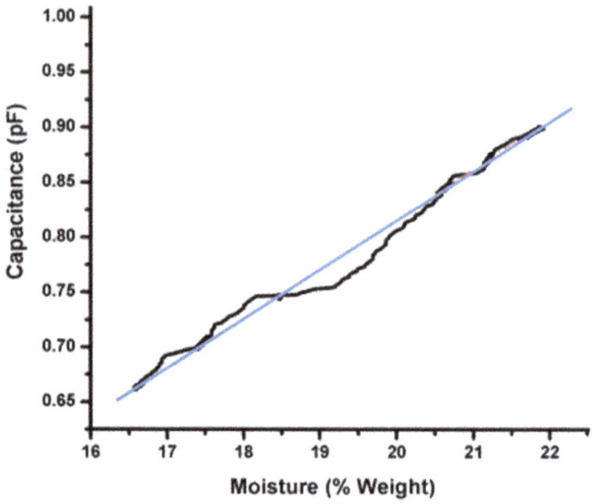

(b)

Figure 3.12. Experimental arrangement. (a) Photograph of the setup interfaced to the data-acquisition board; (b) capacitance variation with moisture content in the cement slab.

hydrophobic film decreases but the response becomes more linear. The gap between the electrodes is now filled with hydrophobic material, so there is a less possibility of water vapor condensation. Two dielectric layers are formed: one the hydrophobic layer of PDMS and another the cement slab. As such, the total capacitance with hydrophobic film is the series combination of two individual capacitances formed by the PDMS layer and the dielectric slab, respectively. The PDMS has a linearizing effect on the total capacitance response with moisture variation.

Moisture measurement of a concrete slab using a basic coplanar structure is encouraging and effective. The capacitance change is significant and repeatable.

3.6 Conclusion

This chapter reviews some electrical methods of SHM. The importance of SHM is discussed. Some of the electrical methods for the determination of initial and final setting time, hydration monitoring and moisture ingress are discussed. It also explains the importance of non-invasive or minimally invasive methods for health monitoring applications.

References

[1] Liu F 2014 Early-age hydration studies of portland cement *PhD Dissertation* Department of Civil and Environmental Engineering, University of Louisville, Kentucky, USA

[2] Akram K J, Islam T and Ahmed A 2018 A simple method on transformer principle for early age hydration monitoring and setting time determination of concrete materials *IEEE Sensors J.* **18** 7266–72

[3] Branco F, Mendes P and Mirambell E 1992 Heat of hydration effects in concrete structures *ACI Mater. J.* **89** 139–45

[4] Jansen D and Goetz-Neunhoeffer F *et al* 2012 The early hydration of Ordinary Portland Cement (OPC): an approach comparing measured heat flow with calculated heat flow from QXRD *Cem. Concr. Res.* **42** 134–8
[5] Morsy M S 1999 Effect of temperature on electrical conductivity of blended cement pastes *Cem. Concr. Res.* **29** 603–6
[6] Kim S G 2010 Effect of heat generation from cement hydration on mass concrete placement *PhD Thesis* Iowa State University
[7] Cataldo A, De Benedetto E and Cannazza G Feb. 2015 Hydration monitoring and moisture control of cement-based samples through embedded wire-like sensing elements *IEEE Sensors J.* **15** 1208–15
[8] Sun M, Staszewski W J and Swamy R N 2010 Smart sensing technologies for structural health monitoring of civil engineering structures: a review *Adv. Civil Engg Eng.* **2010** 724962
[9] Norris A, Saafi M and Romine P 2008 Temperature and moisture monitoring in concrete structures using embedded nanotechnology/microelectromechanical systems (MEMS) sensors *Cons. Building Mat.* **22** 111–20
[10] Baroghel-Bouny M M, Lassabatere T and Coussy O 1999 Characterization and identification of equilibrium and transfer moisture properties for ordinary and high-performance cementitious materials *Cem. Concr. Res.* **29** 1225–38
[11] Khan S, Lorenzelli L and Dahiya R S June 2015 Technologies for printing sensors and electronics over large flexible substrates: a review *IEEE Sensors J.* **15** 3164–85
[12] Zardetto V, Brown T M, Reale A and Di Carlo A 2011 Substrates for flexible electronics: a practical investigation on the electrical, film flexibility, optical, temperature, and solvent resistance properties *J. Polm. Sci. Part B, Polym. Phys.* **49** 638–48
[13] Yu Z, Xia Y, Du D and Ouyang J 2016 PEDOT:PSS Films with metallic conductivity through a treatment with common organic solutions of organic salts and their application as a transparent electrode of polymer solar cells *ACS Appl. Mater. Interfaces* **8** 11629–38
[14] IS 5513 *Specification for VICAT apparatus* (New Delhi, India: Bureau of Indian Standards;) 1996
[15] Kruschwitz S, Bischof E and Taffe A 2015 Multi-sensor investigation of concrete moisture using ultrasound, radar and microwave *Int. Symp., Non-Destructive Testing in Civil Engineering (NDT-CE) (Berlin, September 15–17, 2015)*
[16] Bentz D P, Snyder K A and Ahmed A 2015 Anticipating the setting time of high volume fly ash concretes using electrical measurements: feasibility studies using pastes *J. Mater. Civil. Eng.* **27** 04014129
[17] Luo D, Ibrahim Z, Yang H and Ismail Z 2013 Use of tapered optical fiber sensors in study of the hydration process of cement paste *IEEE Sensors J.* **13** 3415–20
[18] Pessiki S P and Carino N J 1988 Setting time and strength of concrete using the impact-echo method. *ACI Mater. J.* **85** 389–99
[19] Giannelli P, Bulletti A and Capineri L 2017 Multifunctional piezopolymer film transducer for structural health monitoring applications *IEEE Sensors J.* **17** 4583–6
[20] Xiao L Z, Li Z J and Wei X S 2007 Selection of superplasticizer in concrete mix design by measuring the early electrical resistivities of pastes *Cem. Concr. Compos.* **29** 350–6
[21] Mindess S, Young J F and Darwin D 2003 *Concrete* (Upper Saddle River, NJ: Pearson Education)
[22] Salmaz U, Islam T and Suhail S 2020 A novel linear capacitive temperature sensor using polydimethylsiloxane *IEEE Trans. Instrum. Meas.* **69** 7887–94

[23] Lanza V L and Herrmann D B 1958 The density dependence of the dielectric constant of polyethylene, *J. Polym. Sci.* **28** 622–5
[24] Shih H and Flory P J 1972 Equation-of-state parameters for poly(dimethylsiloxane), *Macromolecules* **5** 758–61
[25] Langsdorf A S 1974 *Theory of Alternating Current Machinery* (New Delhi: Tata McGraw Hill Education) 8–20
[26] Woo L, Wansom S and Ozyurt N *et al* 2005 Characterizing fiber dispersion in cement composites using AC-impedance spectroscopy *Cem. Concr. Compos.* **27** 627–36
[27] Liao Y and Wei X 2014 Penetration resistance and electrical resistivity of cement paste with superplasticizer *Mater. Structure* **47** 563–70
[28] Song G 2014 Equivalent circuit model for AC electrochemical impedance spectroscopy of concrete *Cem. Concr. Res.* **30** 1723–730
[29] Weiss J, Snyder K, Bullard J W and Bentz D P 2013 Using a saturation function to interpret the electrical properties of partially saturated concrete *J. Mater. Civil. Eng.* **25** 1097–106
[30] Kong Q and Song G 2017 A comparative study of the very early age cement hydration monitoring using compressive and shear mode smart aggregates *IEEE Sensors J.* **17** 256–60
[31] Mamishev A V, Sundara-Rajan K, Yang F, Du Y and Zahn M May 2004 Interdigital sensors and transducer *Proc. IEEE* **92** 808–46
[32] Akram K J, Ahmed A and Islam T May 2019 Fringing field impedance sensor for hydration monitoring and setting time determination of concrete material *IEEE Trans. Instrum. Meas.* **69** 2131–8
[33] Cao H, Thakar S K, Lee Oseng M and Nguyen C M *et al* 2015 Development and characterization of a novel interdigitated capacitive strain sensor for structural health monitoring *IEEE Sens. J.* **15** 6542–8
[34] Analysis of Multi conductor Transmission Lines, 2nd Ed., Clayton R. Paul, New York, NY, USA: Wiley, 1994.
[35] Bhuiyan R H, Alam M N, Caicedo J M and Ali M 2015 Real-time wireless moisture sensing in concrete using interdigitated stick-on sensors *Sens. Transdu.* **195** 30–8
[36] Siddiqui A Md, Mahboob R and Islam T 2017 A passive wireless humidity tag with digital readout unit for trace moisture measurement *IEEE Trans. Instrum. Meas.* **66** 1013–020

IOP Publishing

Systems for Printed Flexible Sensors
Design and implementation
Tarikul Islam, Subhas Chandra Mukhopadhyay and Boby George

Chapter 4

Resistive sensor interface

Kaushal Kishore and S A Akbar

Resistive sensors are a commonly used sensor type in a wide range of applications such as industrial, commercial and scientific. Such a diverse range of uses also requires innovative and appropriate interface circuits, suitable for the sensors. Sensor interface design is an active area dedicated towards this task, as is fairly evident by the number of research articles published in this domain. The work presented here augments this research area. This chapter focuses on a sensor interface design based on auto-balancing circuits and explores the analysis, design methodology and experimentation with real fabricated sensors to verify the working of interface circuits. As the name suggests, auto-balancing sensor interface circuits are a closed-loop system design for sensor signal conditioning. The work in this chapter targets mid-range resistive and capacitive sensors. A novel auto-balancing, dual-differential architecture has been identified for single-element resistive sensors. The interface circuit calculates a difference of the differential signal across the sensor and reference resistance and generates a control voltage that balances the interface circuit. A detailed analysis of the same is presented here along with applications in gas sensing and temperature sensing. With few circuit modifications, the design is also shown to work for a single-element resistive sensor array. A direct application of the array interface circuit is presented for a silicon nanowire array.

4.1 Introduction

Resistive elements are among the most commonly used sensor types in industrial, scientific and commercial applications. They are inexpensive, can be easily modeled and hence are found in various applications for sensing temperature, humidity, pressure, strain, displacement, gas, etc [1]. Resistive transducers are electrically modeled as $R_x = \rho l / A$, where ρ is the material's resistivity, l is the effective length of the resistor and A is the area of cross section of the resistor. For a typical single-element resistive sensor, this can be re-written as $R_x = R_o(1 \pm kx)$, where k is the transformation factor and R_o is the base value of the resistance under no excitation from the measurand, x. Under an inverse relation, the same can be written as $R_x = \frac{R_O}{(1 \pm kx)}$. The behavior of a resistive sensor in terms of its dynamic range, baseline resistance and linearity highly rely on the fabrication process and interaction of different chemical, biological and physical parameters.

The signal conditioning circuits for resistive sensors are mostly based on bridge-based resistance-to-voltage converters, relaxation-oscillator-based resistance-to-time converters (RTCs) and direct converters [1–6], which have been widely used across many applications. Wheatstone-based signal conditioning circuits are mostly used for interfacing resistive sensors because of their ability to provide highly precise outputs. However, the wheatstone bridge is not preferable for single-element resistive sensors due to their high non-linearity. Further, the measurement range and resolution of bridge-based circuits are limited due to power supply limitations and the high baseline resistance of the sensor. The sensitivity of bridge-based circuits also suffers because half of the signal from the resistance change is attenuated by the adjacent bridge arms. The measurement range and accuracy of bridge-based circuits are enhanced by auto-nulling techniques, as reported in [7–11], using a feedback loop. To cover a wider measurement range, oscillator-based RTC circuits are used. These circuits offer architectural simplicity along with providing said wider measurement range. The variation in the time period of an RTC-based circuit is proportional to the absolute value of the sensor resistance and thus is not suitable for sensors with large and/or variable baseline resistance. Also, oscillator-based circuits show long measurement time and low accuracy for higher resistance values ($\approx$MΩ) and gain-bandwidth product limitations for low resistance values. Under a quarter-bridge configuration, the output response of the bridge is non-linear [1, 12]. This chapter presents instrumentation modules which can deliver a linear output over a wide dynamic range for single-element resistive sensors with a baseline resistance either unknown or known within an estimated range. The circuit provides provision for in-circuit resistance measurement by incorporating a closed-loop feedback system that generates a proportional control voltage to the sensor resistance, thus keeping the circuit in a balanced condition. The circuit delivers a precise, linear and remote measurement capability and has been experimentally verified with commercial and in-house fabricated sensors. Extending the dual-differential architecture, a resistive sensor array interface circuit is designed. The circuit scans the array by selecting sensor elements simultaneously and generates a linear control signal response proportional to the resistive element. Users can choose a hard scan time or an auto-trigger mode based on the in-built trigger signal generated by the interface circuit.

Detailed working and analysis of both the developed interface circuits are expanded upon in subsequent sections.

4.2 Auto-balancing interface circuits

The presented auto-nulling, dual-differential subtractor-based circuit observes potential differences between the sensor and a reference both carrying the same current. The difference of these voltages is used as an error signal to drive a feedback circuit, which ultimately tunes a voltage-controlled resistance (VCR). The technique provides an inherent linear output voltage to change in resistance. The architecture also helps in noise reduction since the voltage drop across the sensing resistor is differentially measured. Moreover, the dual-differential subtraction and auto-nulling ensure that the sensitivity of the developed circuit is almost constant for the entire range of operation, thus giving a linear input–output relation. Details of the circuit design and its operation are described in the following section.

4.2.1 Circuit architecture and operation

For single-element resistive sensors, a signal conditioning circuit based on an auto-nulling, dual-differential subtractor is illustrated in figure 4.1. The sensor resistance, R_s, can be written as $R_o + \Delta R$, where R_o is the nominal resistance and ΔR is the variation in resistance due to the measurand. The base resistance value, R_o, is either unknown or known within an acceptable estimated range. The circuit is excited with a stable current source, represented as I_{ex}. The working of the circuit is explained in

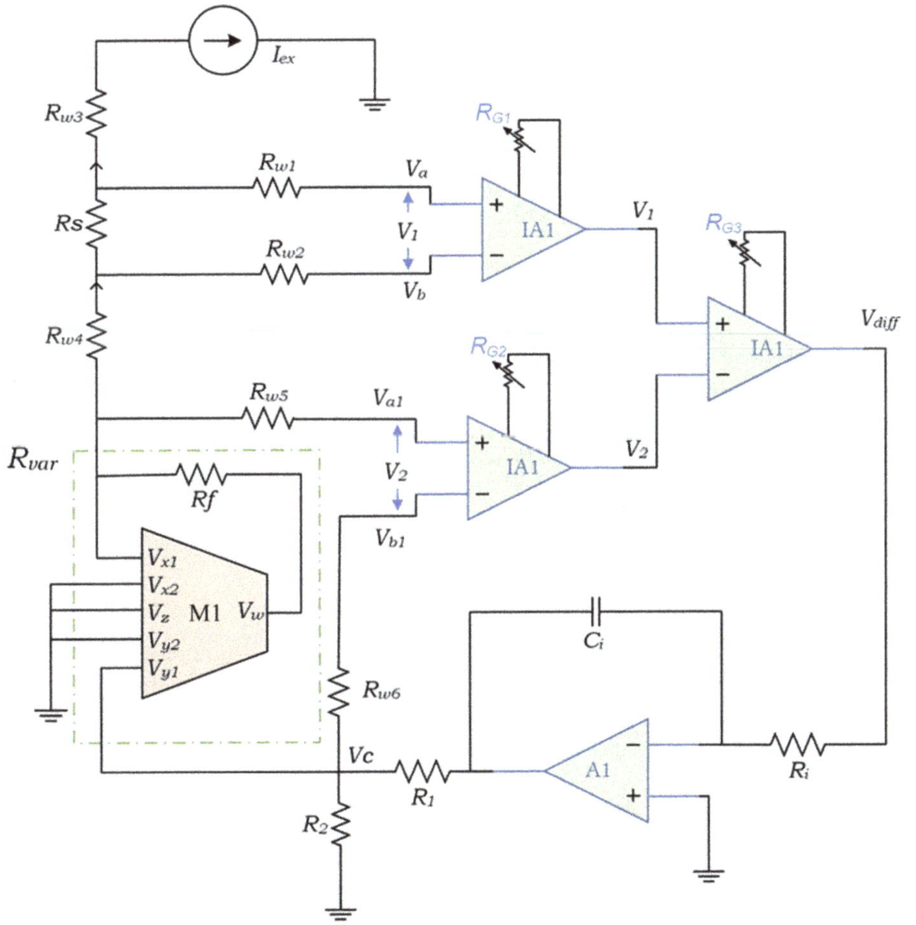

Figure 4.1. Auto-balancing dual-differential resistive sensor interface.

two different functional blocks, namely a voltage-controlled resistor (R_{var} in figure 4.1) and a double-differential subtractor with an auto-nulling feedback loop.

The sensor voltage drop forms the first channel, resulting in a voltage, V_1. The null-detection at the output of the dual-differential subtractor (V_{diff}) adjusts the output control voltage (V_c) of the tunable resistance to ensure that the effective signal at the terminal (V_2) matches the voltage of the terminal (V_1). The tuning control voltage is taken as the output of the circuit, which varies linearly with the sensor resistance change. The following sections describe each functional block in detail.

4.2.1.1 Voltage-controlled resistor (VCR)

An important component of the interface circuit introduced above is the voltage-controlled resistor (VCR). The key device for realizing a VCR is an analog multiplier. Here a reference resistance, R_f, is connected to a commercial four-quadrant multiplier (AD633 [18]), as shown in the schematic (figure 4.1). Considering the properties of the AD633, the transfer function of the multiplier, M1, is expressed as follows:

$$V_w = \frac{(V_{x1} - V_{x2})(V_{y1} - V_{y2})}{10} + V_z. \tag{4.1}$$

The terminals V_{x2}, V_{y2} and V_z are connected to ground. Following the notation shown in figure 4.1, the terminal voltage, V_w, is expressed as

$$V_w = \frac{(V_{x1})(V_{y1})}{10} = \frac{V_{a1}V_c}{10}. \tag{4.2}$$

Considering the input excitation current entering the analog multiplier, the equivalent resistance formed will be

$$I_{ex} = \frac{V_w - V_{a1}}{R_f} = \frac{-1}{R_f}\left[V_{a1} - \frac{V_{a1}V_c}{10}\right], \tag{4.3}$$

$$R_{var} = \frac{V_{a2}}{-I_{ex}} = \frac{R_f}{1 - \frac{V_c}{10}}. \tag{4.4}$$

For the multiplier integrated circuit (IC), the control voltage can swing between ±10 V, but towards the upper limits there is dominant non-linearity. Hence, practically, the device is used till ±9.8 V. Using the practical limits of the control volatge in equation (4.4), the tunable range of voltage-controlled resistance, R_{var}, is

$$R_{var}(v_c) = \begin{cases} 0.505R_f & ; V_c = -9.8V \\ R_f & ; V_c = 0V \\ 50R_f & ; V_c = +9.8V \end{cases}. \tag{4.5}$$

R_f is a constant resistance whose value act as a reference resistance in the voltage-controlled resistance, R_{var}. The value of R_f dictates the range and baseline compensation offered by the circuit and hence also affects the percentage error in the working range of the sensor. Details on this will be discussed in the subsequent section.

4.2.1.2 Auto-balancing dual-differential subtractor

The instrumentation amplifiers IA1, IA2 and IA3 form a dual-differential architecture, each with its own gain resistance R_{G1}, R_{G2} and R_{G3}, respectively. The IA1 and IA2 pair buffers the differential voltages V_1 and V_2 from the sensor and reference resistance, R_{var}. The instrumentation amplifiers offer a high-input impedance and hence the lead wires, R_{w1}, R_{w2}, R_{w5} and R_{w6}, offer the least effects on the reading. In fact, the architecture shown in figure 4.1 represents a four-wire Kelvin measurement. Assuming no loss in power, the voltage across the sensor resistor, R_s, and the tunable resistor, R_{var}, is subtracted by the instrumentation amplifier IA3, generating a difference of the differential voltages V_1 and V_2:

$$V_{diff} = V_1 - V_2 = (V_a - V_b) - (V_{a2} - V_{b2}). \tag{4.6}$$

The differential voltage, V_{diff}, is used as a feedback signal in the auto-balancing loop. The integrator formed by amplifier A1, resistor R_i and capacitor C_i generates a control voltage signal which tunes the voltage V_2 to drive the voltage V_{diff} to zero, thus auto-balancing the circuit. From figure 4.1, equation (4.6) can be extended as

$$\Delta V = (V_1 - V_2) = [(-IR_s)G_1 - (-V_{a1} - V_c)G_2]G_3, \tag{4.7}$$

where $[G_1, G_2, G_3]$ are the gain provided by IA1, IA2 and IA3, respectively. Using $\frac{V_{a1}}{-I_{ex}} = \frac{R_f}{1-\frac{V_C}{10}}$, from equation (4.4) we get

$$R_s = \frac{R_f G_2}{(1 - V_c 10)} + \frac{V_c G_2}{I_s G_1} - \frac{V_{diff}}{I_s G_1 G_3}. \tag{4.8}$$

The above equation suggests that the value of the sensor's resistance can be calculated by measuring the control voltage, V_c, and error signal, V_{diff}. Considering equation (4.8), the un-normalized sensitivity of the interface circuit is calculated as the partial change in control voltage response to the partial change in sensor resistance. Rearranging the sensor resistance equation (equation (4.8)) leads to

$$R_s = \frac{I_s R_f G_1 G_2 G_3 + V_c G_2 G_3\left(1 - \frac{V_c}{10}\right) - V_{diff}\left(1 - \frac{V_c}{10}\right)}{I_s G_1 G_3\left(1 - \frac{V_c}{10}\right)}. \tag{4.9}$$

The sensitivity delivered by the interface circuit is given as

$$S_{R_s}^{V_c,\,v} = \frac{\partial V_c}{\partial R_s} = \frac{I_s G_1 G_3 (1 - V_c 10) + \frac{\partial V_{diff}}{\partial R_S}(1 - V_c 10)}{\frac{I_S R_S G_1 G_3}{10} + G_2 G_3 - \frac{2V_C G_1 G_3}{10} + V_{diff} 10}. \tag{4.10}$$

The feedback action of the auto-balancing network will integrate the error voltage, V_{diff}, and the output of the integrator will approach $V_c = R_2/(R_1 + R_2)\int_0^t V_{diff}dt$, thus adjusting the value of R_{var} using the control voltage until the error diminishes to zero. Under the zero-error condition, the circuit is said to be at a steady state (or balanced state) and the value for the sensor resistance, R_s, is expressed as

$$R_s\,|_{V_{diff}=0} = \frac{R_f G_2}{(1 - V_c 10)} + \frac{V_c G_2}{I_s G_1}. \tag{4.11}$$

Under the balanced condition, the in-circuit resistance of the sensor can be calculated only by measuring the control voltage, V_c, since the error signal is reduced to zero. The sensitivity of the circuit under the balanced condition ($V_{diff} = 0$) is derived using equation (4.11) as

$$S_{R_s}^{V_c}\,|_{V_{diff}=0} = \frac{10 I_s(1 - V_c 10)}{G_1(I_s R_s - 2V_c) + 10G_2}. \tag{4.12}$$

The above equation suggests that the gain, G_1 and G_2, provided by the IA1 and IA2 instrumentation amplifiers, respectively, can be used to adjust the circuit's sensitivity. The optimum sensitivity can be achieved with $G_1 = G_2 = 1$. Keeping $[G_1, G_2, G_3] = 1$, a typical SPICE simulation with a reference resistance, $R_f = 8.1$ kΩ, and an excitation current, $I_s = 10$ μA, shows that the circuit works effectively, giving a linear output for sensor values varying from 15 to 1100 kΩ with a relative error of less than ±0.8%, as shown in figure 4.2. The indicated measurement range covers the balanced state and will be extended if an unbalanced state is also used for measurement using the equation (4.9).

The sensitivity equation (4.12) describes that sensitivity of the circuit can be linearly adjusted with the excitation current, I_{ex}. This effect can be seen by varying the excitation current, and is shown in figure 4.3. Practically, an increase in excitation current is dependent on the input power of the current source and the

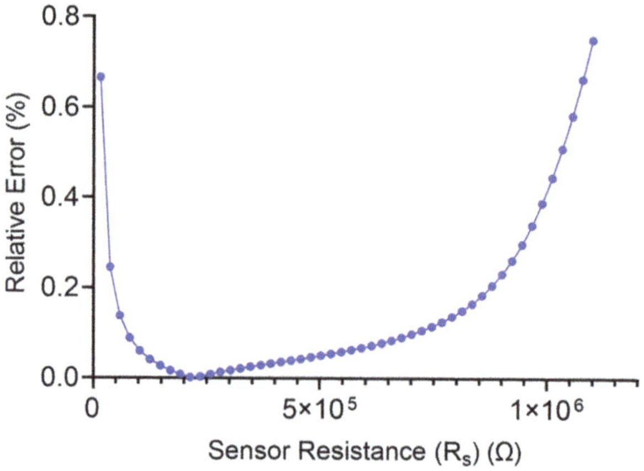

Figure 4.2. Relative error plot for varying sensor resistance, R_s.

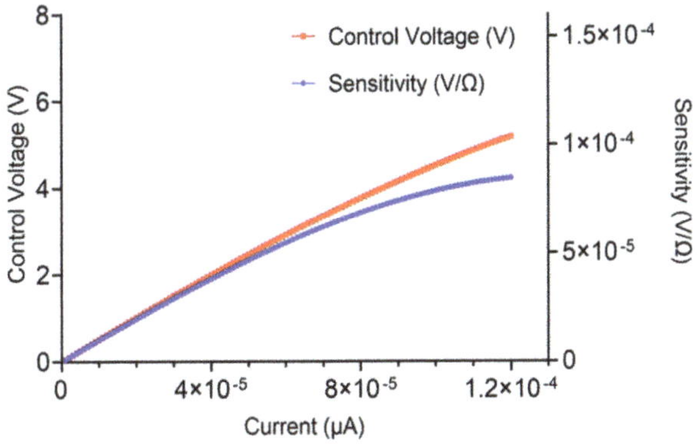

Figure 4.3. Sensitivity variation with respect to varying input excitation current.

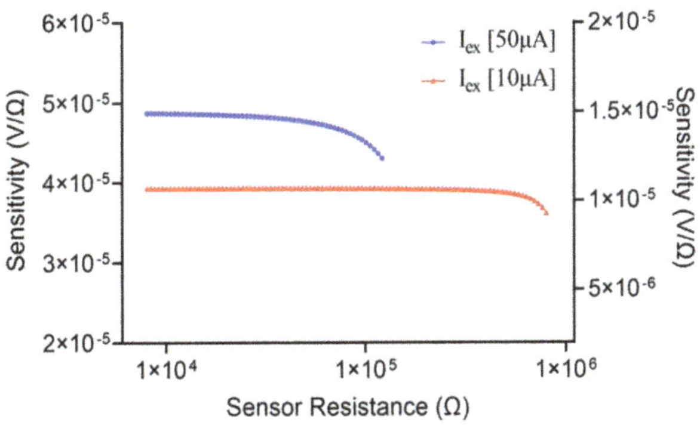

Figure 4.4. Sensitivity plot for excitation current of 50 and 10 μA.

range of the sensor resistance. This in particular is presented in the circuit analysis in section 4.2.2. For a particular excitation current, the sensitivity offered by the circuit is also relatively stable and does not vary much for a wide range of sensor variation. This can be seen in figure 4.4. The circuit delivers an average sensitivity of 9.9 μV Ω^{-1} for an excitation current of 10 μA over a range of 15–1100 kΩ, thus ensuring a linear output. Similarly, for 50 μA of current, an average sensitivity of 48 μV Ω^{-1} is achieved over a resistance range of 8–230 kΩ.

4.2.2 Circuit analysis

4.2.2.1 Linearity analysis

The output equation (equation (4.11)) under a balanced condition and with unity gain is expressed as $\frac{R_f}{1-\frac{V_C}{10}} + \frac{V_C}{I_{ex}} - R_s$. The unknown sensor resistance, R_s, is expressed in terms of a controlled voltage, V_c, which is linearly inseparable and hence represents a coupled equation. Numerical methods, such as bisection, give a better way of solving such an equation to analyze the input–output relation. Rearranging equation (4.8) into a homogeneous form, we get

$$\frac{R_f}{1-\frac{V_C}{10}}+\frac{V_c}{I_{ex}}-R_s=0. \tag{4.13}$$

Solving recursively for V_c to find roots for each sensor resistance and sweeping for the full range of sensor resistance results in a response curve representing the solution for the equation (4.13) and is shown in figure 4.5. For purposes of analysis, let us denote this solution set as $f(R)$. The measure of non-linearity is the average deviation of a function from a linear best fit within the bounded domain of the function [14]. For a given point in the curve, deviation from the linear best fit is given by $\Delta=[f(R)-g(R)]$, where $g(R)$ is the linear best fit of the solution curve. The root-mean-square of the deviation (Δ) along the bounded domain gives a measure for the non-linearity in the curve. This is represented as follows:

$$nL=\sqrt{\frac{\int_{R_l}^{R_u}[f(R)-g(R)]^2\ dR}{R_u-R_l}}, \tag{4.14}$$

where R_l and R_u are the lower and upper limits for the sensor resistance, respectively. The relative non-linearity (λ) with respect to the full-scale output is given as a ratio of nL to output span. For the circuit under consideration, $V_{c|l}$ and $V_{c|u}$ are the lower and upper limits of the control voltages, respectively; hence, the relative non-linearity will be

$$\lambda=\frac{nL}{V_{c|u}-V_{c|l}}. \tag{4.15}$$

Fixing the circuit parameters with $I_{ex}=10\ \ \mu A$ and $R_f=8.1\ k\Omega$, a calculation based on equation (4.14) for a resistance range from 1 to 1000 kΩ results in a

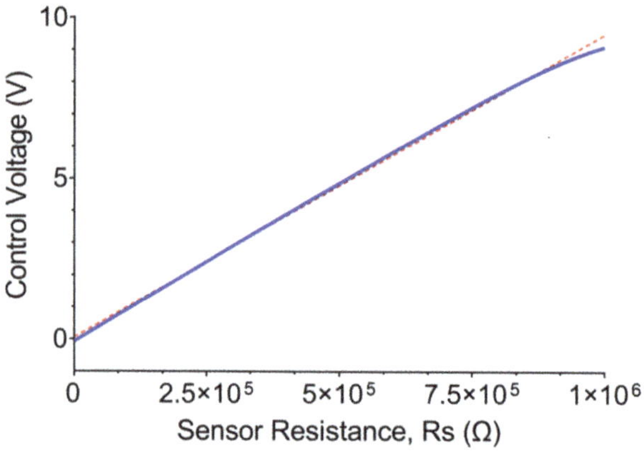

Figure 4.5. Control voltage (V_c) response for different values of sensor resistance, R_S.

non-linearity of 0.15% in the output control voltage, V_c. With such a low level of non-linearity, the circuit is almost linear for a wide working range.

4.2.2.2 Transient analysis

An auto-balancing circuit is a negative-feedback system, trying to minimize a particular quantity with respect to the reference. Considering figure 4.1, the circuit can be analyzed with voltage V_1 as an input signal and V_2 as feedback. A simplified closed-loop model of the circuit, where the gain of $IA3$ ($G_{IA3} = 1$), the voltage divider formed by R1 and R2 (A_d) and the feedback gain $\left(G_{Fb} = 10.\ R_f.\ I_{ex}\right)$, is shown in figure 4.6. The integrator formed by A1, R_i, and C_i can be considered as an integral controller with gain $I = \frac{K_I}{s}$, where $K_I = \frac{1}{\tau} = 1/R_iC_i$.

The desired transient response from the circuit can be tuned by varying the time constant of the circuit. An input perturbation voltage, V_1, in the form of a step change, is applied and the output is measured at V_c in MATLAB, resulting in the response shown in figure 4.7. It can be noted that, for a fast response, the integral gain K_I has to be large. Note that this value is limited by the bandwidth of the operational amplifier (op-amp), A1. For the given circuit, the op-amp, OPA177FP, has $A_o = 130$ dB and $\omega_p = 0.6$ MHz. The value of the RC can be selected such that the system remains stable. For the prototype circuit, the values of R and C are selected as 220 kΩ and 100 pF, respectively, delivering a phase margin of 89.99°, $K_I = 45454.5$ and $\tau = 22$ μs, which confirms that the system is stable. Further, with increasing R_iC_i, a reduction in phase margin will be observed.

The transient response for varying time constant (τ) is shown in figure 4.7 with varying input perturbations. The voltage offset given by the integrator will increase with increase in R_i because of the bias current of the op-amp. Also, the response of the integrator is dependent on the R_iC_i time constant and is limited by the gain-bandwidth product of the op-amp.

Following the error and dynamic response constraints, the parametric values of R_i and C_i will range in a few hundred kilo-Ohm and pF to nF, respectively, and can be chosen by the designer as per the desired transient response.

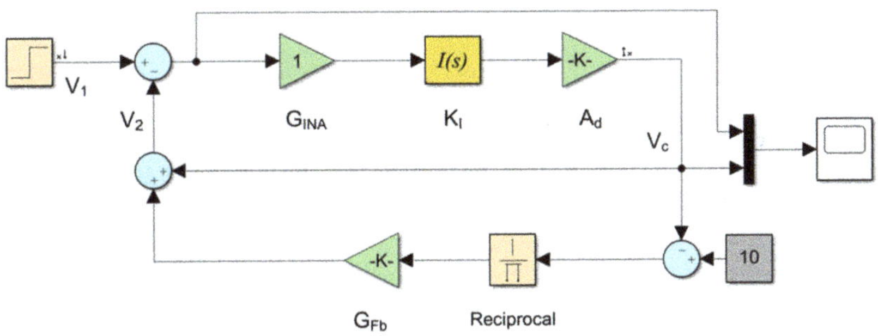

Figure 4.6. System-level block diagram of the interface circuit.

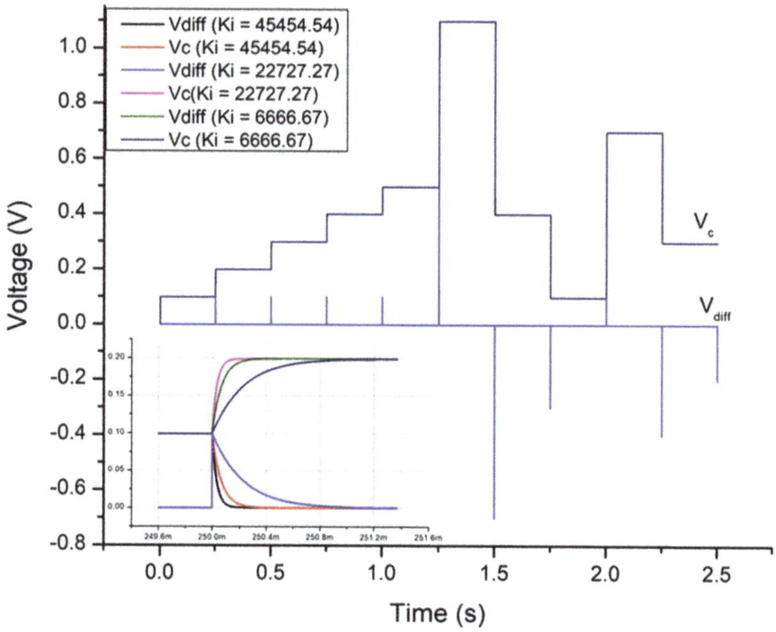

Figure 4.7. Dynamic response effect for varying time constant.

4.2.2.3 Small signal stability

The stability of a system is determined by its ability to contain errors within justified limits when subjected to small disturbances. In more familiar words, a system is stable when it can deliver a bounded output for a bounded input and does not result in an oscillatory response. For an interface circuit, small changes in the sensor's resistance should produce a linear and exponentially stable response. For analysis of the small signal stability of the circuit, let a small change in sensor resistance be r_s, which results in a small response in control voltage, u_c. Introducing a small change in equation (4.13), we get

$$\begin{aligned} R_s + r_s &= \frac{R_f}{1 - \frac{(V_c + u_c)}{10}} + \frac{V_c + u_c}{I_{ex}} \\ r_s &= \frac{R_f}{1 - \frac{V_c}{10}}\left[-1 + \frac{\left(1 - \frac{V_c}{10}\right)}{1 - \frac{(V_c + u_c)}{10}}\right] + \frac{u_c}{I_{ex}}. \end{aligned} \tag{4.16}$$

Neglecting such a small change in response, u_c, with respect to the DC working point, V_c, we get

$$r_s = u_c\left[\frac{\frac{R_f}{10}}{\left(1 - \frac{V_c}{10}\right)^2} + \frac{1}{I_{ex}}\right]. \tag{4.17}$$

The above equation (4.17) shows that the control voltage signal varies linearly with sensor resistance for small perturbations. Also, the proportionality constant is dependent on the operating point $\left[V_c, R_f, I_{ex}\right]$. Further, the expression for the control voltage is given by

$$u_c = -\frac{R_2}{R_1 + R_2}\frac{1}{R_iC_i}\int_0^t \Delta V_{diff}dt. \tag{4.18}$$

The term ΔV_{diff} represents the change in error between the sensor response V_1 and feedback V_2. Calculating the error introduced by the perturbation, r_s, we get

$$\begin{aligned} \Delta V_{diff} &= -I_{ex}(R_s + r_s) - \left[\frac{I_{ex}R_f}{(1 - V_c/10)} - (V_c + u_c)\right] \\ \Delta V_{diff} &= u_c\frac{R_f}{10}\frac{1}{(1 - V_c/10)^2}. \end{aligned} \tag{4.19}$$

Equation (4.19) can be modified using equation (4.18) to calculate u_c as

$$u_c = -\left[\frac{R_2}{R_1 + R_2}\frac{R_f}{10}\frac{1}{\left(1 - \frac{V_c}{10}\right)^2}\right]\frac{u_c(s)}{R_iC_is}. \tag{4.20}$$

Representing in state space form, $\dot{u}_c = -Au_c$

Where $A = \frac{1}{RC}\left[\frac{R_2}{R_1+R_2}\frac{R_f}{10}\frac{1}{\left(1-\frac{V_C}{10}\right)^2}\right]$ is a strictly positive value. This suggests that small changes in the sensor resistance will produce a linear and exponentially stable response, u_c. The decay rate is defined by the $[R_iC_i]$ time constant of the integrator in the feedback loop.

4.2.2.4 Effect of the input offset voltage

The input offset voltage of the instrumentation amplifier will introduce an error in the sensor resistor's measurement. Considering the input offset voltage due to IA1, IA2 and IA3 as V_{os1}, V_{os2} and V_{os3}, respectively, the expressions for the output of the IAs, including the input offset voltage, can be derived as follows:

$$V_1 = -I_{ex}R_s + V_{os1}, \tag{4.21}$$

$$V_2 = -\left(\frac{R_f}{1 - \frac{V_C}{10}}\right)I_{ex} - V_c + V_{os2}, \tag{4.22}$$

$$V_{diff} = -I_{ex}R_s + V_{os1} + \left(\frac{R_f}{1 - \frac{V_C}{10}}\right)I_{ex} + V_c - V_{os2} + V_{os3}. \tag{4.23}$$

Under the balanced condition, the above equation gives an estimate of the sensor resistor, R_s, as

$$R_s = \frac{R_f}{1 - \frac{V_c}{10}} + \frac{V_c}{I_{ex}} + \frac{V_{os1} - V_{os2} + V_{os3}}{I_{ex}}. \tag{4.24}$$

Another offset source is the op-amp, A1, with input offset voltage (e_{os}) and bias current (i_{os}). The total offset introduced by the A1 op-amp is $V_{os} = (e_{os} + i_{os}R)$. Considering these added terms, the output response of the circuit will be further extended to include the integrator offset voltage as

$$R_s = \frac{R_f}{1 - \frac{V_c}{10}} + \frac{V_c + V_{os}}{I_{ex}} + \frac{V_{os1} - V_{os2} + V_{os3}}{I_{ex}}. \tag{4.25}$$

This effect can be further reduced by using the right off-the-shelf commercial IC. Careful selection of an instrumentation amplifier with a low input offset will result in reduced offset effects. For prototyping, a INA114 with an offset voltage of 50 V is used. Designers can choose commerical ICs such as LTC2053 and AD8421, which have offset voltages as low as 10 and 25 μV, respectively.

4.2.2.5 Gain error of instrumentation amplifier

Deviation from the ideal gain equation of the instrumentation amplifier results in a gain error. $G = 1 + \left(\frac{R_{fo}}{R_g}\right)$ represents the ideal gain of an IA, where R_{fo} and R_g are the internal and external resistor to adjust the gain. Assuming a variation in gain as $G_t = G_o + \Delta G$, where G_o and ΔG are the theoretical gain and the gain variation, we can rewrite the gain equation as $G_t = G_o(1 + G_\epsilon)$, where $G_\epsilon = \Delta G / G_o$ is the relative gain error.

For the prototype circuit, the theoretical gain, G_o, for all the IAs is equal to unity. The effective gain for the circuit can be written as $G_t = (1 + G_\epsilon)$. Considering the gain errors $G_{\epsilon 1}$, $G_{\epsilon 2}$ and $G_{\epsilon 3}$ for IA1, IA2 and IA3, respectively,

$$V_1 = -I_{ex}R_s \quad (1 + G_{\epsilon 1}), \tag{4.26}$$

$$V_2 = \left[-\left(\frac{R_f}{1 - \frac{V_c}{10}}\right)I_{ex} - V_c\right](1 + G_{\epsilon 2}), \tag{4.27}$$

$$V_{diff} = \left[\left(-R_s(1 + G_{\epsilon 1}) + \frac{R_f}{1 - \frac{V_c}{10}}(1 + G_{\epsilon 2})\right)I_{ex} + V_c(1 + G_{\epsilon 2})\right](1 + G_{\epsilon 3}). \tag{4.28}$$

Under the balanced condition, the estimated sensor resistance from the above equations is given as

$$R_s = \frac{(1 + G_{\epsilon 2})}{(1 + G_{\epsilon 1})}\left(\frac{R_f}{1 - \frac{V_c}{10}} + \frac{V_c}{I_{ex}}\right). \tag{4.29}$$

The commercial instrumentation amplifier ICs used to implement IA1, IA2 and IA3 are the same and with unity gain. The gain error with unity gain is minimum and thence this effect will be negligible. The gain error variation between the same commercial ICs will be less, and hence the ratio in equation (4.29) will turn out to be one. A suitable method to completely remove this error would be to use a single-package, quad-instrumentation amplifier IC with matched devices.

4.2.2.6 Error due to analog multiplier

Variation in multiplier gain and voltage offset are two main sources of error presented by the analog multiplier. For the AD633, the internal gain is V/10. Considering the shift in the gain of an analog multiplier is $1/(10 \pm G_{am})$ and the offset is V_{ofm}, the expression for the voltage, V_{a1}, from equation (4.4) can be written as follows:

$$V_{a2} = -I_{ex}\frac{R_f}{1 - \frac{V_C}{10(1 \pm G_{\epsilon am})}} \pm V_{ofm}, \tag{4.30}$$

where $G_{\epsilon am} = \Delta G/10$. The expression for sensor resistance, R_s, now will be modified as follows:

$$R_s = \frac{R_f}{1 - \frac{V_C}{10(1 \pm G_{\epsilon am})}} - \frac{V_C \pm V_{ofm}}{I_{ex}}. \tag{4.31}$$

A precalibration of the multiplier IC will reduce this error. The effect of the offset voltage, V_{ofm}, can further be reduced by adjusting the offset pin of the AD633. The designer can choose altogether from various other commercially available ICs suited for a particular design, such as the MPY633, from Texas Instruments, with an accuracy of 0.5%.

4.2.2.7 Effect of excitation current variation

One of the most crucial components in the interface circuit design is the excitation current source. It is required that the current source is DC and free from variations. For analysis, let ΔI_{ex} be the variation in the excitation current. Under this variation, the sensor resistance expression will change to

$$R_s = \frac{R_f}{1 - \frac{V_C}{10}} + \frac{V_c}{I_{ex}(1 \pm \epsilon_{I_{ex}})}, \tag{4.32}$$

where $\epsilon_{ex} = \frac{\Delta I_{ex}}{I_{ex}}$. Static DC level shifts can be compensated by initial calibration and hence do not create a challenge. However, varying AC components must be filtered.

4.2.2.8 Effect of reference resistor, $\boldsymbol{R_f}$

R_f is a constant resistance whose value acts as a reference resistance in the voltage-controlled resistance, R_{var}. The resolution of the circuit at a constant excitation current is dependent on the slope of the control voltage response. The higher the slope, the higher the sensitivity, and thus a better resolution is delivered. According

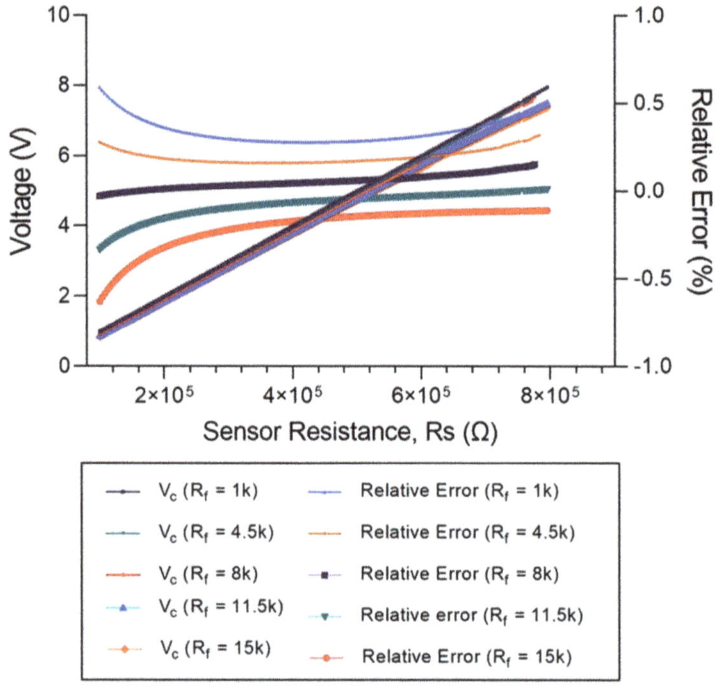

Figure 4.8. Effect of reference resistance, R_f, on the control voltage response and circuit error.

to equation (4.13), the response curve of the control voltage shifts downwards with a relatively constant slope, indicating that the sensitivity of the response is not changing. Minor variation in the slope and sensitivity is seen only towards the higher extremes of the measurement range. This effect is shown in figure 4.8.

4.2.2.9 Error due to variation in reference resistor, $\boldsymbol{R_f}$

A stable value of the resistance R_f is also crucial for accurately estimating the sensor resistance. Any variation in the reference resistor R_f due to variation in temperature or any other environmental factor introduces an error in the measurement. Considering ΔR_f as the variation in the R_f due to external disturbances, the expression for the sensor resistance can be written as follows:

$$R_S = \frac{R_f(1 \pm \epsilon_{R_f})}{1 - \frac{V_c}{10}} + \frac{V_c}{I_{ex}}, \tag{4.33}$$

where $\epsilon_{rf} = \frac{\Delta R_f}{R_f}$.

4.2.2.10 Effect of input power on measurement range

For a given power, $P_{ex} = V_{ex}I_{ex}$, of the excitation current loop, the input power delivered by the source must be greater than or equal to the total power dissipated across the loop [15, 16] and is given as

$$V_{ex}I_{ex} \geqslant I_{ex}^2 R_s + I_{ex}^2 R_{w3} + I_{ex}^2 R_{var}. \tag{4.34}$$

Using equations (4.4) and (4.9) in equation (4.35), under null-condition, we get

$$V_{ex} \geqslant I_{ex}R_s + I_{ex}R_{w3} + I_{ex}\left(R_s - \frac{V_c}{I_{ex}}\right)R_s \leqslant \frac{V_{ex} + V_c}{2I_{ex}} - \frac{R_{w3}}{2}. \tag{4.35}$$

Clearly, the range of operation for the circuit is dependent on the input excitation power. The feedback control enhances the range by an additional term of V_c and minimizes the lead wire effect by half. Assuming a negligible contribution of the lead wire, the limiting range of operation for an excitation current of 10 μA powered through a 15 V power supply would be around 1.2 MΩ.

4.2.3 Prototype design and experimental results

Experimental validation of the interface circuit shown in figure 4.1 is carried out by choosing commercially available IC and off-the-shelf components. A complete list of the commercial ICs and component values are listed in table 4.1. A precision instrumentation amplifier, INA114, is used to implement IA1, IA2 and IA3.

The circuit is supplied with ±15 V connecting all active devices. The output of the integrator can reach ±15 V, which will exceed the maximum voltage allowed at the V_{y1} terminal of the multiplier, which is 10 V. To avoid this situation, a voltage divider configuration with R_1 = 50 kΩ and R_2 = 105 kΩ is formed at the output of the integrator. The resulting divider scales the output voltage of the integrator by 0.677, ceiling the control voltage V_c at the V_{y1} terminal of multiplier M1 within ±10 V. The reference resistance, R_f = 8.1 kΩ, is used between the V_{x1} and V_w terminals of the multiplier. All the passive components are verified with an LCR meter (Keysight, E4980AL) for a tolerance within ±1% for resistive elements and ±5% for capacitive elements. The excitation current is provided by a 2450 Source Meter, from Keithley, and the sensor resistance is swept using a calibrated decade resistance box. With changing input resistance, the control voltage response is acquired by the

Table 4.1. Component Model/Value

Components	Model/Value
Instrumentation Amplifier IA1, IA2, IA3	INA114
Amplifier A1	OPA177
Analog Multiplier M1	AD633
R_i	220 kΩ
C_i	100 pF
R_1	50 kΩ
R_2	105 kΩ
R_f	8.1 kΩ

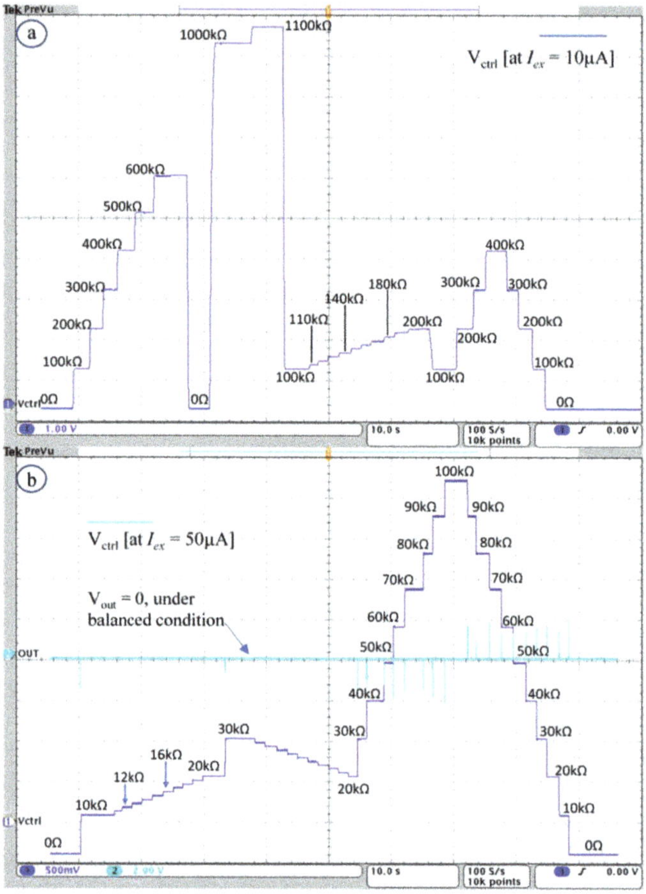

Figure 4.9. (a) Control voltage, V_c, variation with different sensor resistance values with an excitation current of 10 μA. (b) Control voltage variation with different sensor resistance values with an excitation current of 50 μA. Each step change corresponds to 10 kΩ.

oscilloscope (MDO4034-B-3) and is shown in figure 4.9(a) and (b) for a 10 and 50 μA excitation current, respectively. By varying the resistance from the decade box, we obtained two different sets of outputs for different excitation currents. Using these data points, we can now compare and analyze the hardware level output with the analytical result obtained in section 4.2.2.

4.2.3.1 Linearity test

Using the experimental data, figures 4.10 and 4.11 show the response of control voltage with changing sensor resistance and the corresponding relative error. The experimental evaluation indicates that the circuit can follow sensor variations within an interval of 20 kΩ–1.1 MΩ with a relative error of less than 0.78% for an excitation current of 10 μA.

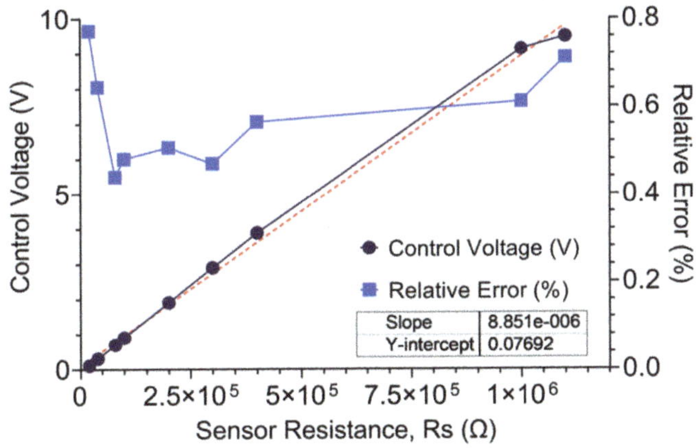

Figure 4.10. Control voltage response at an excitation current of 10 μA with the relative error. (Inset shows the coefficient of linear equation.)

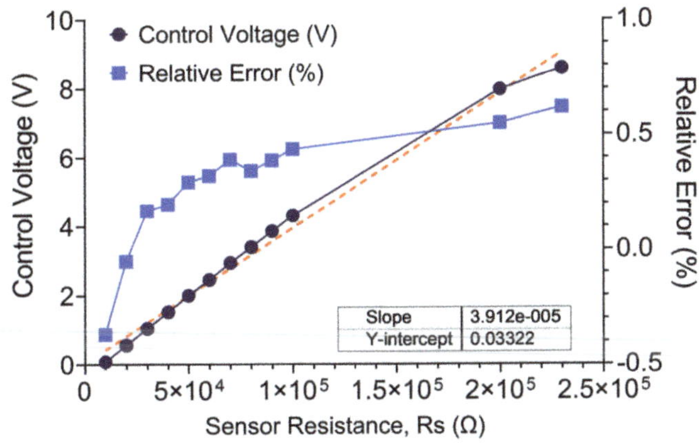

Figure 4.11. Control voltage response at an excitation current of 50 μA with the relative error. (Inset shows the coefficient of linear equation.)

The higher relative error towards the start and end of the resistor variation can be attributed to the non-linear behavior of the multiplier IC AD633 as it reaches the extremes of the control voltage. The observed behavior of the control voltage and relative percentage error as a function of sensor resistance is depicted in figure 4.12.

With three different cases of excitation current, –1, 10, and 50 μA, it can be noticed that the error assumes a very low value in each of the cases and the control voltage varies linearly in that region. However, the range can be extended at the cost of relative error and can practically reach 30 kΩ–10 MΩ (2.5 decades) for 1 A with an input excitation power of 20 W.

4.2.4 Experimental result with sensor

The dynamic response for the interface circuit was tested with an in-house fabricated gas sensor to detect ammonia. The sensor has an interdigitated microheater with a

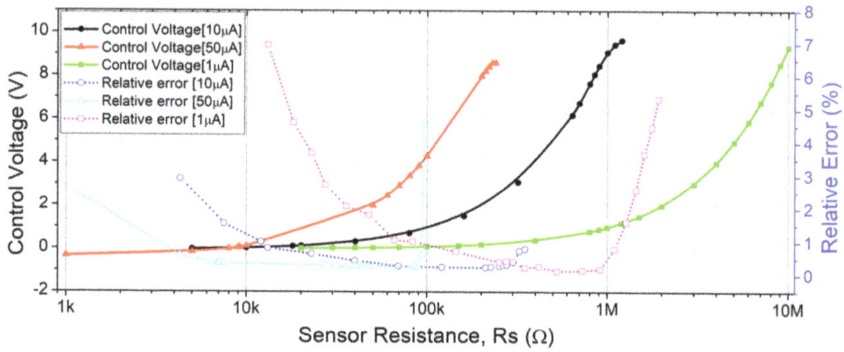

Figure 4.12. Experimental output control voltage and relative error vs. sensor resistance for different excitation currents.

gas-sensing membrane on top of the heater film, which shows a negative response to increasing ammonia concentration [17].

For proper testing, a gas sensor testing and characterization facility is used. The setup consists of a four-channel gas line with mixer and mass flow controller (MFC), a glass chamber sealed with an o-ring, a power supply, a controlled heater and a multichannel data-acquisition system. The MFC has a gas line connection to the pressurized gas cylinders used as source gases. A zero-air gas line is also driven to the MFC for generating the desired gas concentration. To achieve the right gas concentration of any gas, the MFC controller mixes that particular gas with zero air and passes it into the gas chamber at a fixed flow rate. In this case, ammonia gas with 99.995% purity is mixed with zero air to generate the experimental test concentrations. The sensor is placed inside a glass chamber and sealed with the o-ring arrangement for experimenting. The temperature of the heater film is raised to 220 °C for ammonia sensing and zero air is purged inside the chamber through the MFC. The sensing layer resistance is measured using the developed interface circuit with an excitation current of 10 μA. As expected, with increasing ammonia concentration, the sensor film resistance drops; the resulting sensor resistance values are shown in figure 4.13 at different concentration levels of ammonia. The sensor response to ammonia concentrations of 0 to 100 ppm is from 360 to 270 kΩ. For the stated sensor range, the experimental result shows that the interface circuit delivers a relative error of less than 0.78%.

4.2.4.1 Experimental result with negative temperature coefficient thermistor sensor

The experiment was also conducted with a commercial negative temperature coefficient (NTC) thermistor temperature sensor using a standard temperature controller setup. The NTC temperature sensor was placed in a glass chamber with a bottom glass plate in contact with a heater, from Bio-Rad. The sensor was coated with a thermal paste to enhance the thermal conductivity and sealed inside a glass chamber. Leads from the sensor element were connected to the circuit. The response of the sensor is continuously measured from the interface circuit for varying

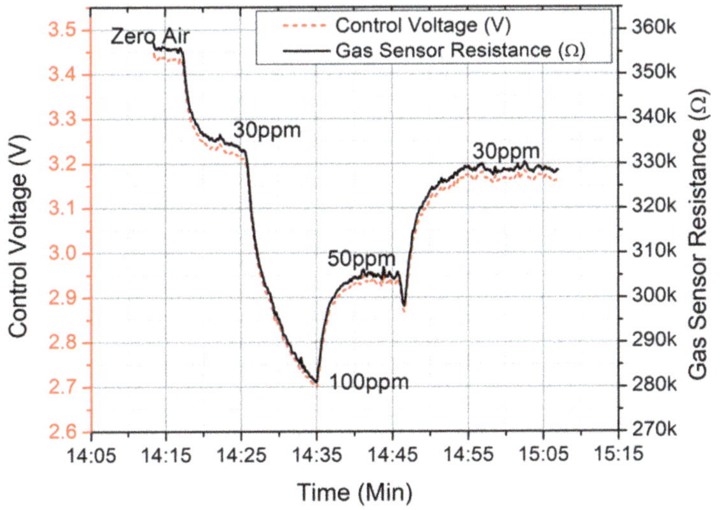

Figure 4.13. Response of the interface circuit with changing ammonia gas concentrations.

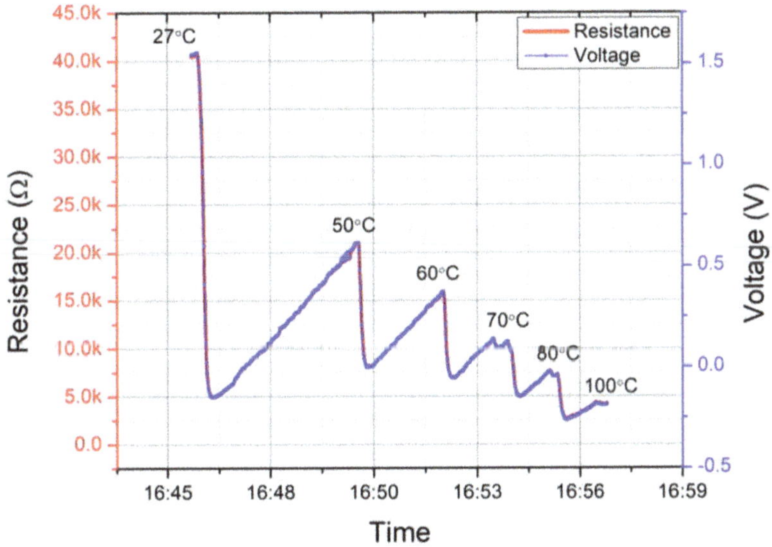

Figure 4.14. The response of the interface circuit with changing temperature for a 47k NTC.

temperature values. A resistance plot for changing temperature is shown in figure 4.14. The sensor was first kept at room temperature, of 27 °C, for some time and then the set temperature of the temperature controller was changed to 50 °C. The controller triggers to the set point and slowly reaches the set temperature. The same can be seen as a ramping voltage response, shown in figure 4.14. Similarly, the set points were varied up to 100 °C and the entire response of the interface circuit was recorded. From the experiment, we can see that the circuit could track the transient variations in sensor resistance with good accuracy.

4.2.5 Discussion

The auto-balancing resistive bridge configuration reported in [7, 9] delivers a wide dynamic range for resistance measurement for both floating and grounded resistances. This corresponds to an automatic tuning range of 5.6–170 kΩ for the first bridge configuration and a range of 6.2–194 kΩ for the second bridge configuration. For both the mentioned circuits, the circuit resolution is dependent on the range of the sensor resistance and keeps on decreasing with an increase in sensor resistance; hence, the output response is non-linear. A similar effect is seen in [9], as well. Moreover, the reported work presents an auto-balancing bridge with dual-loop control, hence increasing the complexity. On the other hand, an oscillator-based design for RTC reported in [18] delivers a wide measurement range but at the expense of increased relative error.

Moreover, the implementation also suffers from varying sensitivity with respect to the sensor resistance, although the reported work suggests that the sensitivity decrease beyond 1 GΩ achieves a time compression for limiting the output pulse width within a specified range. However, this introduces linearity error and loss in measurement resolution.

Compared to state-of-the-art circuits, the developed circuit is packed with some of the key features needed for dynamic sensor resistance measurement. The circuit features in-built four-wire Kelvin measurement with the instrumentation amplifiers, IA1 and IA2, acting as differential voltage sensors with high input impedance. Hence, the excitation current will not enter the input terminal of the instrumentation amplifier. This makes R_{w1}, R_{w2}, R_{w5} and R_{w6} impart a negligible effect on the circuit performance, and hence the circuit is also suitable for remote sensor measurements. Further, the auto-balancing loop delivers a linearly varying output control voltage, thus ensuring a low linearity error. Considering the stated features, the developed circuit is a promising candidate for a high-precision, linear and low-cost sensor system for both dynamic and static resistive sensors such as gas sensors, temperature sensors and flow sensors, which essentially follow a single-element sensor topology.

The presented design can also be extended for a resistive sensor array within the same input power. This can be done by implementing a parallel channel using an analog multiplexer switch. The subsequent section describes a multisensor array interface based on the developed auto-balancing, dual-differential architecture.

4.3 Auto-balancing interface for resistive sensor array

The previous section introduced a dual-differential-architecture-based auto-nulling interface circuit for single-element resistive sensors. The presented architecture delivered high precision, linear response, and a resistive gas sensor and temperature sensor. One of the features described earlier in the chapter was the extension of the circuit to handle a resistive sensor array. In this section, an auto-balancing interface circuit based on dual-differential architecture for a resistive sensor array is presented. The interface circuit provides a linearly varying control voltage output for linear change in input resistance. The circuit sweeps across the sensor array and calculates the difference of the differential voltage between the sensor element and a reference sensor formed by a VCR. Experimentally, the efficacy of the sensor interface circuit is tested on a silicon nanowire (SiNWs) array intended as a sensing platform for various biochemical sensing applications. The design, analysis and testing of the interface circuit are discussed in the following sections.

4.3.1 Circuit architecture and operation

A schematic diagram of the proposed signal conditioning circuit is shown in figure 4.15. Functionally, the circuit can be explained in three different sections: (a) a sensor array with a switch matrix, (b) a voltage-controlled resistance (R_{var}), and (c) a dual-differential subtractor with auto-balancing feedback. The operation of the circuit is based on a dual-differential, subtractor-based auto-balancing work presented in our previous research for single-element resistive sensors, reported in [19]. The sensor array, R_1, R_2, ..., R_n, is connected at terminal V_a using a multiplexer switch with a 1 to n decoder. The select line for the multiplexer is represented by S_0, S_1, S_2, ..., S_n, which can be connected to any microcontroller to select the desired element in the sensor array. The sensor array is excited using a constant-current source, I_s, with a direction away from the node V_a. The circuit analysis of the array

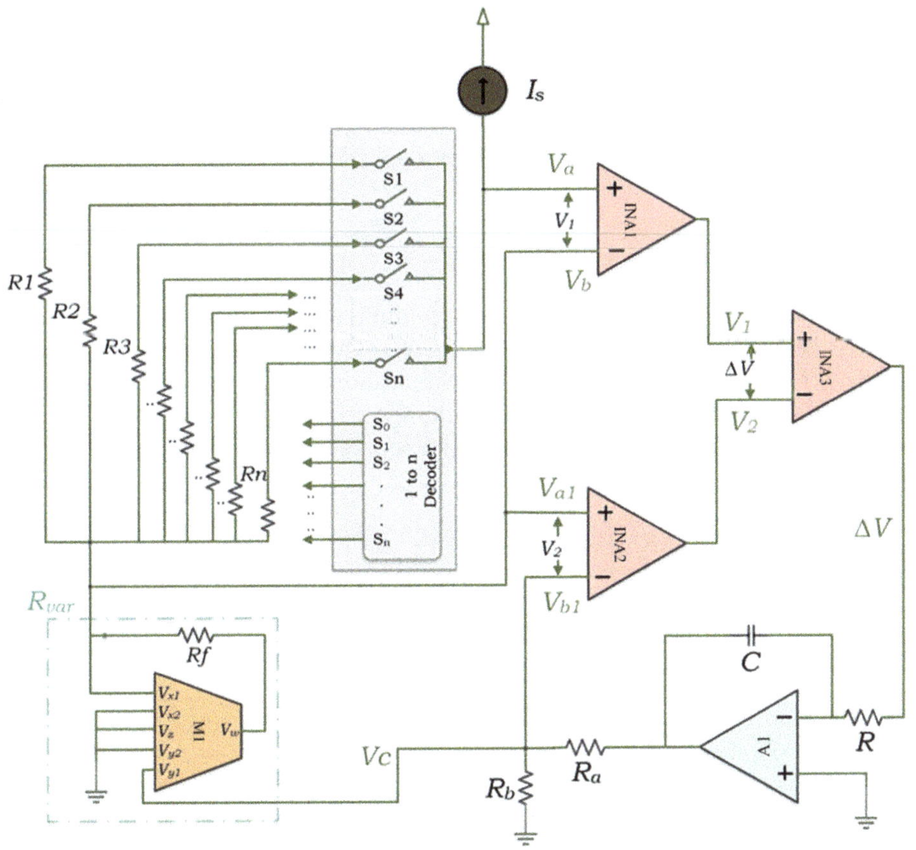

Figure 4.15. Proposed interface circuit for resistive sensor array.

interface will be similar to that discussed in section 4.2.2 with each channel acting as an independent dual-differential subtractor circuit. For each channel, the control voltage will be generated based on the sensor value being measured. The auto-balancing of the circuit is achieved through a tunable VCR, represented in figure 4.15 as R_{var}, formed using a four-quadrant analog multiplier, M_1. For design purposes, a commercially available analog multiplier, AD633, is considered, the transfer function of which is expressed as

$$V_w = \frac{(V_{x1} - V_{x2})(V_{y1} - V_{y2})}{10} + V_z. \tag{4.36}$$

Under the condition of V_{x2}, V_{y2} and V_{z2} connected to ground and with an input current of I_s, equation (4.36) is modified to give

$$\begin{aligned} I_s &= \frac{V_w - V_{a1}}{R_f} = \frac{-1}{R_f}\left[V_{a1} - \frac{V_{a2}.\ V_{c,\,n}}{10}\right] \\ R_{var} &= \frac{V_{a1}}{-I_s} = \frac{R_f}{1 - V_{c,\,n}/10}. \end{aligned} \tag{4.37}$$

To understand the functionality of the proposed circuit, let us assume that the select line $[S_0, S_1, ..., S_n]$ is at S_n and the output of the instrumentation amplifier IA3 is at a non-zero value, ΔV_n volts, when the circuit is switched on. Considering the notations given in figure 4.15 with a unity-gain instrumentation amplifier, and following the analysis presented in section 4.2.2, the output equation of the circuit will be

$$R_s = \frac{R_f}{(1 - V_{c,\,n}10)} + \frac{V_{c,\,n}}{I_s} - \frac{\Delta V}{I_s}. \tag{4.38}$$

The above equation suggests that the value of the sensor resistance can be calculated by measuring the control voltage ($V_{c,\,n}$) and ΔV for the particular sensor channel, n. Considering equation (4.38), the un-normalized sensitivity of the interface circuit is calculated as the partial change in control voltage response with respect to the partial change in sensor resistance. Rearranging the sensor resistance equation (equation (4.38)) as

$$R_s = \frac{I_sR_f + V_{c,\,n}(1 - \frac{V_{c,\,n}}{10}) - \Delta V(1 - \frac{V_{c,\,n}}{10})}{I_s(1 - \frac{V_{c,\,n}}{10})}, \tag{4.39}$$

the sensitivity for the interface circuit is defined as

$$S_{R_s}^{V_{c,\,v}} = \frac{\partial V_{c,\,n}}{\partial R_s} = \frac{I_s(1 - V_{c,\,n}10) + \frac{\partial \Delta V}{\partial R_S}(1 - V_{c,\,n}10)}{\frac{I_SR_S}{10} + -\frac{2V_{c,\,n}}{10} + \Delta V10}. \tag{4.40}$$

The feedback action of the auto-balancing network will integrate the error voltage, ΔV_n, and the output of the integrator will approach $V_{cn} = R_b/(R_b + R_a)\int_0^t \Delta V_n dt$, thus adjusting the value of R_{var} using the control voltage until the error diminishes to zero. Under the zero-error condition, the circuit is said to be at a steady state (or balanced state) and the value for sensor resistance, R_s, is expressed as

$$R_s\,|_{\Delta V=0} = \frac{R_f}{(1 - V_{c,\,n}10)} + \frac{V_{c,\,n}}{I_s}. \tag{4.41}$$

Under the balanced condition, the in-circuit resistance of the sensor can be calculated only by measuring the control voltage, $V_{c,\,n}$. Similarly, the sensitivity of the circuit under a balanced condition ($\Delta V = 0$) is derived using equation (4.39) as

$$S_{R_s}^{V_{c,\,n}}\,|_{\Delta V=0} = \frac{10I_s(1 - V_{c,\,n}10)}{G_1(I_sR_s - 2V_{c,\,n}) + 10G_2}. \tag{4.42}$$

An example SPICE simulation with a reference resistance, $R_f = 10.5$ kΩ, and an excitation current, $I_s = 50$ nA, shows that the circuit works effectively, giving a linear output for sensor values varying from 4 to 190 MΩ with a relative error less than ±1%, as shown in figure 4.16. This measurement range only includes the balanced state and will be extended if the unbalanced state is also used for measurement using the equation (4.38).

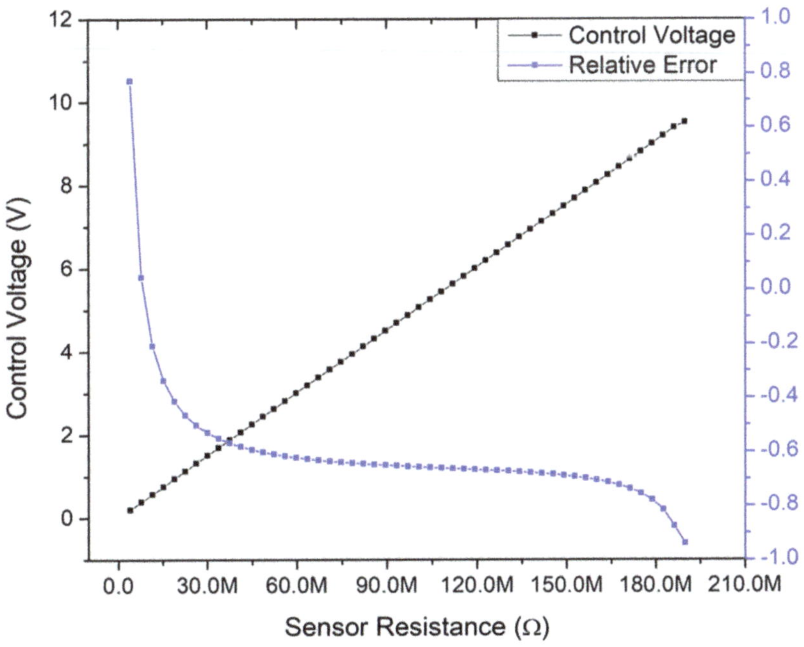

Figure 4.16. Output equation response with different values of the sensor along with percentage relative error.

The design can be easily scaled up to incorporate a large sensor array by selecting a switch matrix with a higher number of channels. This is possible because the sensor elements are not excited in series. Hence the voltage drop across the sensor is only dependent on the selected sensor and not on the combined resistance value of the array.

4.3.2 Circuit analysis

4.3.2.1 Effect of multiplexer non-idealities

The DC performance of a multiplexer switch is mainly affected by its ON resistance (r_{on}) and leakage current (I_{LKG}), while the dynamic performance is majorly affected by the switch charge injection (Q_J). When the switch is ON, the ON resistance of the multiplexer switch forms a series resistance with the sensor element. For a $1 \times n$ analog multiplexer, the sensor resistance seen by IA1 will be ($R_s + r_{on}$). The effective resistance value can be written as $R'_s = R_s(1 + r_\epsilon)$, where $r_\epsilon = \frac{r_{on}}{R_S}$. A modified expression for sensor resistance under ON resistance consideration can be given by

$$R'_s = \left[\frac{R_f G_2}{(1 - V_{c,\,n}10)} + \frac{V_{c,\,n} G_2}{I_s G_1}\right]\frac{1}{1 + r_\epsilon}. \tag{4.43}$$

The relative error introduced due to the ON resistance of the multiplexer is expressed as

$$E_{r_{on}} = \frac{R_s - R'_s}{R_s} = \frac{r_\epsilon}{1 + r_\epsilon}. \tag{4.44}$$

The leakage current of the multiplexer switch affects the DC performance by forming a circuit combination with $[R_s - r_{on} - I_s]$, as shown in figure 4.17.

The effect of r_{on} is dominant for low-resistance circuits, while high-resistance circuits are affected mainly by the leakage current. Therefore, a switch with a low leakage current is preferred. The effect can be analytically expressed as

$$V_a = \frac{I_s R_f}{(1 - V_{c,\,n}10)} - I_s(R_s + r_{on}) + I_{LKG}(R_s + r_{on}). \tag{4.45}$$

During switching, the control input voltage is applied at the gate of the CMOS switch. This fast change in voltage injects a charge into the switch output through gate-drain capacitance, C_Q, as shown in figure 4.17. As a result of this charge, a step change in output voltage, ΔV_a, is observed, which is expressed as $\Delta V_a = Q_J C_L$ with C_L being the load capacitance and Q_J being the charge injected. Since the design presented here does not drive any capacitive load, the only parasitic capacitance will act as a load. Considering the design parameters, the proposed circuit utilizes an analog multiplexer IC, MUX506, with a low ON resistance of 125 Ω and a low leakage current of 1 pA with $Q_J = 0.13$ pC at 1 nF of capacitive load. Under these design considerations, the error introduced by the analog multiplexer is negligible.

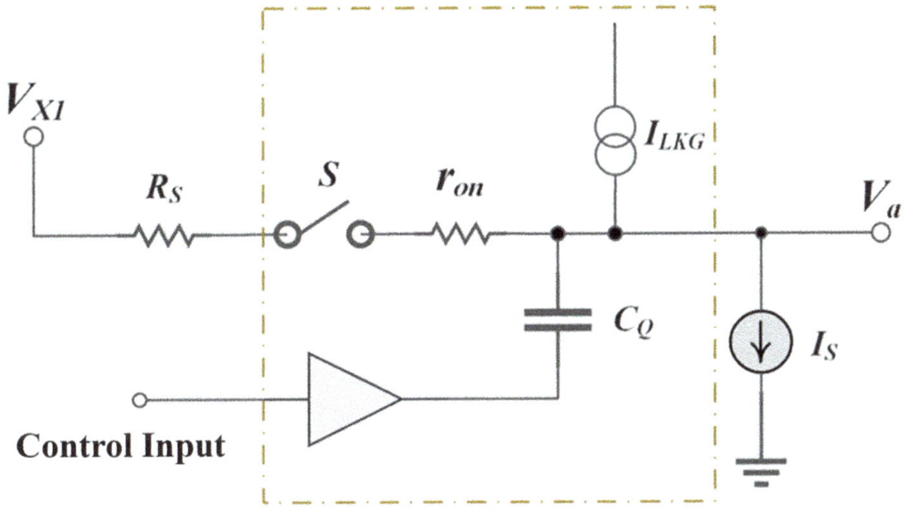

Figure 4.17. Real model of analog multiplexer switch with leakage current and ON resistance (r_{on}).

4.3.2.2 Effect of reference resistance

R_f is a constant resistance whose value acts as a reference resistance in the voltage-controlled resistance, R_{var}. The resolution of the circuit at a constant excitation current is dependent on the slope of the control voltage response. The higher the slope, the higher the sensitivity, and thus a better resolution is delivered. According to equation (4.41), the response curve of the control voltage shifts downwards with a relatively constant slope, indicating that the sensitivity of the response is not changing. Minor variation in the slope and sensitivity is seen only towards the higher extremes of the measurement range. This effect is shown in figure 4.18.

For a reference resistance of 10.5 kΩ, the relative error curve is lowest and within 0.8% for the given sensor resistance range.

4.3.2.3 Stability and loop-gain analysis

The loop gain of a feedback circuit is the amplification of a test signal in one loop of the circuit. This can be calculated by setting all the independent input signal sources to zero and inserting a test signal into the loop. Shown in figure 4.19, a test voltage source, v_t, at the high-impedance node of INA3 is inserted and a voltage, v_r, at the output of INA2 is measured. The small-signal model of the op-amp in the integrator is utilized to derive the loop gain. Considering the small-signal model of the op-amp, the gain of the loop can be written as follows:

$$H(s) = -\frac{v_r(s)}{v_t(s)} = -\frac{R_b}{R_b + R_a}\left[\frac{A(s)}{1 + sRC(1 + A(s))}\right], \tag{4.46}$$

where $A(s)$ is the open loop-gain transfer function of the op-amp, which can be written as $A_0/(1 + s/\omega_1)$, where A_0 is the DC gain of the op-amp and ω_1 is the unity-gain frequency. Considering this, equation (4.46) can be modified as follows:

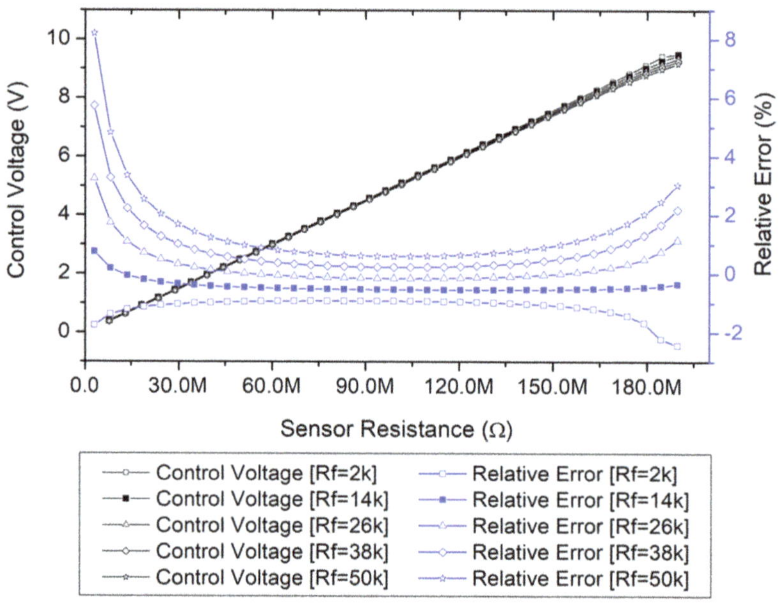

Figure 4.18. Effect of reference resistance, R_f, on the control voltage response.

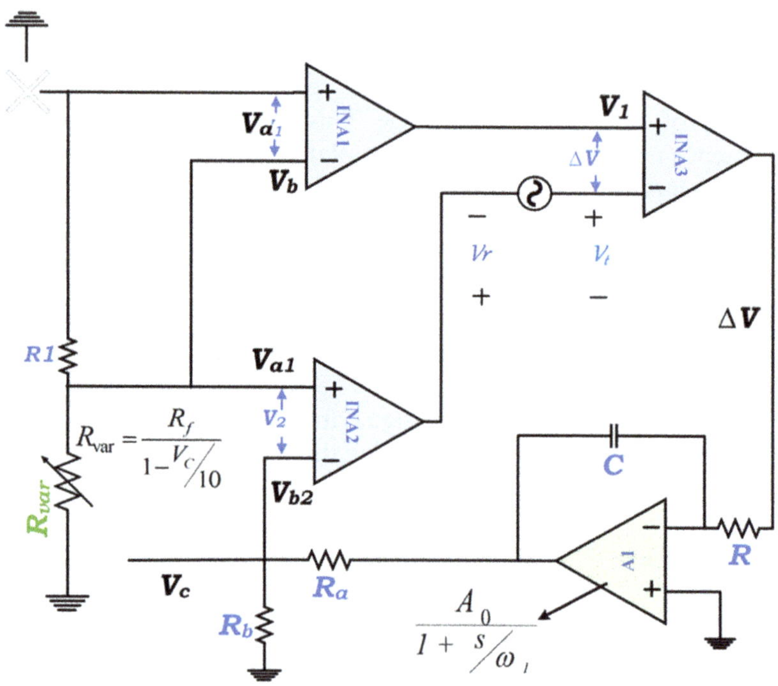

Figure 4.19. Small signal analysis of the interface circuit.

$$H(s) = \frac{R_b}{R_b + R_a}\left[\frac{A_0}{s^2\frac{RC}{\omega_1} + s\left(\frac{1}{\omega_1} + RC(1 + A_0)\right) + 1}\right]. \tag{4.47}$$

Equation (4.47) suggests that the loop gain for the feedback circuit is positive, thus confirming the stability of the designed interface circuit. Furthermore, for a second-order system, the value of the phase margin of the system to verify the stability can be given as follows:

$$PM = \tan^{-1}\left[\sqrt{\frac{2}{\sqrt{(1+4Q^4)} - 1}}\right], \tag{4.48}$$

where $Q = \frac{\frac{1}{\omega_1} + RC(1 + A_0)}{\sqrt{\omega_1 / RC}}$.

For the given circuit, the TL082 op-amp has $A_0 = 100$ dB and $\omega_1 = 4$ MHz. The value of the RC can be selected as per equation (4.46) such that the system remains stable. For the interface circuit, the values of R and C are chosen as 100 kΩ and 1000 pF, respectively. From these values, the system gives a phase margin of 90°, which confirms that the system is stable.

4.3.3 Prototype design and experimental results

To carry out experimental tests, a prototype of the circuit is fabricated using commercial, discrete off-the-shelf components. INA333, a precision instrumentation amplifier, is used to realize INA1, INA2, and INA3. A junction field effect transistor (JFET) op-amp, TL082, is used along with a resistance $R = 100$ kΩ and $C = 1000$ pF to form an inverting integrator. The active components are provided with a supply voltage of ±15 V. The analog multiplier, M1, combined with reference resistance, $R_f = 20$ kΩ, forms the voltage-controlled resistance, controlled by the signal, V_{cn}, at the V_{y1} node. Since the control voltage can swing between ±10 V, the integrator's output is scaled by the resistance divided by R_a and R_b, scaling the output of the integrator by 0.67 times. The sensor array is connected to the interface circuit using a 16-channel analog multiplexer, ADG706BF. Each sensor element is connected to the input channel of the multiplexer and select lines are connected to the GPIO of a microcontroller (Atmega328) for scanning the sensors' elements using a software logic. The excitation current, I_s, is provided by a source meter (Keithley 2450) and the results are recorded on a digital storage oscilloscope (Keysight DSOS054A). To test the effect of sensor resistance on the circuit's performance, a precision resistance box is used to emulate the values of the sensors. This testing enables us to calibrate the system and validate the circuit functionality before applying it with real sensors. By varying the sensor resistance value, the control output voltage varies to keep the circuit balanced. The response of the control voltage with varying sensor resistance is captured and shown in figure 4.20.

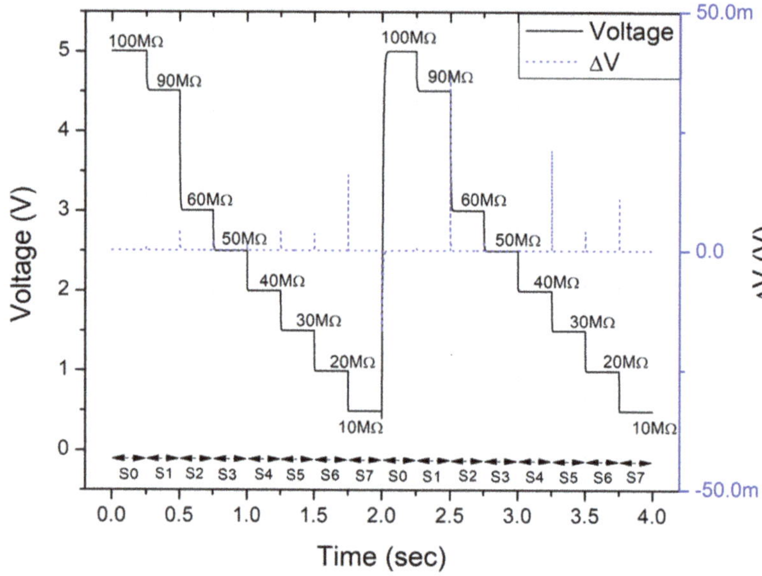

Figure 4.20. The response of interface circuit with changing select line. Each select line (*S*1, *S*2, ..., *S*7) connects to a sensor element.

4.3.4 Experimental measurement with a silicon nanowire array

The final evaluation of the array interface system is done with a fabricated SiNW array platform. The sensor is fabricated in-house and intended as a sensing platform for various biochemical sensing applications. For a current sweep from 10 to 45 nA, the response of the SiNW array is summarized in figure 4.21. As can be seen, the device has nine SiNW arrays with nominal resistances within a range of 29–35 MΩ. The fabricated device is optimized to exhibit resistive-based sensing, which is evident from the I-V characteristics for each sensing element in the array.

These uncalibrated resistive sensor arrays are connected to the input channels of the analog multiplexer, occupying nine channels, as shown in figure 4.22. Each element is then measured by selecting the sensor channel and reading the respective control voltage signal. The channel is shifted to select the next channel when the difference voltage, ΔV, goes to zero. The measured resistance of each sensing element is shown in figure 4.23 along with the relative error. The interface circuit can measure the sensor resistance within a relative error of less than 0.65%, making it suitable for SiNW sensing applications.

4.3.5 Discussion

The developed circuit delivers a linear, accurate output reading for a wide measurement range. The sensor is excited by a current signal forming an in-built four-wire Kelvin measurement setup, making the circuit immune to lead wire effects. With a closed-loop auto-balancing signal approach, the user can scan the sensor array with the error signal being the trigger for channel switching. Since the overall settling time of

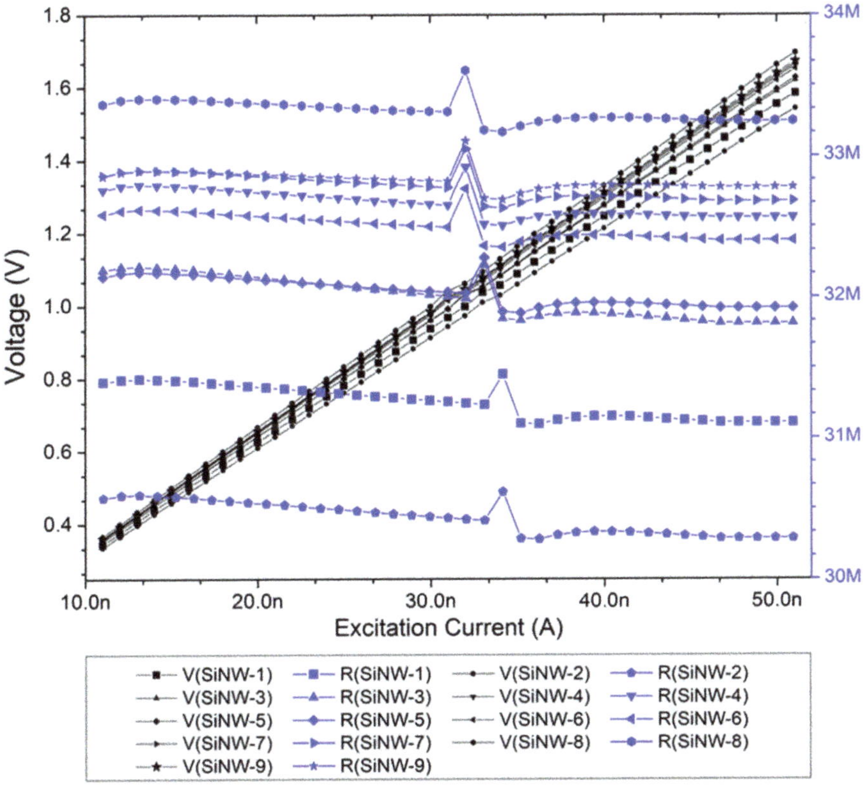

Figure 4.21. Silicon nanowire response for a current sweep from 10 to 45 nA. Each sensor element has a base resistance in the range of 30–34 MΩ.

each sensor signal readout is independent of the sensor's absolute value and depends on the integrator time constant, the scan rate for the developed interface is periodic and can be controlled using the error (ΔV) trigger signal for precise timed measurement.

The measurement range of the proposed interface covers a range of possible applications such as a gas sensor array, carbon nanotubes, nanowires, etc, with a relative error below 1%. For interface circuits based on an RTC, relaxation oscillator or charge integration [20, 21], the settling time is proportional to the absolute value of the sensor resistance. The oscillator-based solution presented in [20] implements resistance-to-frequency conversion for a wide measurement range of 1 MΩ to 1 GΩ. However, this is achieved by using multiple analog switches to select the appropriate integrating capacitor creating a discrete frequency output with respect to the input sensor resistance. For a resistive array application, the scanning trigger becomes aperiodic when the output settling time depends on the sensor value and varies with each element [22], making it difficult to design an auto-trigger scanning and acquisition system. Moreover, these designs tend to have a higher non-linearity error. For the proposed interface circuit, both auto-triggered and external triggered measurement is possible since the ΔV acts as the external trigger for channel selection. The dynamics of the interface are independent of the sensor value.

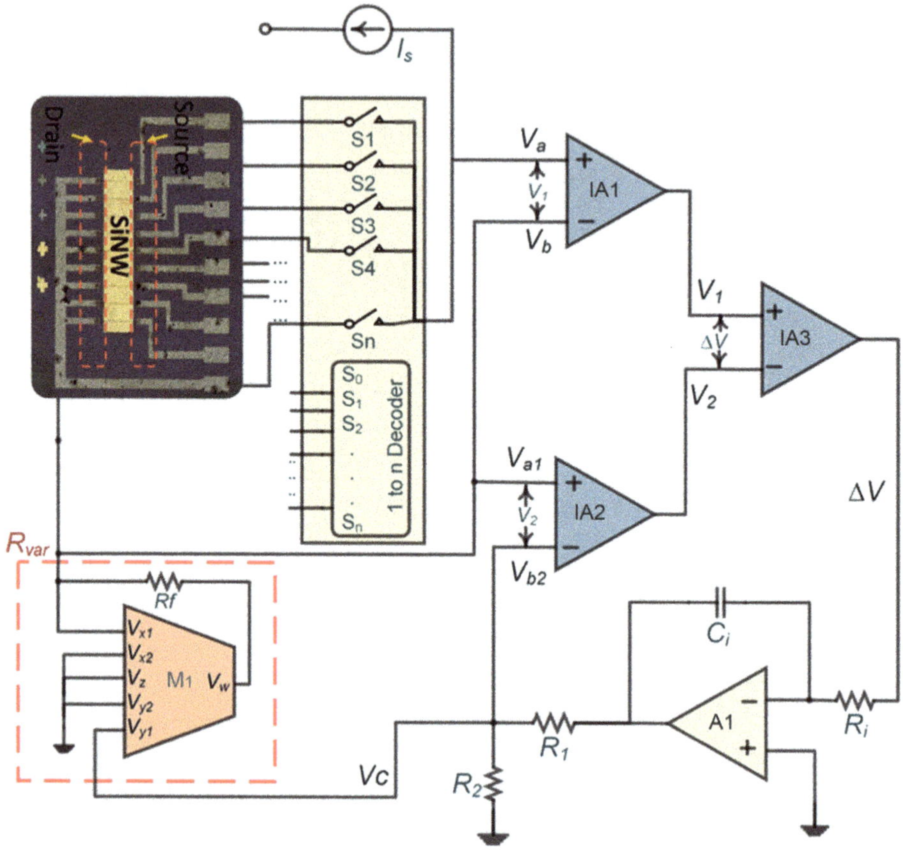

Figure 4.22. Experimental setup with SiNW array.

Such features, combined with independent scan rate and low non-linearity error, prove advantageous for a sensor array interface.

4.3.6 Summary

In this chapter, the design, development and testing of instrumentation modules for single-element resistive sensors have been presented. The interface circuit is based on an auto-balancing, dual-differential architecture designed to adjust a variable voltage controller reference resistor to auto-null the absolute sensor resistance value. The interface design provides a single control voltage output that is immune to lead wire variations and is proportional to the change in sensor resistance, achieved through an auto-nulling technique that enhances the range of the interface circuit while maintaining linearity and stable sensitivity. The circuit is analyzed in terms of linearity, transient response and stability, with separate analysis undertaken to understand the effect of component non-idealities such as offset voltages, gain mismatches, reference resistance variations and loop-gain limitations. Various measurement tests were conducted to verify the measurement range, stability, linearity and sensitivity using a precision resistive decade box. The interface electronics

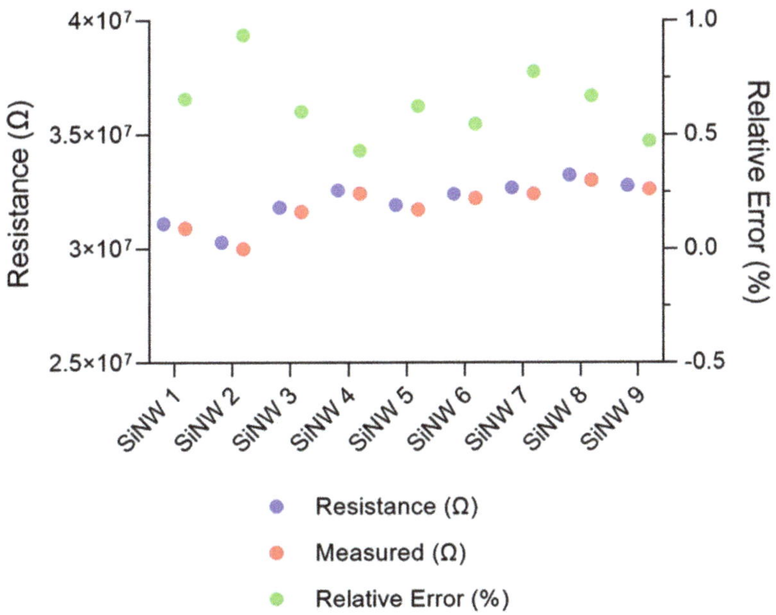

Figure 4.23. Measured resistance of SiNW array with the array interface circuit.

are also simple to configure for the desired range and resolution, which has been shown and experimentally demonstrated. The final verification of the interface circuit was performed using an in-house fabricated resistive gas sensor and a commercial NTC temperature sensor. The results indicate that the interface circuit is a viable candidate for dynamic, high-precision, linear and low-cost sensor systems and can be an efficient alternative to a Wheatstone bridge for single-element resistive sensors.

An extension of the developed instrumentation module for resistive sensor array is also presented in this chapter. The circuit delivers linear and stable sensitivity for the entire working range and has been analyzed in detail by considering the non-ideal behavior of its constituent components. The developed prototype was tested for a range of 4–910 MΩ and the relative error was found to be below 0.95% in the auto-balancing range. The circuit was tested with an in-house fabricated SiNW array consisting of nine sensing nanowires resistors. The circuit could scan the SiNW platform in an automated manner using an auto-nulling signal as a trigger for the next scan. Each scan resulted in a respective control voltage signal proportional to the sensor resistance. The experimental results confirm that the circuit is a good sensor interface system for the nanowire platform and can also be tailored to the desired range and sensitivity of the sensor being interfaced.

4.4 Conclusion and industrial aspects

Resistive gas sensors and nanowires are widely used in industrial processes, environmental monitoring, medical diagnosis, safety and pharmaceuticals. The number of toxic and hazardous gases have significantly increased in recent years due to the increase in automobiles and rapid industrialization, causing severe indoor and outdoor pollution and many adverse health effects. Toxic gases can cause harm at low concentrations over long periods of time (chronic exposure) or at higher concentrations over short periods of time (acute exposure). Gas sensors, such as CO, CO_2, methane, hydrogen, oxygen, etc, are used to monitor and control combustion processes, top gas composition in furnaces to predict end-product quality and process condition, and environmental emission monitoring in mines.

Temperature is one of the most common parameters used to measure, monitor and control metallurgical and mineral process industries. Resistive temperature detectors and thermocouples are perhaps the best solutions to collect accurate temperature measurements over wide ranges and to improve application efficiencies. These sensors are used to monitor and control temperature in operating furnaces, rotary kilns and gas turbine exhausts, as well as for engine condition monitoring and temperature increases in bearings, thrust bearing plates, shafts and motor windings. Tip-sensitive resistive temperature sensors are miniature in size and have fast response and are designed for surface temperature measurement in bearing shoe applications to give reliable indication of bearing wear and oil film breakdown through continuous monitoring of temperature. These are effectively used in many applications in a variety of other industries including medical, chemical, manufacturing and aerospace.

Strain gauge sensors work by measuring the change in electrical resistance across a thin conductive foil. They measure strain directly, which can be used to indirectly determine stress, torque, pressure and deflection. Their applications include the measurement of stress on wet and porous media such as rock, to determine strain in material due to static and dynamic loads, to detect grinding process fluctuations, to determine wing deflection/deformation during flight, to monitor the stress in railway lines, as well as in structural component testing and construction applications.

The presented interface circuits are suited for a wide range of resistive sensing applications. However, they are ideally suited for industrial applications involving remote measurement such as gas plants, chemical industries, engine manufacture, metallurgy, etc. The array interface design presents a compact solution for multi-sensor readout and can be suited for applications such as flexible sensing involving an array of electrodes, environment monitoring, etc. Owing to its simple design, wide range and clear advantages, the interface circuit seems to be a promising candidate for high-precision, linear and low-cost sensor systems and can be an efficient alternative to Wheatstone bridges for resistive sensors.

4.5 Future of sensing technology

Sensors have encapsulated wide aspects of industrial, medical and personal life, creating a cocoon of environment monitoring applications interconnecting humans to the digital world. Their use can be found in virtually all aspects of our daily life, including health monitoring, security, automotion, entertainment and many more. In addition, sensors heavily dominate the scientific and industrial sectors, particularly now with the coalescence of artificial intelligence (AI) and cyber-physical systems centered around sensor systems. A partial reason for this ever-developing sector is the advancement in manufacturing technology, sensor and system design capabilities and novel material development. Further, sensors with integrated energy harvesting from solar and integrated fuel cells [23] are paving the way for self-reliant and robust platforms for environment sensing applications.

Despite the demands of miniaturization, the current generation of sensors are becoming extremely powerful and robust. Every new design targets enhanced accuracy, effective sensor fusion, low failure rate and reduced power consumption, making sensor systems and devices future ready. This new era of industrialization is fueled by rapidly progressing silicon technology and material science innovation [24]. Sensing technology is evolving in line with advancements in process automation, resulting in more capable and better networked sensing devices. A majority of industries have already started automating and upgrading their processes and plants to an Industrial Internet-of-Things platform (IIoT), making it 'Industry 4.0' ready.

The field of biosensing is expanding its capabilities in visual, tactile, olfactory, audio and physical systems. Since the sensing technologies are now utilizing both micro- and nanostructures, transforming the use of sensors from surface devices [25] to in vivo devices [26], these cutting-edge innovations are having a deep impact in drug monitoring, diagnostics and continuous health assessment [27].

Similarly, smart and intelligent sensor technology is looking to revolutionize the future of the world in the coming years. They will be constantly evolving, such as measuring invisible things and adapting to be more wearable. By ensuring the integrity and reliability of the results generated by these sensors, their applications in medicine, consumer products, communication devices, homes and transportation is inevitable. Further advances in sensing technology will not be limited to sensing devices. By converging intelligent sensing with AI and machine learning, it will support agile decision making and drive performance optimization.

References

[1] Fraden J 2010 *Handbook of Modern Sensors* (New York: Springer)

[2] Jain V and George B 2017 An efficient digitization scheme for resistive sensors interfaced through quarter bridge *2017 Eleventh Int. Conf. on Sensing Technology (ICST)* (Piscataway, NJ: IEEE)

[3] George B, Roy J K, Kumar V J and Mukhopadhyay S C (ed) 2017 *Advanced Interfacing Techniques for Sensors* (Cham: Springer)

[4] Malik S, Kishore K, Artee, Akbar S and Islam T 2016 A CCII-based relaxation oscillator as a versatile interface for resistive and capacitive sensors *2016 3rd Int. Conf. on Signal Processing and Integrated Networks (SPIN)* (Piscataway, NJ: IEEE)

[5] Nagarajan P R, George B and Kumar V J 2017 *IEEE Trans. Instrum. Meas.* **66** 2736–44

[6] Kishore K, Akbar S A, Kumar S S, Mukhiya R and Tanwar A 2018 CCII based current signal interface for piezoresistive pressure sensor *2018 15th IEEE India Council Int. Conf. (INDICON)* (Piscataway, NJ: IEEE)

[7] Marcellis A D, Ferri G and Mantenuto P 2013 *Sensors Actuators B: Chem.* **187** 371–8

[8] Mantenuto P, De Marcellis A and Ferri G 2012 *Procedia Eng.* **47** 261–4

[9] Marcellis A D, Ferri G and Mantenuto P 2013 *Sensors Actuators B: Chem.* **189** 130–40

[10] Alnasser E 2018 *IEEE Sensors J.* **18** 635–40

[11] Malik S, Kishore K, Islam T, Zargar Z H and Akbar S 2015 *Sensors Actuators A: Phys.* **234** 248–62

[12] Nagarajan P R, George B and Kumar V J 2017 *IEEE Sensors J.* **17** 1696–705

[13] Devices A and Norwood MA USA, Datasheet. AD633, Low Cost Analog Multiplier http://analog.com/media/en/technical-documentation/data-sheets/AD633.pdf>

[14] Emancipator K and Kroll M H 1993 *Clin. Chem.* **39** 766–72

[15] Wyatt J and Papadopoulos G 1984 *IEEE Trans. Circuits Syst.* **31** 657–61

[16] Penfield P, Spence R and Duinker S 1970 *IEEE Trans. Circuit Theory* **17** 302–5

[17] Prajesh R, Jain N and Agarwal A 2015 *Microsyst. Technol.* **22** 2185–92

[18] Marcellis A D, Ferri G, Depari A and Flammini A 2011 A novel time-controlled interface circuit for resistive sensors *2011 IEEE SENSORS Proc.* (Piscataway, NJ: IEEE)

[19] Kishore K, Malik S, Baghini M S and Akbar S A 2020 *IEEE Sens. J.* **20** 3047–56

[20] Liu H, Tee Hui Teo and Zhang Y P 2009 A low-power wide-range interface circuit for nanowire sensor array based on resistance-to-frequency conversion technique *Proc. of the 2009 12th Int. Symp. on Integrated Circuits* 13–6

[21] Ferri G, Carlo C D, Stornelli V, Marcellis A D, Flammini A, Depari A and Jand N 2009 *Sensors Actuators B: Chem.* **143** 218–25

[22] Depari A and Flammini A 2012 *Procedia Eng.* **47** 148–51

[23] Cui Y, Lai B and Tang X 2019 *Biosensors* **9** 92

[24] Madagalam M, Bartoli M, Tagliaferro A and Carrara S 2021 *IEEE Sens. J.* **21** 11155–62

[25] Criscuolo F, Ny Hanitra I, Aiassa S, Taurino I, Oliva N, Carrara S and De Micheli G 2021 *Sensors Actuators* B **328** 129017

[26] Barbruni G L, Asti F, Ros P M, Ghezzi D, Demarchi D and Carrara S 2021 A 20 mbps, 433 mhz rf ask transmitter to inductively power a distributed network of miniaturised neural implants *IEEE Int. Symp. on Medical Measurements and Applications (MeMeA)* pp 1–6

[27] Wuthibenjaphonchai N, Haruta M, Sasagawa K, Tokuda T, Carrara S and Ohta J 2021 *IEEE Sens. J.* **21** 9402–12

IOP Publishing

Systems for Printed Flexible Sensors
Design and implementation
Tarikul Islam, Subhas Chandra Mukhopadhyay and Boby George

Chapter 5

Interfacing circuit for capacitive sensors

Sougata Kumar Kar and Siddhartha Sen

5.1 Introduction

In the past three decades, the development of micro- and nanosensors has progressed to a great extent to make them an integral part of many products. The fabrication process of these sensors is quite similar to that of integrated circuit (IC) technology. This has helped the growth of this trend of integrated system development. However, there are several issues in their fabrication, such as stiction, residual stress, uniformity, etc, in the realization of a working sensor structure. Proper characterization of the process is required to improve the yield. The fabricated sensors experience considerable variations after fabrication because of the smaller feature size (μm or nm) of the device, the complexity of the sensors' structures and lack of proper standardization. Moreover, the small feature size leads to less change in the sensor output. Detection of this small change with a suitable electronic interfacing circuit is non trivial.

Among the various types of micro- and nanosensors, capacitive sensors have found a wide range of applications. They offer higher sensitivity with negligible temperature coefficients [1]. Figure 5.1 shows a general block diagram of a capacitive sensor system in which the change in capacitance is measured by an electronic interface, providing analogous output. However, the capacitive sensors often suffer from mismatch and parasitic effects [1–3]. In practice, capacitive sensors are configured both as single-ended, or unipolar, and differential modes. In the case of a single-ended capacitive sensor, the capacitance is compared either with respect to a reference capacitor or the absolute value of the capacitor is measured. In the case of a differential configuration, two sensing capacitors change in opposite direction; when one increases the other decreases. The differential configuration offers higher sensitivity and better rejection of common-mode signals as well as interference.

doi:10.1088/978-0-7503-3935-3ch5

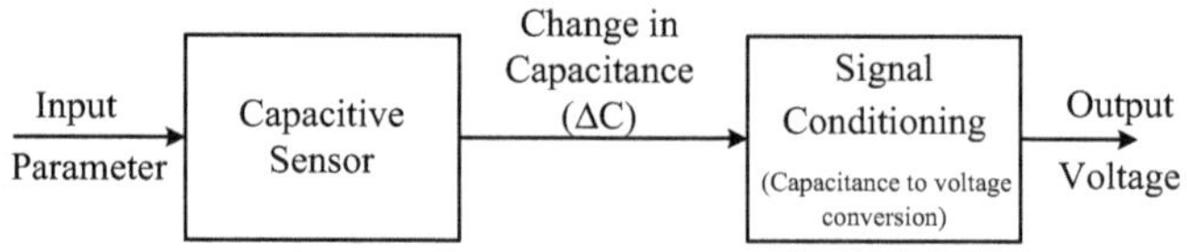

Figure 5.1. Capacitive sensor system.

5.2 Capacitive sensors: basic principles

Technically, capacitive sensing acquires the capacitance between two or more conductive plates separated by a dielectric medium [1], where the sensing principle is one of variable capacitance:

$$C = \frac{\varepsilon_0 \varepsilon_r A}{d}, \tag{5.1}$$

where ε_0 is the free space permittivity (8.854×10^{-12} $\mathrm{Fm^{-1}}$), ε_r is the relative permittivity of the medium sandwiched between the two conductors, d is the space between the plates and A is the cross-sectional area of the plates.

Variation in any of these parameters changes the capacitance. In this capacitive sensing principle the capacitance variation is possible by changing either the gap or overlap area between the parallel plates. Figures 5.2(a) and (b) represent single-ended sensing, where the movable plate provides the change in capacitance by changing the space between two plates or by varying the effective cross-sectional area of the plates.

5.2.1 Variable gap type

In this method, two parallel plates are used to form a capacitor, where the gap between the two plates is varied horizontally by connecting one movable plate to the measuring signal and the other plate as a fixed plate, as shown in figure 5.2(a). In this case the capacitance can be expressed as

$$C = \frac{\varepsilon_0 \varepsilon_r A}{d \pm x}, \tag{5.2}$$

where x is the effective change in the gap between the parallel plates and d is the nominal gap.

5.2.2 Variable area type

Here, the overlap area between the fixed plate and the moving plate is varied by displacing the movable plate parallel to the fixed plate. In this method the effective cross-sectional area varies according to the applied measuring signal. Figure 5.2(b) represents the sensing methodology, where a movable plate moves parallel to the fixed plate due to the physical signal to cause a change in area of overlap. In this case the capacitance can be expressed as

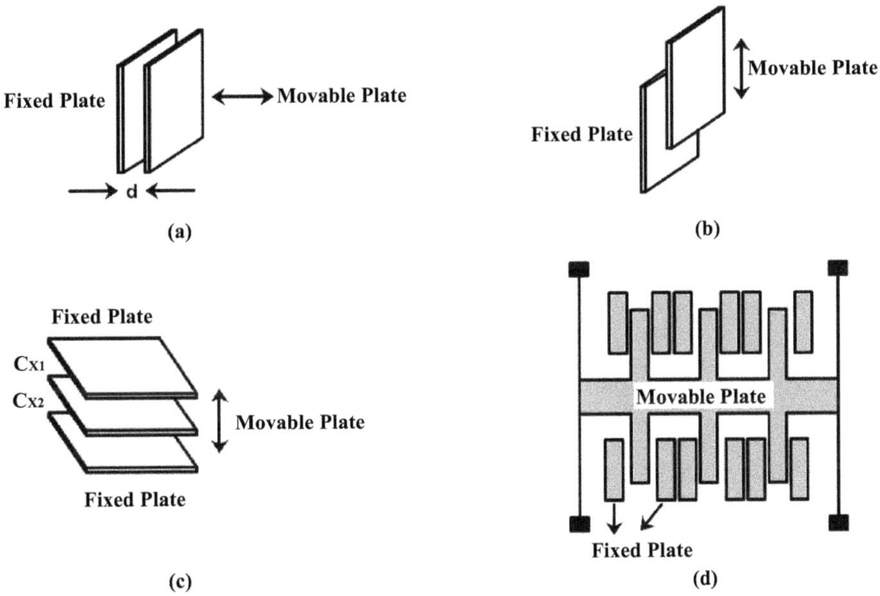

Figure 5.2. (a) Variable gap type, (b) variable area type, (c) and (d) differential configurations.

$$C = \frac{\varepsilon_0 \varepsilon_r (A \pm \Delta A)}{d}, \tag{5.3}$$

where ΔA is the effective change in the area between the plates. Thus the capacitance variation is proportional to ΔA here.

5.2.3 Variable dielectric type

In this type of sensor, both of the two plates are fixed but the relative permittivity between them changes. This type of sensor is particularly suitable for liquid level measurement, where the liquid is non-conduting.

5.2.4 Differential configuration

In this capacitive sensing principle, three or multiple parallel plates are used, which ultimately forms two sense capacitors (C_{X1}, C_{X2}). This arrangement consists of two fixed plates and a movable plate, which then provides a differential capacitance (ΔC). The capacitance variation is obtained by moving the common movable plate towards any of the fixed plates. Figures 5.2(c) and (d) show differential capacitive sensing, where the movable plate is arranged in the middle of two fixed plates. This adjustment creates two capacitors, and their capacitance depends on the movable plate's position.

5.2.4.1 Three-plate system

In this method three parallel plates are used, in which two plates are fixed and one plate is movable. The arranged setup forms two capacitors, and the capacitive sensing is through gap variation. Figure 5.2(c) illustrates this topology, where the common movable plate is displaced according to the applied measuring signal:

$$C_{X1} = \frac{\varepsilon_0 \varepsilon_r A}{d + x} = C_0\left(\frac{1}{1 + \frac{x}{d}}\right), \tag{5.4}$$

where $C_0 = \varepsilon_0 \varepsilon_r A/d$. Now, if we expand equation (5.4) using a Taylor series:

$$C_{X1} = C_0\left[1 - \frac{x}{d} + \left(\frac{x}{d}\right)^2 - \dots\right]. \tag{5.5}$$

Equation (5.5) is non-linear, and can be approximated to

$$C_{X1} \approx C_0\left(1 - \frac{x}{d}\right) \approx C_0 - \Delta C, \quad \text{if} \quad \frac{x}{d} \ll 1, \tag{5.6}$$

where

$$\Delta C = C_0 \times \frac{x}{d}.$$

Similarly, we can write

$$C_{X2} = \frac{\varepsilon_0 \varepsilon_r A}{d - x} = C_0\left(\frac{1}{1 - \frac{x}{d}}\right) = C_0\left[1 + \frac{x}{d} + \left(\frac{x}{d}\right)^2 - \dots\right], \tag{5.7}$$

which can be approximated to

$$C_{X2} \approx C_0\left(1 + \frac{x}{d}\right) \approx C_0 + \Delta C, \quad \text{if} \quad \frac{x}{d} \ll 1. \tag{5.8}$$

From equations (5.5) and (5.7), it is clear that the individual capacitances are non-linear w.r.t. the displacement to be measured. To reduce this non-linearity error a differential configuration is preferred, where we can get the change in capacitance as output instead of absolute capacitance. The differential configuration is shown in figure 5.2(c). The difference in capacitance between C_{X1} and C_{X2} can be expressed as

$$\Delta C = C_{X2} - C_{X1} = \frac{\varepsilon_0 \varepsilon_r A}{d - x} - \frac{\varepsilon_0 \varepsilon_r A}{d + x} \tag{5.9}$$

$$= \left(\frac{\varepsilon_0 \varepsilon_r A}{d^2 - x^2} \times 2x\right) = C_0\left(\frac{2x}{d}\right)\left[\frac{1}{1 - \left(\frac{x}{d}\right)^2}\right], \tag{5.10}$$

which can be expanded using a Taylor series:

$$\begin{aligned} \Delta C &= C_0\left(\frac{2x}{d}\right)\left[1 + \left(\frac{x}{d}\right)^2 + \left(\frac{x}{d}\right)^4 \dots\right] \\ &= C_0\left(\frac{2x}{d}\right) + C_0 2\left(\frac{x}{d}\right)^3 + C_0 2\left(\frac{x}{d}\right)^5 + \dots \end{aligned} \tag{5.11}$$

If $\frac{x}{d} \ll 1$, then

$$\Delta C \approx C_0 \times \frac{2x}{d}. \tag{5.12}$$

If we look at equations (5.11) and (5.12), the sensitivity is doubled and the non-linearity is less in the differential capacitance measurement as compared to the single-capacitance sensor. Apart from this, as d decreases, the sensitivity increases, thus reducing the size of the sensor. Any parasitic effect common to the sensing capacitors will be rejected as in any differential measurement.

5.2.4.2 Interdigitated comb structure

Here, multiple parallel plates are used in which a movable plate is arranged between fixed plates to form differential capacitance, as shown in figure 5.2(d). Here the movable plate is displaced according to the applied measuring signal. In this method, the developed differential capacitance is directly proportional to the displacement occurring due to the physical signal, as shown in (5.12). In the case of single-ended capacitive sensing, the capacitance change is inversely proportional to the displacement. As a conclusion, differential capacitive sensing is preferred in most applications because it offers higher linearity and sensitivity.

5.3 Sensor interfacing techniques

The basic purpose of a sensor interface is to convert the change in capacitance (ΔC) to an equivalent electrical signal. They are broadly classified into continuous-time and discrete-time methods. The interfaces can provide an analog or digital output. For digital outputs, one can have an analog-to-digital converter (ADC) at the output of an analog-output sensor interface. Some interfaces can deliver a digital output directly, which are called capacitance-to-digital converters (CDC). The most commonly used types of interface are discussed below.

5.3.1 Capacitance-to-frequency converter

The capacitance-to-frequency conversion technique is one of the simplest interfacing schemes. Here, the capacitance variation is converted to frequency variation, thus the measuring scheme is simple. The basic configuration is shown in figure 5.3. It is similar to a simple RC oscillator circuit that generates a square waveform. Normal RC oscillators use a resistive-type potential divider in the feedback path, whereas, in the present case, a differential capacitance arrangement is used as a potential divider [1].

Considering the feedback voltage to be $\boldsymbol{V_F = xV_o}$, the output of the oscillator will be a square wave with time period

$$T = 2RC \ln\left(\frac{1 + x}{1 - x}\right). \tag{5.13}$$

However, the expression is not linear. Besides, variation of R and C will cause a drift in the output.

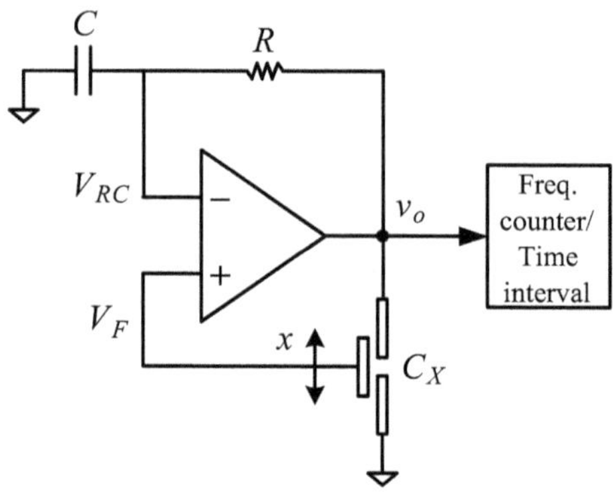

Figure 5.3. Capacitance-to-frequency conversion.

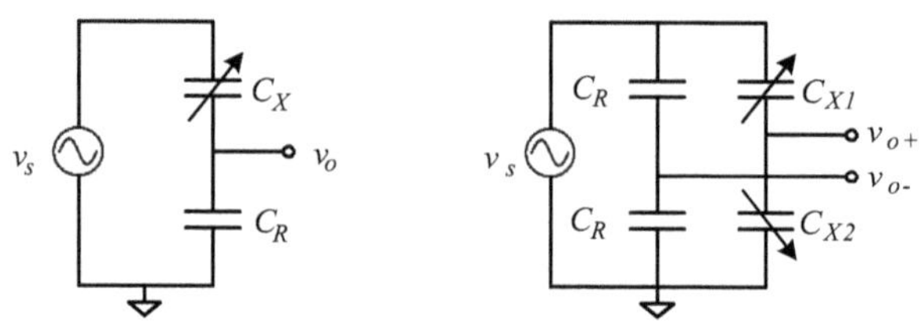

(a) (b)

Figure 5.4. Continuous-time interface for (a) single/unipolar sensor, (b) differential sensor.

5.3.2 Continuous-time interface

Continuous-time interfaces are simple in structure and preferred in many applications. Two basic continuous-time interfaces are shown in figures 5.4(a) and (b). The first is a potential divider with a single sensor, C_X, and a reference capacitor, $C_{R,}$ and which is excited by an AC signal source, $v_{s.}$ Considering C_X to be a variable-gap-type sensor, the output of this potential divider can be expressed as

$$v_o = \frac{C_X}{C_R + C_X} v_s = \frac{C_0}{C_0 + C_R\left(1 \pm \frac{x}{d}\right)} v_s. \tag{5.14}$$

Equation (5.14) represents a non-linear relationship between the output voltage and the displacement. The second configuration in figure 5.4(b) is an AC bridge with differential capacitance arrangement. The bridge output can be calculated as

$$\Delta v_o = v_{0+} - v_{0-} = \left[\frac{C_{X1}}{C_{X1} + C_{X2}} - \frac{1}{2}\right] v_s \tag{5.15}$$

$$= \frac{1}{2}\left[\frac{C_{X1} - C_{X2}}{C_{X1} + C_{X2}}\right] v_s = \frac{\pm x}{2d} \times v_s. \tag{5.16}$$

The output voltage of the bridge is linearly proportional to the displacement irrespective of the amount of displacement. The bridge output will be applied to a differential amplifier to amplify the signal for further processing. This bridge configuration suffers from a mismatch of the reference capacitors and parasitics from the sensing capacitors.

An operational amplifier (op-amp)-based constant current configuration is shown in figure 5.5(a) [4]. Due to the virtually shorted inverting and non-inverting terminals of the op-amp, the current is constant through C_R. The output of this configuration can be expressed as

$$v_o = -\frac{C_R}{C_X}v_s = -\frac{C_R}{C_0}\left(1 \pm \frac{x}{d}\right)v_s \tag{5.17}$$

$$= -\frac{C_R}{C_0}v_s \mp \frac{C_R}{C_0} \times \frac{x}{d}v_s. \tag{5.18}$$

The first part of the output is constant, but the second part is proportional to the displacement. If C_R and C_0 are made to be equal, then this output needs to be added with v_s to get the effective voltage due to the change in displacement only. This configuration is also affected by the sensor's parasitic capacitance but not affected by the parasitic capacitance in between the inverting node and ground. The feedback resistor sets the DC bias at the inverting node of the op-amp. The configuration in figure 5.5(b) is called a charge amplifier configuration [4]. A variation in charge across the sensing capacitor develops an equivalent voltage at the output across C_F.

Assuming that the value of R_F is very high compared to the impedance of the feedback capacitance, the output of the charge amplifier can be obtained as

$$v_o = -\frac{C_X}{C_F}v_s = -\frac{C_0}{C_F\left(1 \pm \frac{x}{d}\right)}v_s \tag{5.19}$$

$$\approx -\frac{C_0}{C_F}\left(1 \mp \frac{x}{d}\right)v_s = -\frac{C_0}{C_F} \times v_s \pm \frac{C_0}{C_F} \times \frac{x}{d} \times v_s. \tag{5.20}$$

The non-linearity error depends on the relative displacement, but the effect of parasitics is less in this configuration because of the virtually shorted inverting node.

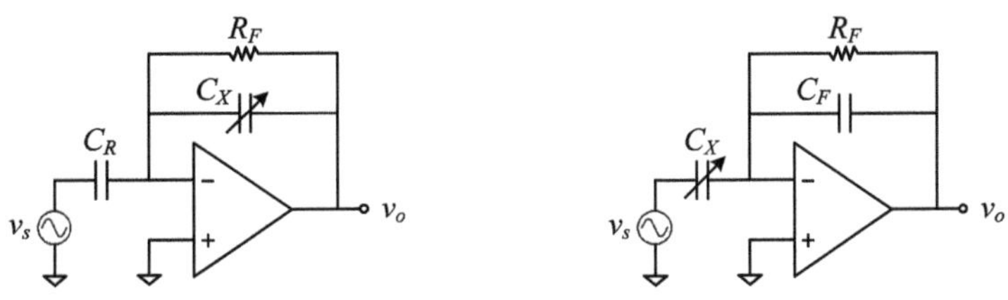

Figure 5.5. (a) Constant current and (b) charge amplifier configurations.

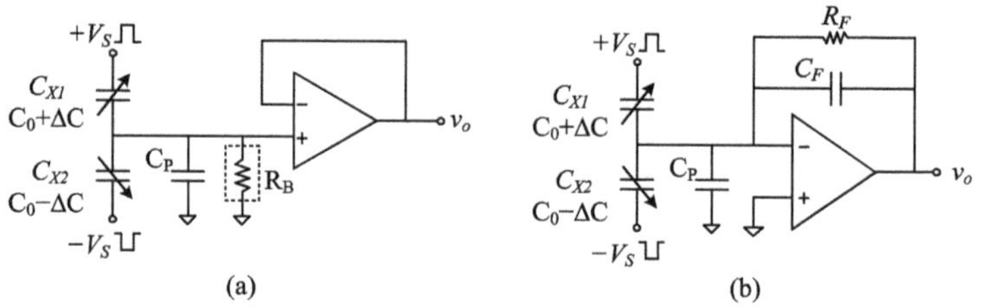

Figure 5.6. Input interfaces for a differential sensor: (a) voltage-mode and (b) charge amplifier configuration.

The continuous-time configurations in figures 5.6(a) and (b) are mostly used for differential capacitive sensor applications [3]. They can also be used for single, or unipolar, sensors along with a reference capacitor. The first configuration in figure 5.6(a) is called half-bridge voltage-mode interfacing, in which the half-bridge voltage is buffered for further signal conditioning. The half-bridge is excited with out-of-phase signals, either sinusoidal or square wave.

The bridge output voltage at the non-inverting node of the op-amp can be derived as

$$v_x = \frac{\Delta C}{2C_0 + C_P} \times V_{S-P-P}. \tag{5.21}$$

The output is proportional to the change in capacitance ΔC and $\Delta C = C_{X1} - C_{X2}$. The parasitic capacitance, C_P, at the sensing node reduces the sensitivity and increases the non-linearity. $V_{S\text{-}P\text{-}P}$ is the peak-to-peak excitation voltage. R_B is the bootstrap resistor and is used to bias the sensing node. Without R_B, this node is a floating node for DC signals. This must be of a much high value (~MΩ) so that it draws much less current from the sensing node.

The configuration in figure 5.6(b) is a charge amplifier interfaced with a half-bridge differential sensor. The feedback maintains proper DC bias at the inverting node of the op-amp. Therefore, the effect of C_P is negligible. R_F must be much higher compared to the impedance of the feedback capacitance. The output volage, v_o, can be expressed as

$$v_o = \frac{2\Delta C}{C_F} \times V_{S-P-P}. \tag{5.22}$$

Both configurations in figure 5.6 require high-value resistors to bias the sensing node. This can be avoided in discrete-time implementations.

5.3.3 Discrete-time interface

A basic discrete-time interface is shown in figure 5.7 [5, 6]. It has two phases, a reset phase and an amplification phase. In the reset phase, the reset switch is closed and the op-amp is in a unity-gain feedback configuration. This sets the inverting node voltage to the same as the non-inverting node voltage. The difference between the two node voltages is the offset of the op-amp. In this phase, the charge stored across C_F is zero. In the next phase, that is, the amplification phase, the reset switch is opened and the charge stored across the sense capacitors C_{X1} and C_{X2} is distributed.

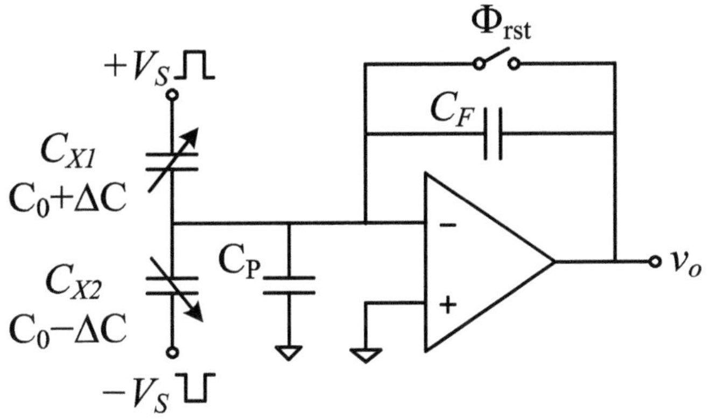

Figure 5.7. Discrete-time input interface for a differential sensor.

This develops a voltage across the feedback capacitor as well as the output of the op-amp.

Using the charge balance equation at the sensing node, the output voltage at the amplification phase can be written as

$$v_o = \frac{2\Delta C}{C_F} \times V_{S-P-P}. \tag{5.23}$$

The effect of C_P is negligible here, but the op-amp offset voltage introduces error in the output. By tuning the feedback capacitor, the gain can be changed.

The configurations discussed until now have drawbacks. They are affected by the non-idealities of the op-amp such as offset voltage, noise and finite gain. In general, a low-pass filter is used to remove the high-frequency noise. But the low-frequency noise, in which the flicker noise is significant, degrades the performance of the interfacing circuits by limiting the minimum detectable change in the input parameter. Moreover, the offset and noise of the op-amp are also amplified along with the input signal. The most common techniques which are used to take care of these issues are chopper stabilization (CHS), auto-zeroing (AZ) and correlated double-sampling (CDS), which will be discussed here one by one.

5.3.4 Chopper stabilization

This technique is used in continuous-time systems as well as for discrete-time systems. It operates based on the principle of modulation and demodulation. This method is also known as synchronous modulation and demodulation, chopper modulation, etc. Figure 5.8 shows a basic modulation and demodulation scheme implemented with a differential capacitive sensor. The half-bridge differential capacitive sensor is excited with two out-of-phase square-wave signals, which acts as a modulator and provides a modulated signal. The change in capacitance (ΔC) is the modulating or input signal and the square-wave excitation is the carrier signal. The effect of offset and low-frequency noise is eliminated by shifting them to higher frequencies and then removing them by low-pass filtering. Here, the input signal is

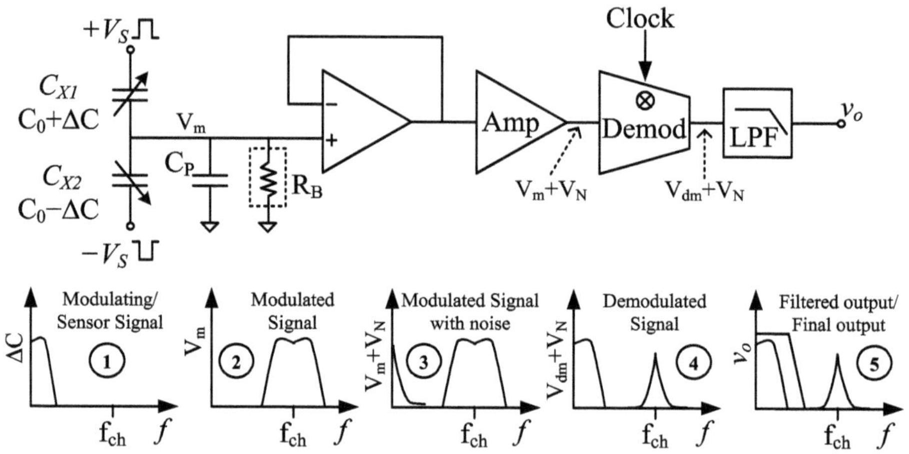

Figure 5.8. Chopper stabilization in a continuous-time system.

modulated and demodulated twice whereas the offset and noise of the op-amp is modulated once. This method removes the DC offset and 1/*f* noise [7–9] of the op-amps.

Frequency spectra of the signals at different stages of the signal interfacing circuits are also shown in figure 5.8. The first spectrum in figure 5.8 shows the original input signal band of ΔC, the signal of interest. We represent two signals, the modulating (ΔC) and the carrier signal (V_S) in (5.24). The carrier signal (V_S) is the square-wave excitation which contains all the odd harmonics of the fundamental frequency, as shown in (5.25):

$$\Delta C = \Delta C_0 \cos(\omega_m t) \quad \text{and} \quad V_S = V_P Sq(\omega_{ch} t), \tag{5.24}$$

where

$$Sq(\omega_{ch} t) = \frac{4}{\pi}\left(\sin\omega_{ch}t + \frac{1}{3}\sin 3\omega_{ch}t + \frac{1}{5}\sin 5\omega_{ch}t + \dots\dots\dots\dots\right). \tag{5.25}$$

The modulating signal is modulated within the bridge by the square-wave excitation signal, V_S. Equation (5.26) gives the expression of the modulated waveform:

$$V_m = V_{cm} + \frac{\Delta C_0}{C_0} V_P \cos(\omega_m t) \times Sq(\omega_{ch} t). \tag{5.26}$$

where V_{cm} is the common mode signal. If we consider only the fundamental frequency of the square wave in (5.25) for simplicity, then (5.26) can be expressed as

$$V_m \approx V_{cm} + \frac{\Delta C_0}{C_0} V_P \times \frac{2}{\pi}\{\sin(\omega_{ch} + \omega_m) + \sin(\omega_{ch} - \omega_m)\}. \tag{5.27}$$

The original signal is shifted to the carrier frequency. This is shown in the second spectrum in figure 5.8. After the modulation, noise is added from the buffer op-amp.

Both the noise and signal are then amplified by the amplifier, and the amplifier noise is also added. The total noise is considered as V_N. This is shown in the third spectrum. Only the low-frequency noise spectrum, mainly $1/f$ noise, is considered here. It is also assumed that the amplifier does not add any harmonics in the spectrum. The demodulated signal is represented by (5.28):

$$V_{dm} = V_{cm} + \left[\frac{\Delta C_0}{C_0} V_P \times \frac{2}{\pi}\{\sin(\omega_{ch} + \omega_m) + \sin(\omega_{ch} - \omega_m)\} + V_N\right] \times Sq(\omega_{ch} t). \tag{5.28}$$

In the demodulation, if we consider the same fundamental frequency of the square wave in (5.25), then (5.28) can be approximated as

$$V_{dm} \approx V_{cm} + \frac{\Delta C_0}{C_0} V_P \frac{8}{\pi^2} \cos(\omega_m t) + \frac{4}{\pi} V_N \sin(\omega_{ch} t) + \text{higher freq. terms.} \tag{5.29}$$

From (5.29), it is clear that, after demodulation, the original signal is back to its initial frequency band whereas the noise is shifted to the carrier frequency, which is shown in fourth diagram. This also occurs for the op-amp DC offset, as well. A low-pass filter with a cut-off frequency slightly higher than the signal bandwidth removes the noise and offset shifted to chopping or carrier frequency and the other higher frequency terms, as shown in the fifth spectrum. If we consider other odd harmonics of the excitation signal, those will be at much higher frequencies and will be removed by the low-pass filter, as well.

The demodulator in figure 5.8 is also known as a phase-sensitive demodulator (PSD). The reason for this is that it provides a phase-sensitive output. It is required in many applications such as capacitive sensors, linear variable differential transformers (LVDTs), etc, where the direction of the input change (positive and negative or up and down) changes the phase of the output, keeping the amplitude the same. In most cases they are 180° out of phase. Therefore, to sense the direction of the input, phase detection is necessary, and hence the PSD plays an important role. The operation of a PSD is explained in figure 5.9 with a simplified voltage-mode sensing scheme.

In figure 5.9, V_X is the sensing node. The waveforms are shown for DC change of ΔC. If $C_{X1} > C_{X2}$, ΔC is positive, and for $C_{X1} < C_{X2}$, ΔC is negative. For positive ΔC, V_X will be in phase with V_S; or else, in phase with $-V_S$ means 180° out of phase with V_S. The amplitude of the square wave at the sensing node will be as per (5.21). For same values of ΔC, whether positive of negative, the amplitude will be same; only difference will be in phase. To detect this, a phase-sensitive detector is required which is implemented by two switches and one inverting amplifier with unity gain. The switches are driven by Clk and inverted Clk. Clk is in phase with V_S. The output of the demodulator will be a DC voltage, greater or less than V_{CM} for $C_{X1} > C_{X2}$ and $C_{X1} < C_{X2}$, respectively. It also contains the switching spikes, which can be filtered out by a low-pass filter after the demodulator. This is how the PSD provides a direction-sensitive output to detect the direction of the input change.

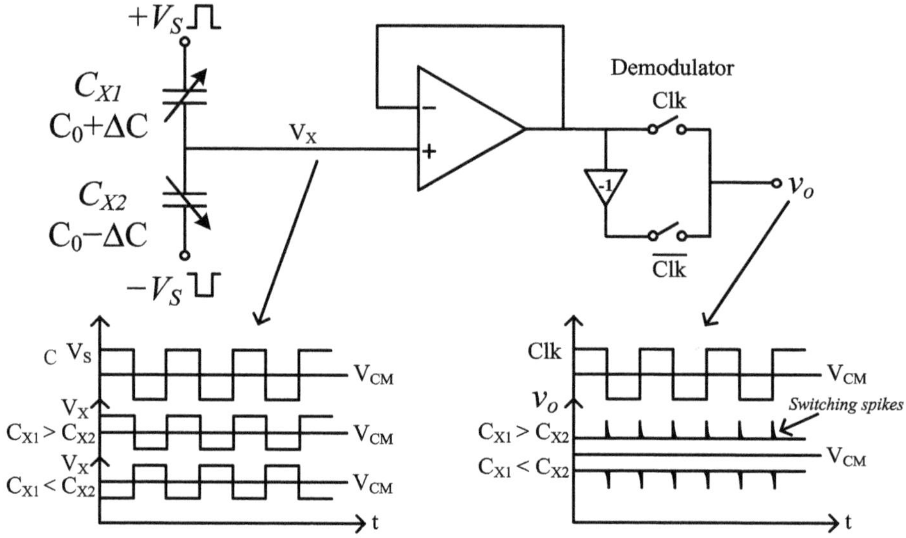

Figure 5.9. Illustration of modulated and demodulated waveforms for DC +ve and −ve ΔC.

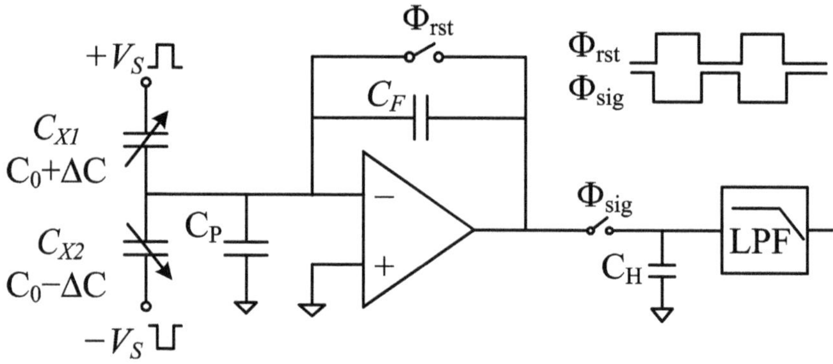

Figure 5.10. Chopper stabilization in a discrete-time system.

Figure 5.10 shows the implementation of CHS in a discrete-time system. The modulation is the same as the previous scheme, but the demodulation is implemented with a sample and hold circuit. In the reset phase the sampling switch is open, but in the amplification phase signal is passed to the holding capacitor. In the reset phase it holds this voltage of the amplification phase. The low-pass filter after this sample and hold removes the high-frequency spikes and noise.

5.3.5 Auto-zeroing (AZ) technique

This technique removes offset and low-frequency noise automatically by using a sample and hold technique, as shown in figure 5.11. This technique requires two clock phases. In the first phase, the instantaneous noise and offset are sampled and stored in the auto-zero capacitor, C_{AZ}, and then in the second phase it is subtracted

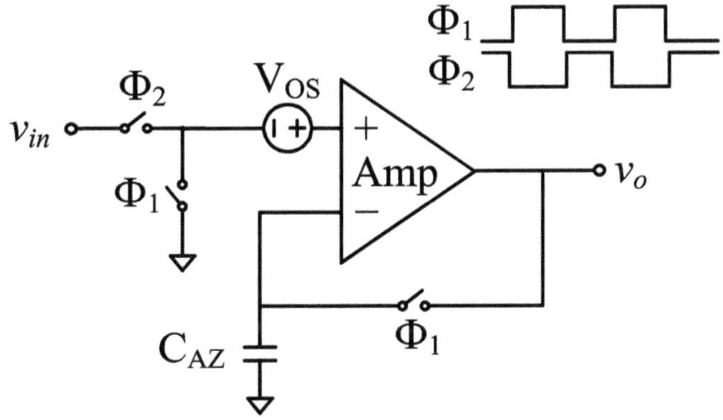

Figure 5.11. Implementation of auto-zeroing scheme.

from the signal along with the instantaneous unwanted noise and offset. The auto-zeroing process exhibits high-pass characteristics, which reduces the low-frequency noise [7, 8]. The output voltages at two clock phases can be written as follows.

First phase/sampling phase Φ_1:

$$v_o = v_{C_{AZ}} = V_{OS} + V_N. \tag{5.30}$$

Second phase/amplification phase Φ_2:

$$v_o = A(v_{in} + V_{OS} + V_N - v_{C_{AZ}}) = Av_{in}. \tag{5.31}$$

5.3.6 Correlated double-sampling

CDS (figure 5.12) is an extension of the auto-zeroing technique and operates on the same principle of auto-zeroing [7]. This means it samples the error voltage in a first phase and the signal with error in a next phase. In this process the correlated offset and noise are removed in the second phase. In the first sampling phase, a capacitor, C_{CDS}, stores the error voltage such as offset and instantaneous low-frequency noise.

In the first phase Φ_{rst} is closed and the charge stored in capacitor C_{CDS} w.r.t. the V_{cds} node is

$$Q_{rst-C_{CDS}} = -C_{CDS}(V_{OS} + V_N). \tag{5.32}$$

In the next phase Φ_{rst} is opened and the output of the op-amp voltage becomes

$$v_{out-op-amp} = \frac{2\Delta C}{C_F}V_{S-P-P} + V_{OS} + V_N. \tag{5.33}$$

By using the principle of charge conservation, the output voltage at the v_{cds} node can be written as

$$v_{cds} = \frac{2\Delta C}{C_F}V_{S-P-P}. \tag{5.34}$$

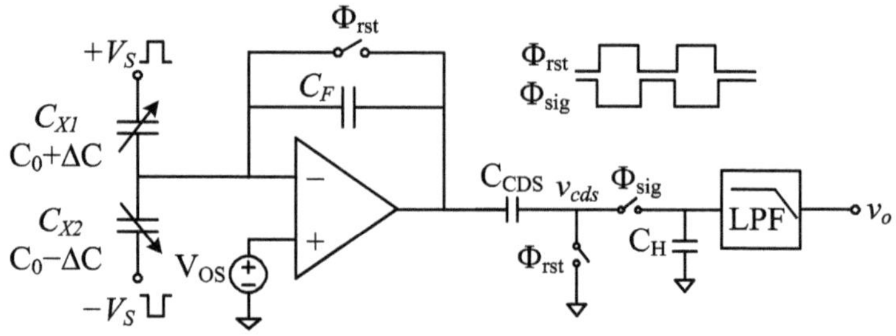

Figure 5.12. Basic implemention of correlated double-sampling technique.

This voltage is transferred to the holding capacitor C_H in this signal phase, Φ_{sig}. Though this method reduces the offset and $1/f$ noise, it suffers from kT/C noise, charge injection and clock feedthrough. Moreover, the use of C_{CDS} in series in the signal path shows high-pass characteristics, as well.

5.4 Capacitance-to-digital converters

Capacitance-to-digital converters (CDCs) convert capacitance or change in capacitance directly into digital bits without using separate analog-to-digital converters (ADCs). In conventional CDCs, the analog voltage from the signal-conditioning circuits is converted to digital bits using ADCs. In most cases, the capacitive sensor is directly integrated into the ADC architectures. Three commonly used architectures are discussed here one by one [10–14].

5.4.1 Dual-slope CDC

A dual-slope-type CDC operates on a similar principle to a dual-slope ADC. They can be used for both single and differential capacitance sensors. Figure 5.13 shows one such implementation for a differential capacitive sensor [10]. The fixed integrator capacitance in the ADC structure is replaced by the differential sensor capacitances C_{X1} and C_{X2}. A single-pole double-throw switch, S1, is used to select $+V_R$ and $-V_R$. Another switch, S2, a single pole single throw type, is used across C_{X2} to short it when required. The timing is controlled by the timing and control circuit, and the digital output is provided by the counter. The integrator output goes to the comparator and the comparator's output is used to generate the timing signals. In this CDC, if the capacitance change is due to displacement, then the sensor capacitances can be expressed as [10]

$$C_{X1} = C_0(1 \pm Kx) \text{ and } C_{X2} = C_0(1 \mp Kx)\,, \tag{5.35}$$

where x is the parameter to be measured and K is the transformation factor.

The output voltage of the CDC can be written as

$$Kx = \frac{C_{X1} - C_{X2}}{C_{X1} + C_{X2}} = \frac{T_1 - 2T_2}{T_1}, \tag{5.36}$$

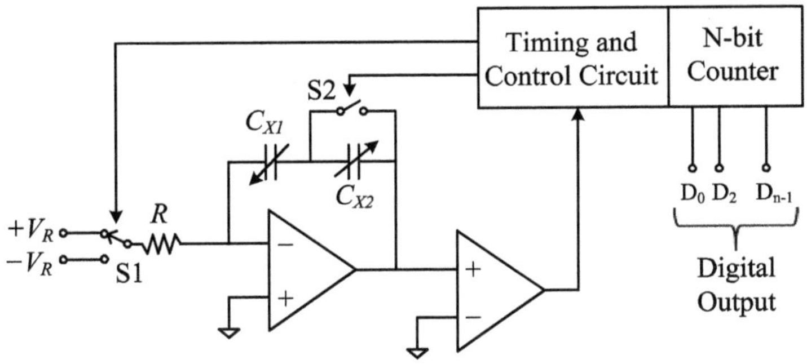

Figure 5.13. Dual-slope CDC for differential capacitive sensor.

where T_1 and T_2 are the charging and discharging time periods, respectively. At the start of the conversion phase, switch S1 is connected to $-V_R$ and S2 is shorted. Then, the output of the integrator ramps up with a slope, V_R/RC_{X1}, for a fixed time period, T_1. In this time capacitor C_{X1} is charged to the voltage $V_R T_1/RC_{X1}$. Now, S1 is connected to $+V_R$ and S2 is opened. Then, the output of the integrator decreases with a slope, $V_R(C_{X1}+C_{X2})/RC_{X1}C_{X2}$. T_2 is the time period in which the output reaches zero. Time periods T_1 and T_2 are counted in terms of clock cycles. For example, let T_1 be set equivalent to 1000 clock pulses, then T_2 is obtained by counting down from 1000 at twice the clock frequency. The final counting value will represent *Kx*. This CDC provides a linear relationship for inverse sensor characteristics, also [10].

5.4.2 Succesive approximation register type

A successive approximation register (SAR)-type CDC is shown in figure 5.14 with a single capacitive sensor along with a reference capacitor [11, 12]. It operates on two phases, a sampling (Φ_1) phase and a conversion (Φ_2) phase. In the sampling phase V_{CC} is sampled across C_X and C_R. At the same time the N-bit capacitive digital-to-analog converter (DAC) is reset. At the end of the sampling phase, the connections of C_X and C_R are reversed. The net charge to be transferred to the DAC is proportional to $(C_X - C_R)$. In the starting of the conversion phase, the voltage difference $(V_{CM} - V_X)$ at the input of the comparator is

$$\Delta V = V_{CC}\frac{C_X - C_R}{C_T} + V_{OS}, \tag{5.37}$$

where $C_T = C_X + C_R + C_{DAC-T}$, C_{DAC-T} is the total DAC capacitance and V_{OS} is the comparator offset.

Now, the SAR quantifies this charge by connecting the appropriate DAC capacitance to V_{CC}. It starts with the MSB and ends in the LSB. A particular bit is kept at '1' (connected to V_{CC}) if $V_X < V_{CM}$ or at '0' if $V_X > V_{CM}$. During the conversion process the modified difference in voltage can be expressed as

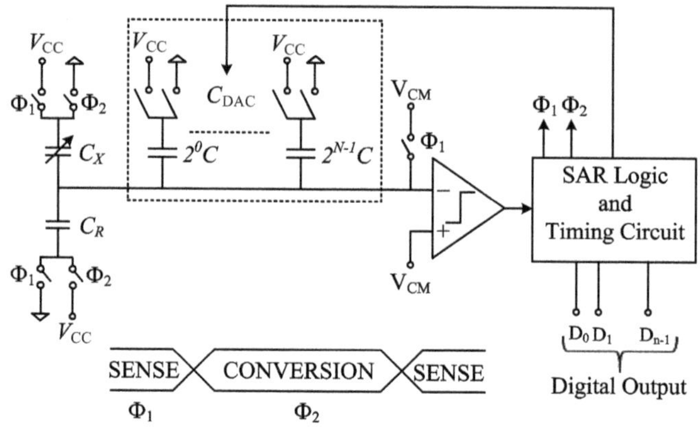

Figure 5.14. SAR-type CDC for a single capacitive sensor.

$$\Delta V_{\text{conv}} = V_{\text{CC}} \frac{(C_X - C_R) - C_{\text{DAC}}}{C_T} + V_{\text{OS}}. \tag{5.38}$$

The conversion process continues till C_{DAC} matches with $(C_X - C_R)$ within the resolution of the capacitive DAC. The error due to the offset of the comparator is

$$C_{\text{error}} = \frac{C_T \quad V_{\text{OS}}}{V_{\text{CC}}}. \tag{5.39}$$

5.4.3 Sigma delta modulator-type

A sigma-delta ($\Sigma\Delta$)-type CDC offers high resolution compared to previous architectures. This is due to oversampling and noise shaping [13, 14]. A first-order $\Sigma\Delta$ CDC with a unipolar/single sensor, C_X, is shown in figure 5.15 along with a reference capacitor, C_R. The CDC operates on an oversampling principle with logic feedback to balance the charge flow from the sensing and reference capacitor to the feedback capacitor C_F. C_F forces the averaged charge to be zero. Let it take N clock cycles to complete one measurement. Then, the total amount of charge transferred from C_X to C_F is $-NC_XV_{\text{REF}}$ and the charge from C_R to C_F is nC_RV_{REF}, where n is the number of '1's in the output bitstream 'Y'. Here, 'Y' is fed back and 'AND'ed with $\Phi_{2\text{D}}$. Then, we can write

$$-NC_XV_{\text{REF}} + nC_RV_{\text{REF}} = 0. \tag{5.40}$$

The average value of Y (n/N) gives the ratio of the two capacitors:

$$Y_{\text{avg}} = \frac{n}{N} = \frac{C_X}{C_R}. \tag{5.41}$$

The main sources of error are clock feedthrough and charge injections from the MOS switches. A differential sensor can also be integrated in this configuration by replacing C_R with C_{X2}. Then, based on the same principle, the average value of Y can be calculated as [14]

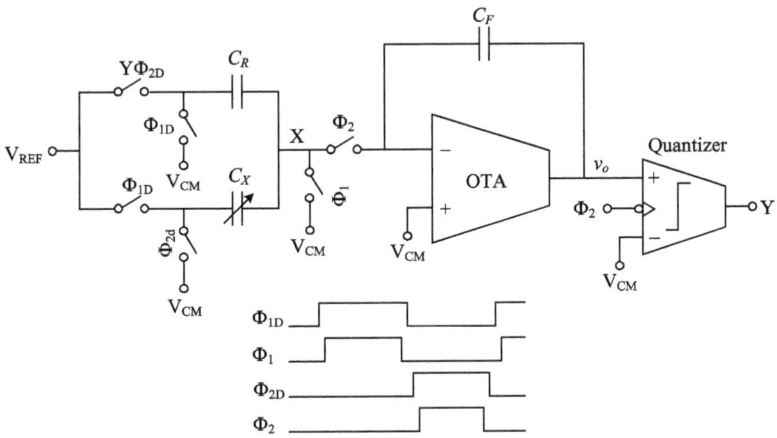

Figure 5.15. ΣΔ-type CDC for a unipolar/single capacitive sensor.

$$Y_{avg} = \frac{n}{N} = \frac{C_{X1}}{C_{X1} + C_{X2}}. \quad (5.42)$$

5.5 Sensor mismatch cancellation

Up to now we have discussed various interfacing techniques for single/unipolar and differential capacitive sensors. Though a differential sensor configuration offers several advantages, it suffers from individual capacitance mismatch. Without any input signal, the two capacitances, C_{X1} and C_{X2}, must be the same and equal to C_0. However, due to fabrication inaccuracy and process variations, they will be different by some amount. This mismatch introduces offset in the measured output, termed as zero offset. To overcome this problem one commonly used method is to add capacitance with the lower value sense capacitor to make it equal to the other capacitor [15, 16]. But this is limited by the resolution of the added capacitance. In [15] an on-chip capacitance array was included within a signal-conditioning ASIC for this type of compensation. The second method is the charge compensation method, in which a reference voltage is applied to compensate for the charge imbalance between the sensing capacitors [16]. This method is also limited by the resolution of the reference voltage.

5.5.1 Capacitance array

The basic structure of the capacitance array is shown in figure 5.16. The value of the added capacitance can be selected through external pins followed by a decoder [15] or by internal shift registers which can be programmed externally. In most cases a thermometric decoder is used to reduce the area consumed by the capacitor array. Only one array of capacitors is sufficient, which can be selected in parallel with either C_{X1} or C_{X2}. Before starting the compensation process, first the smaller capacitor is detected by checking the output of the interfacing circuit without any input. If $C_{X1} > C_{X2}$,

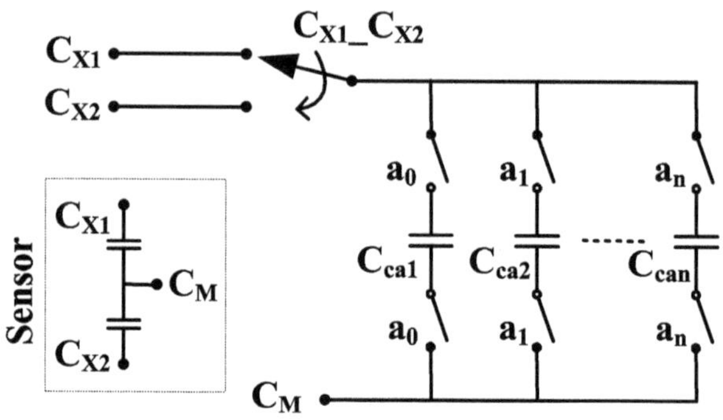

Figure 5.16. Capacitance array for sensor mismatch cancellation.

the output will be greater than the output common-mode voltage or vice versa. In general, the output common-mode voltage is zero for a bipolar supply voltage or $V_{DD}/2$ for a unipolar supply. Once the smaller capacitor is detected, then capacitance is added in parallel starting from the smallest value in the array until the output of the interface changes its state from the previous one. Here, the state of output means either above or below the common-mode voltage. This process is mostly done manually before starting measurement. An automatic compensation scheme based on the SAR principle is demonstrated in [15].

5.5.2 Voltage-mode charge balance method

The voltage-mode charge balance method is based on using the voltage as the calibration knob [16]. This calibration voltage controls the charge to be transferred, which is equal to the charge generated due to the mismatch. A schematic is shown in figure 5.17. This configuration is fully differential with two differential capacitive sensors, C_{X1P}, C_{X1N} and C_{X2P}, C_{X2N}. C_{OS} is the OTA offset compensation capacitor and C_{OFF} is for sensor offset compensation. In the Φ_1 phase, the system is reset and charge equivalent to $C_{OFF}(V_{DD} - V_{CAL})$ is stored across the C_{OFF} capacitor. In the Φ_2 phase, this charge is transferred to the feedback capacitor, which compensates the charge mismatch in the sensing capacitors in the reset phase. The output voltage can be expressed as

$$\Delta V_{OUT} = \frac{0.5V_{DD}}{C_F}(\Delta C_{sense} + \Delta C_{mismatch}) + \frac{C_{OFF}}{C_F}(0.5V_{DD} - V_{CAL}), \qquad (5.43)$$

where ΔC_{sense} is the capacitance change due to applied input and $\Delta C_{mismatch}$ is the mismatch in the sensing capacitors without any external input. This method of compensation offers better resolution compared to the capacitance array but is still limited by the resolution of the DAC that is used to generate V_{CAL}.

5.5.3 Auto-cancellation of sensor mismatch

As discussed before, the cancellation of sensor mismatch can be done both manually and automatically. An automatic calibration method is discussed here based on the

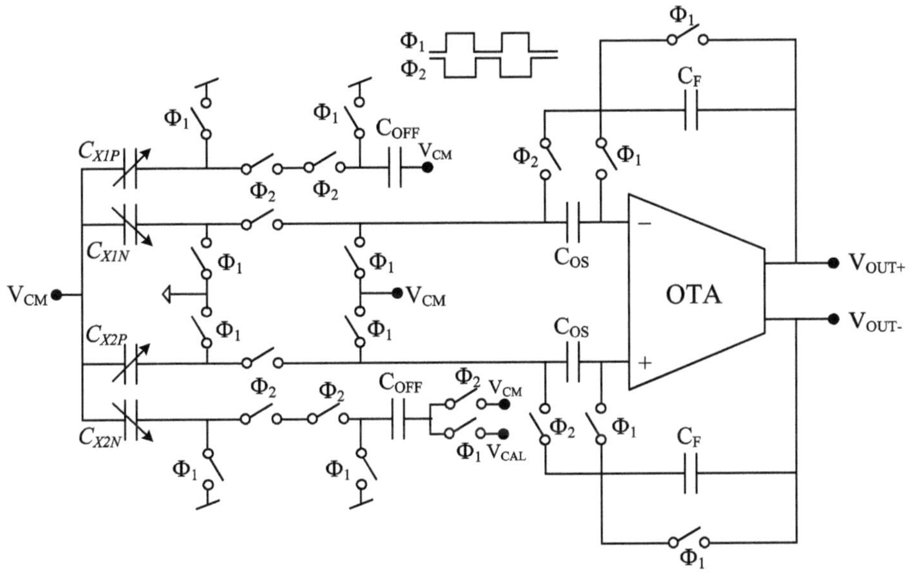

Figure 5.17. Charge balance method for sensor mismatch cancellation.

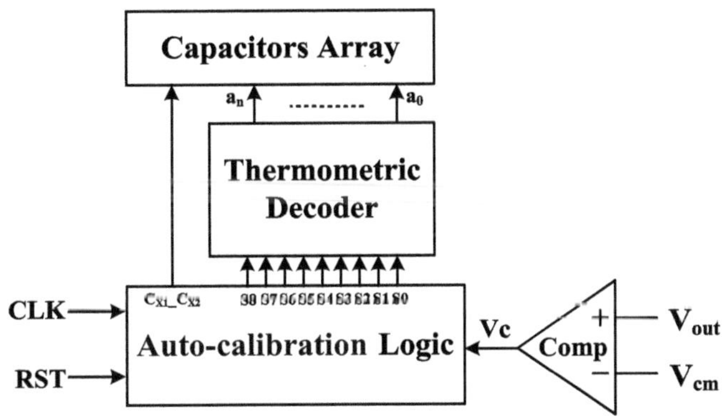

Figure 5.18. Block diagram for automatic cancellation of sensor mismatch.

SAR principle [15]. The compensation is done as mentioned before through a capacitor array, and the selection of the proper capacitance value is done by control bit selection from the MSB to LSB. The auto-calibration logic uses the comparator output, as shown in figure 5.18, which compares V_{out} with V_{cm} and, based on the logic discussed afterwards, provides the control signal $C_{X1}\_C_{X2}$ and 'n' number of control bits to the on-chip capacitor array. The logic requires 'n+2' clock pulses to provide proper selection bits. The first clock pulse resets all the flip-flops and the second clock pulse enables and decides the compensation capacitance, either C_{X1} or C_{X2}, based on the comparator output. From the third clock pulse the SAR logic starts by setting the MSB logic '1' for a predefined period; later a combinational logic and the comparator output decides the state of the bit. Details of this logic is provided below.

- Clock pulse 1 resets the flip-flops.
- Clock pulse 2 enables and selects either C_{X1} or C_{X2} by comparing the output voltage with the common-mode voltage. If the output voltage is more than the common-mode voltage, then it represents $C_{X1} > C_{X2}$, so capacitance C_{X2} will be selected by the control bit $C_{X1}\_C_{X2}$ and vice versa.
- Clock pulse 3 sets the MSB bit as logic 'HIGH' and remains 'HIGH' for a predefined time period. If V_{out} remains in its previous state (either $> V_{cm}$ or $< V_{cm}$), then there is no change in comparator state also and the bit is kept '1', else it is set to '0'. Then at each clock pulse the subsequent bits are set in the similar fashion.
- Clock pulse 4 sets the next bit (S_{n-1}).
- This goes on until the LSB is set at (n+2) clock pulse.

This auto-calibration logic can be implemented by using any field programmable logic array. It can also be integrated within the ASIC itself to make it fully integrated.

5.6 Case study: integrated capacitance measurement system

In this section the development of an integrated capacitance measurement system is discussed to gain a complete system overview. The complete system is shown in figure 5.20 [15]. Throughout this chapter we have mainly concentrated on the different types of input interfaces along with a capacitance array for sensor mismatch compensation. But, in practice, a complete sensor interface includes many other blocks such as demodulators, filters, clock generators, reference voltage generators, serial input parallel output (SIPO) registers, etc, for proper functioning of the circuit. In that case, a capacitance sensor is to be integrated with the interface to form a complete system. The integration can be in three different ways. The first such way is PCB integration, in which the sensor and the circuit are fabricated and packaged separately and integrated on a PCB. The second method is hybrid integration, in which the two fabricated dies are placed on a single package and wire bonded. The third integration is monolithic integration, in which the sensor and the circuit are fabricated on the same wafer and connected together. This last method of integration offers the best performance compared to the first two methods, but it requires a dedicated process for fabrication of the sensor and circuit. In this work a spring-mass-damper-based acceleration sensor is used as a differential capacitive sensor; it is shown in figure 5.19. Due to applied acceleration, the mass is displaced and the capacitance changes. This change in capacitance is measured by the interfacing circuit.

5.6.1 Circuit components

Figure 5.20 shows a fully differential capacitive sensor interface with most of the associated circuit blocks [15]. The first interface block is the charge amplifier, which is implemented with a fully differential op-amp. Several architectures are available

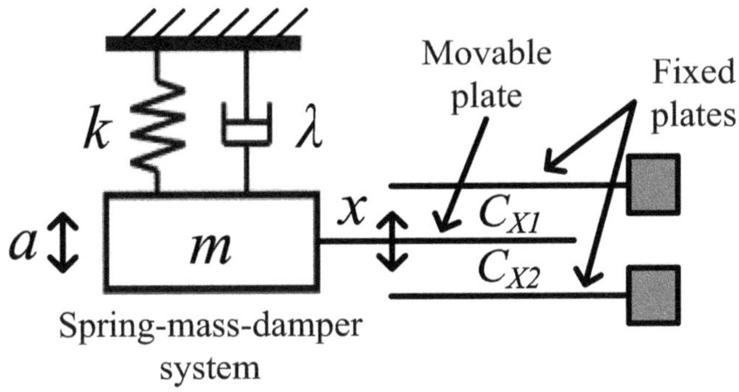

Figure 5.19. A differential capacitive acceleration sensor.

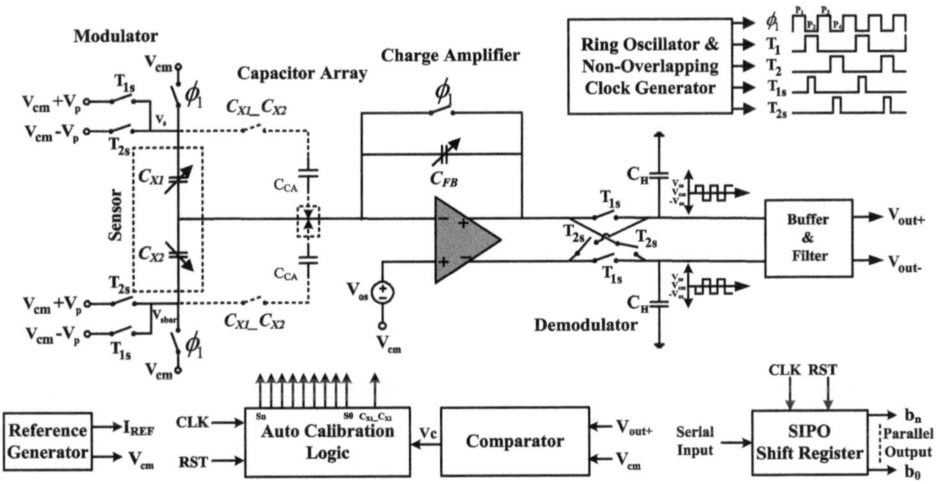

Figure 5.20. An integrated capacitive sensor interface.

for fully differential op-amps like cascode, folded cascode, telescopic, etc. The open-loop gain should be >60 dB to keep the offset less and the phase margin should be >60° for closed loop stability. A common-mode feedback circuit is a must to fix the output common-mode voltage at the desired level. In the charge amplifier gain can be tuned by tuning the feedback capacitor, C_{FB}. In general there are two feedback loops but in this configuration only one is used, which reduced the gain mismatch in between the two loops; further, only one differential sensor configuration is required for differential output.

After the charge amplifier there is a cross-coupled demodulator, which is implemented with the transmission gate switches. Non-overlapping clocks are required to drive these switches. This particular demodulator configuration removes the offset efficiently without using any switched-capacitor-based technique. The offset waveforms are shown in the figure, which are averaged out by the filter. Clock generation is done by a ring oscillator first, then, by using certain combinational logic, the required non-overlapping clock pulses are generated [15], as shown in figure 5.21. Non-overlapping clocks also reduce the charge loss in the capacitors as well as the charge injection during the transition of the switches. It also includes an on-chip capacitance array, which was discussed before. The demodulator output is buffered and filtered. The filter is a second-order low-pass filter, which can be implemented with different topologies. In this design a Sallen–Key architecture is used. This removes the high-frequency noise and spikes mainly the low-frequency noise, which is shifted to higher frequency due to chopper modulation and demodulation.

A reference generator is also an integral part of any interfacing circuit, in which the required current and voltage references are generated. In general, a bandgap reference is used for both supply and temperature-independent biasing voltage and current. In this design the reference current is 50 μA and the reference voltage is 900 mV. The reference generator also includes tuning options to fine-tune the generated current and voltage.

A serial input parallel output (SIPO) shift register is utilized to reduce external pin count and to program the different control bits used in this interface for tuning. Blocks such as the capacitor array, reference generator, charge amplifier, etc, require multiple control bits to set the desired value. These bits are set by the SIPO. This SIPO can be programmed from outside with the help of any microcontroller. The SIPO shift register is designed and implemented with positive edge-triggered D-flip-flops with a reset option, as shown in figure 5.22.

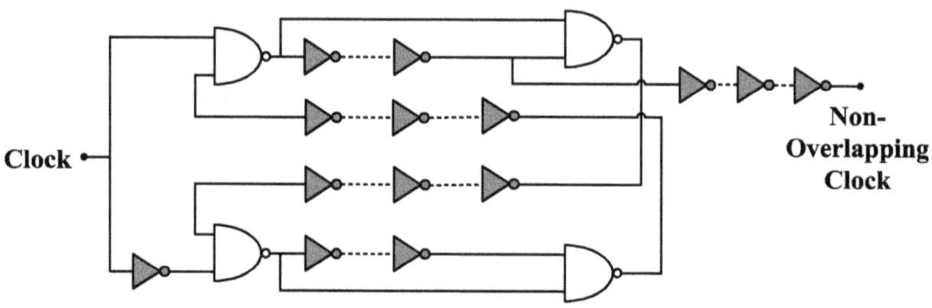

Figure 5.21. Non-overlapping clock generator.

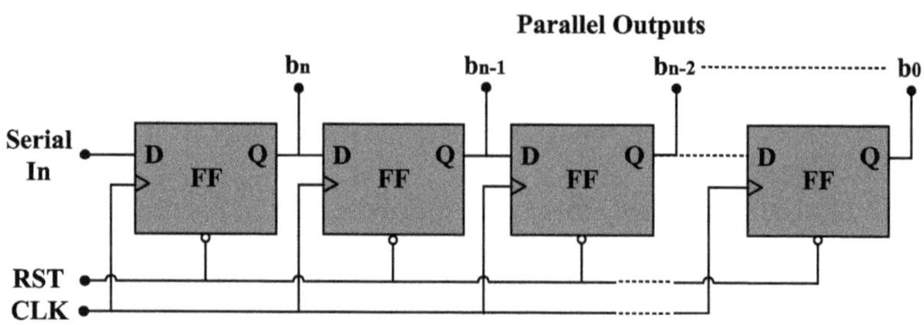

Figure 5.22. Serial in parallel out (SIPO) register.

5.7 Conclusion

In this chapter, various sensor interfaces for capacitive sensors are discussed. A sensor can be single or differential based on the requirements and application. Variable-gap-type capacitive sensors are the most common, although variable-area and dielectric-type capacitive sensors are also popular. Sensor interfaces convert the change in capacitance to an equivalent electrical output. This conversion process is affected by many factors, not only from the sensor but also from the interface itself. The main motivation is to improve the accuracy and resolution. The accuracy is improved by reducing the non-linearity and resolution by minimizing the input referred noise. The micro- and nanosensors used today impose more constraints on the design of sensor interfaces. The reduction of power for battery-operated systems is one such major requirement in today's market. This power requirement is also applicable for IoT sensor nodes, which mainly operate on renewable energy sources. Therefore, though many circuit interface topologies are available, still significant effort is required to address current requirements.

References

[1] Baxter L K 1997 *Capacitive Sensors Design and Applications.* (New York: IEEE Press)

[2] Amini B V and Ayazi F 2004 A 2.5-v 14-bit Δ CMOS SOI capacitive accelerometer *IEEE J. Solid-State Circuits* **39** 2467–76

[3] Kar S K, Swamy K M, Mukherjee B and Sen S 2015 Systematic development of integrated capacitance measurement system with sensitivity tuning *IEEE Trans. Instrum. Meas.* **64** 2738–46

[4] Pallas–Areny R and Webster J G 1991 *Sensors and Signal Conditioning.* (New York: Wiley–Blackwell)

[5] Yazdi N, Kulah H and Najafi K 2004 Precision readout circuits for capacitive micro-accelerometers *Sensors. IEEE* pp 28–31

[6] Kar S K, Chatterjee P, Mukherjee B, Swamy K B M M and Sen S 2018 A differential output interfacing asic for integrated capacitive sensors *IEEE Trans. Instrum. Meas.* **67** 196–203

[7] Enz C C and Temes G C 1996 Circuit techniques for reducing the effects of op-amp imperfections: autozeroing, correlated double sampling, and chopper stabilization *Proc. IEEE* **84** 1584–614

[8] Razavi B 2002 *Design of Analog CMOS Integrated Circuits* (New Delhi: Tata McGraw-Hill Education)

[9] Yazdi N, Mason A, Najafi K and Wise K D 2000 A generic interface chip for capacitive sensors in low-power multi-parameter microsystems *Sens. Actuators* A **84** 351–61

[10] Mohan N M, Shet A R, Kedarnath S and Kumar V J Nov. 2008 Digital converter for differential capacitive sensors *IEEE Trans. Instrum. Meas.* **57** 2576–81

[11] Tanaka K, Kuramochi Y, Kurashina T, Okada K and Matsuzawa A 2007 A 0.026 mm^2 capacitance-to-digital converter for biotelemetry applications using a charge redistribution technique *2007 IEEE Asian Solid-State Circuits Conf.* pp 244–47

[12] Sanyal A and Sun N 2016 A 55 fJ/conv-step hybrid SAR-VCO ΔΣ capacitance-to-digital converter in 40 nm CMOS *ESSCIRC Conf. 2016: 42nd European Solid-State Circuits Conf.* pp 385–8

[13] Alhoshany A and Salama K N Feb. 2019 A precision, energy-efficient, oversampling, noise-shaping differential SAR capacitance-to-digital converter *IEEE Trans. Instrum. Meas.* **68** 392–401

[14] Li B, Sun L, Ko C, Wong A K and Pun K July 2014 A high-linearity capacitance-to-digital converter suppressing charge errors from bottom-plate switches *IEEE Trans. Circuits Syst.* I **61** 1928–41

[15] Tirupathi R and Kar S K 2021 On-chip implementable autocalibration of sensor offset for differential capacitive sensor interfaces *IEEE Trans. Instrum. Meas.* **70** 1–9

[16] Jeong Y and Ayazi F Nov. 2013 A novel offset calibration method to suppresscapacitive mismatch in MEMS accelerometer *Proc. 9th Best PaperAward Samsung Electro-Mech.* pp 1–4

IOP Publishing

Systems for Printed Flexible Sensors
Design and implementation
Tarikul Islam, Subhas Chandra Mukhopadhyay and Boby George

Chapter 6

Interface electronics and conditioning circuits for triboelectric flexible sensors

Chunkai Qiu and Mehmet Rasit Yuce

Triboelectric-based sensors have shown great potential in applications such as sensing environmental parameters, biomedical monitoring, chemical sensing, and human–machine interaction. The analog front-end circuits should be designed to process the real-time signal generated by triboelectric-based transducers, thereby enhancing the functional applicability of triboelectric sensor interfaces. However, the high intrinsic impedance of triboelectric sensors means that the transducers are inherently low power and susceptible to external noise. Therefore, techniques for buffering high-impedance input and minimizing circuit noise should be investigated for processing the signal of triboelectric sensors. This chapter introduces analog interface circuits for signal conditioning of triboelectric sensors. The configuration of the sensing circuit of triboelectric-based sensors is described, as well as the issues that should be addressed in the design.

6.1 Introduction

Triboelectric nanogenerators (TENGs) have become a powerful manner in which to collect ambient mechanical energy as well as sense the applied physical and chemical changes [1, 2]. TENGs can convert mechanical energy into electrical energy, relying on the coupling of the triboelectrification and electrostatic induction effects [3, 4]. Under the context of this new technology, numerous types of TENGs with diverse functionalities have been proposed. Generally, the working mechanism of TENGs can be classified into four modes: a contact-separation mode, a linear-sliding mode, a single-electrode mode and a freestanding triboelectric layer mode [5, 6]. Each mode has its unique advantages and characteristics, and should be appropriately chosen or combined to operate in different application circumstances [7–9]. Due to

doi:10.1088/978-0-7503-3935-3ch6

the superiority of the self-generated electrical output, as well as its easy fabrication and low cost, triboelectric-based sensors are well received in Internet-of-Things (IoT) applications [10, 11].

A TENG-based sensor can transform an applied mechanical force into a self-generated electrical signal. By analyzing the electrical signal (e.g., amplitude, frequency, period and number of peaks), information about the mechanical motion can be retrieved [12, 13]. Therefore, a TENG can act as a self-powered sensor to detect physical parameters such as human interaction [14, 15]. By leveraging a microcontroller unit (MCU) or a computer to analyze the generated output signal, useful information related to human activity or ambient environment can be obtained. This valuable information can be used to control various external objects, exhibiting promising potential applications in human–machine interfaces (HMIs) [16, 17]. With its advantages of being self-powered, cost effective, having a simple device structure and easy fabrication processes, environmental friendliness, and high reliability [18–20], various triboelectric sensors have been explored and applied in a variety of applications, including vibration sensing [21], pressure sensing [22], chemical sensing [23], biomedical monitoring [24] and displacement tracking [25].

To achieve the practical implementation of these sensors, a portable and small-sized electronic signal-conditioning circuit is necessary for real-time signal processing and data computation. The output characteristics of a TENG sensor can be described by voltage, current, power density and the amount of generated charges. Due to the small sensor size and high-output impedance attribute, the output power of a TENG sensor is typically relatively low. This poses a challenge in the design of the signal-conditioning circuits. A general block diagram of a signal-processing circuit for triboelectric transducers is presented in figure 6.1. The interface electronics consist of three parts: (1) a preamplifier with high input impedance; (2) filtering circuits (e.g., low-pass, high-pass or notch filters) and gain amplifiers; (3) an MCU with built-in analog-to-digital converters (ADCs) for signal sampling and data processing. According to the specific application and properties of the transducer, the design of the circuit electronics should be adjusted accordingly. This chapter explores practical solutions to process the AC signal generated by triboelectric sensors, including several design considerations.

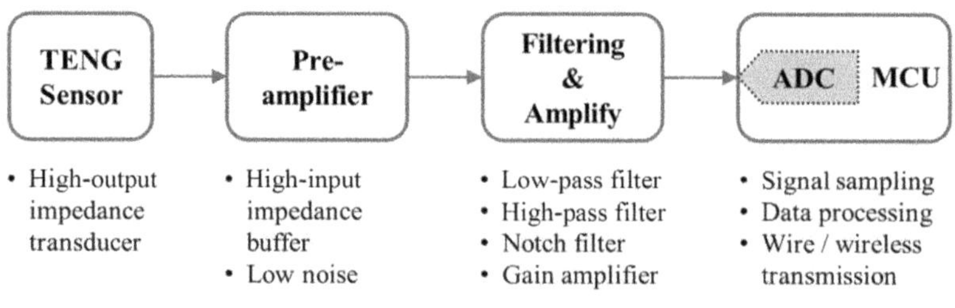

Figure 6.1. Block diagram of a signal-conditioning circuit for triboelectric sensors.

6.2 Characteristics of triboelectric sensors

Before designing the circuit, it is necessary to investigate the properties of the triboelectric sensors, because these properties will directly affect the circuit design. TENGs performance can be influenced by factors such as friction materials, device structure, applied mechanical stimulus, load resistance and environmental conditions [35, 36]. Besides these influences, TENGs usually exhibit high-output impedance characteristics [26, 37]. Figure 6.2 illustrates the theoretical output performance (e.g., peak voltage, current and power) of a vertical contact-separation TENG under a constant velocity motion with varying load resistances. The TENG has three regions of operation depending on the load resistance value, as shown in figure 6.2(a). In region I (<1 kΩ), where the load resistance is relatively low, the TENG behaves similarly to a current source. In this region, the peak output current remains approximately the same as the short-circuit current, and the voltage is kept at a relatively low level. Conversely, in region III (>1 GΩ), the TENG performs like a voltage source. Since the load resistance is high, the TENG shows similar properties as in the open-circuit condition. The output voltage in this region saturates at the open-circuit voltage, and the current remains relatively low. On the other hand, region II is the transitional region in which the TENG transforms from a short-circuit condition to open-circuit condition. In this area, with increasing load resistance, the output current of the TENG decreases dramatically. In contrast, the output voltage increases from the minimum level towards the maximum level.

Based on the impedance-matching effect, the TENG delivers maximum output power when the resistive load is equal to the source impedance [38, 39]. Therefore, the intrinsic impedance of the TENG can be determined by investigating the relationship between the output power and the load resistance. Typically, the maximum instantaneous power of a TENG is reached in region II, as depicted in figure 6.2(b). Some examples of the measured TENG output impedances from existing works are summarized in table 6.1. Information analysis of the triboelectric sensors mainly relies on the characteristics of the output voltage or current signals, including amplitude, frequency and the number of peaks. Therefore, rational design

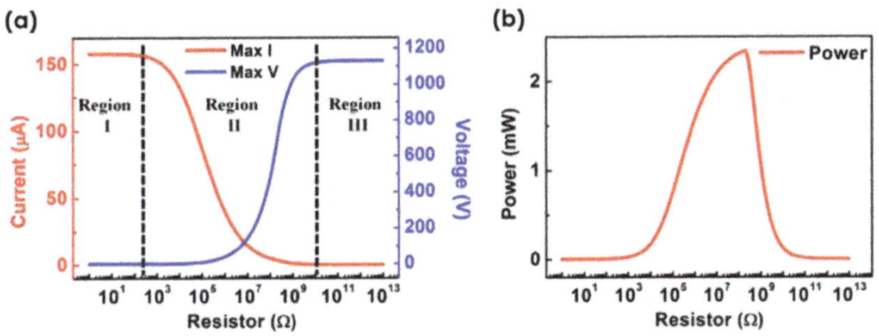

Figure 6.2. Theoretical output performance of contact-separation-mode TENGs with different load resistances. (a) The influence of load resistance on the output voltage and current; (b) the output power against the load resistance. Reproduced with permission from [26]. Copyright 2013, Royal Society of Chemistry.

Table 6.1. Comparison of TENG output impedance.

Reference	[27]	[28]	[29]	[30]	[31]	[32]	[33]	[34]
TENG mode[1]	CS	CS	LS	LS	SE	SE	FT	FT
Internal Resistance[2]	23 MΩ	100 MΩ	50 MΩ	10 MΩ	70 MΩ	3 MΩ	80 MΩ	80 MΩ

[1] CS: contact separation; LS: lateral sliding; SE: single electrode; FT: freestanding triboelectric layer.
[2] The optimum load resistance where maximum power is reached.

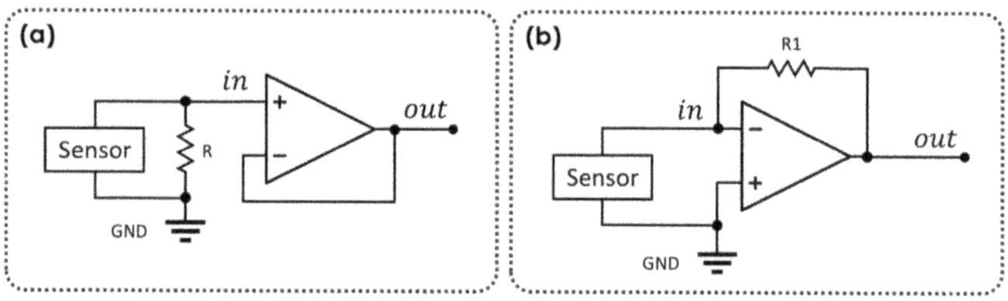

Figure 6.3. Circuit diagram of preamplifiers: (a) voltage follower; (b) transimpedance amplifier.

of interfacing electronics with suitable input impedance to interact with the TENG is critical [40, 41]. For instance, if the input impedance of the electronic circuit is very low, the voltage signal will be very weak; while if the input impedance of the electronic circuit is very high, the voltage signal can be saturated.

6.3 Block structure of interface electronics

6.3.1 Preamplifier

The high-output-impedance nature of the triboelectric sensors reveals that a preamplifier should be utilized in order to convert the high-impedance source signal to a low-impedance output signal for the following stages [42]. A voltage follower is one of the simplest configurations of high-input-impedance preamplifier that can be used with TENG sensors, as shown in figure 6.3. Owing to the high-impedance sources, operational amplifiers with low input bias current and low noise features can be employed to minimize the output errors [43]. However, the input bias current of a preamplifier can flow through the external resistance and produce a voltage across the source resistance, which escalates system errors. Several commercially available low-input bias current operational amplifiers suitable for high-impedance sensors are compared in table 6.2.

For example, LTC6240 (from Linear Technology) is a low-noise, low-offset, rail-to-rail output operational amplifier suited for a high-impedance transducer amplifier. The LTC6240 has an input resistance of 10^3 GΩ. The input bias current (I_{BIAS}) is only 0.2 pA and is guaranteed to be 1 pA maximum. The extremely low input bias current enables it to be used with high-impedance transducers and biasing resistors

Table 6.2. Comparison of low-input bias current operational amplifiers.

Part number	Voltage supply	I_{BIAS} (pA)	V_{OS} (μV)	I_{OS} (pA)	e_n (nV/$\sqrt{Hz}$)	i_n (fA/$\sqrt{Hz}$)	GBP (MHz)
LTC6240	2.8–6 V	0.2	50	0.2	7	0.56	18
LTC6081	2.7–5.5 V	0.2	70	0.1	13	0.5	3.6
AD8606	2.7–5.5 V	0.2	80	0.1	8	10	10
AD8616	2.7–5.5 V	0.2	80	0.1	10	50	24
OPA392	1.7–5.5 V	0.01	1	0.01	6	40	13
LMP7707	2.7–12 V	0.2	37	0.04	9	1	14
LMC6001	4.5–15.5 V	0.01	350[1]	0.005	22	0.13	1.3
MAX4476	2.7–5.5 V	1	70	1	4.5	0.5	10

[1] Maximum limit (at 25 °C).

while maintaining high accuracy. The input offset voltage (V_{OS}) is typically just 50 μV, along with a low-input offset current (I_{OS}) of 0.2 pA. Additionally, the input voltage noise (e_n) is guaranteed to be less than 10 nV/$\sqrt{Hz}$ at 1 kHz. The combination of low-input bias current, low offset voltage, low noise and wide bandwidth, along with a wide power supply range, makes these amplifiers perfect for signal conditioning of high-impedance sensors.

Another simple configuration of a preamplifier is a current preamplifier, as shown in figure 6.3(b). This approach is based on a transimpedance amplifier that can convert current input to voltage output. The work in [44] utilized a transimpedance amplifier for sensing the current signal generated by the TENG. A 10 kΩ feedback resistor was used, which resulted in a sensitivity of 10 mV μA^{-1}. The LMC6001 amplifier (from Texas Instruments) is selected to construct the circuit due to its ultra-low input current and low-noise features. The presented circuit successfully detected the output current waveform generated by a contact-separation-mode TENG sensor when pressed and released. Furthermore, in comparison with a commercial current preamplifier system (SR570, from Stanford Research Systems), the difference of measurements is only about 5% [44].

6.3.2 Filtering

Electromagnetic interference and cross talk could deteriorate the reliability and performance of TENG sensors. To suppress these noises, filters such as low-pass filters, high-pass filters, band-pass filters and band-stop filters can be deployed. Some popular types of filters are Butterworth, Chebyshev and Bessel filters [45]. These filters can either be implemented digitally on digital processing units or physically by contrasting analog circuits. With regard to the actual circuit implementation, the Sallen–Key topology is one of the most well-known and popular active-filter configurations with the advantages of being simple to implement, exhibiting good stability, and being least affected by the performance of the operational amplifier [46].

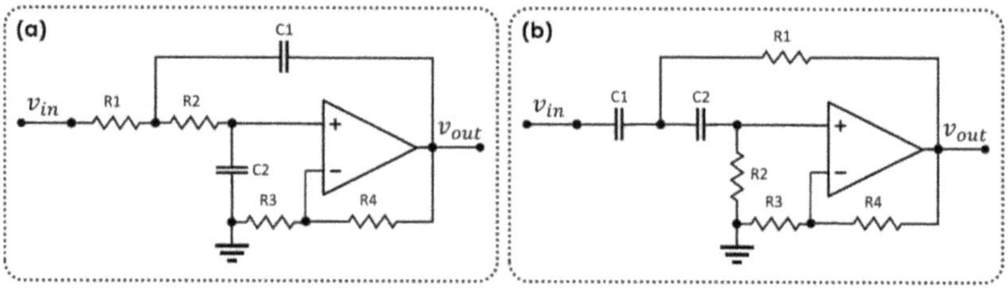

Figure 6.4. Sallen–Key architecture. (a) Low-pass filter; (b) high-pass filter.

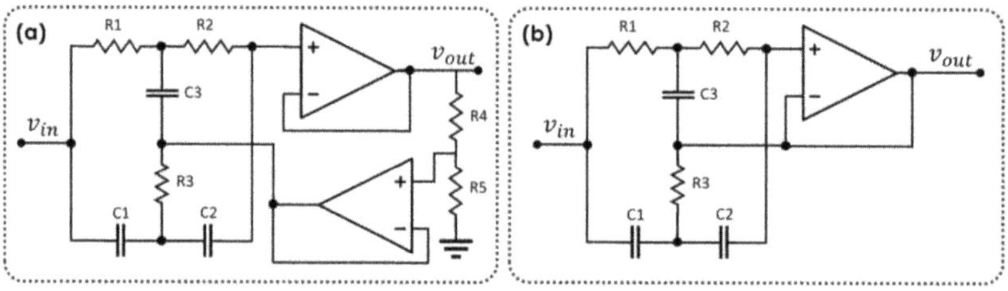

Figure 6.5. Active twin-T notch filter. (a) Notch filter with adjustable Q factor; (b) high-Q-factor notch filter.

The low-pass and high-pass Sallen–Key architectures are shown in figures 6.4(a) and (b), respectively. The cut-off frequency (f_c) of the Sallen–Key filter is determined in (6.1), and the gain of the amplifier is given by (6.2):

$$f_c = 1/(2\pi\sqrt{R_1R_2C_1C_2}), \tag{6.1}$$

$$\text{Gain} = 1 + R_4/R_3. \tag{6.2}$$

The generated signal of a triboelectric-based sensor can be interfered by power line frequency (e.g., 50 Hz). Electromagnetic fields from the environment can induce interference electric and magnetic fields (EMF) when they pass through the sensor electrode. The larger the sensor size, the higher the EMF induced. Besides, triboelectric sensors are often used in HMI applications, and considerable line interference could be induced when the sensor contacts the human body [47, 48]. Therefore, a twin-T notch filter can be implemented to minimize the 50 Hz interference. A widely deployed active twin-T notch filter circuit is shown in figure 6.5(a). This is a notch filter design with an adjustable quality factor (Q factor) [49, 50]. By changing the ratio of R_4/R_5, a fraction of the output is fed back to the twin-T filter by a second voltage buffer. The Q factor is adjusted accordingly due to the Q factor being dependent on the amount of feedback. As a consequence, the notch depth of the filter is adjusted. The 50 Hz twin-T active notch filter can be designed according to (6.3)–(6.6). To obtain the maximum Q factor and reach the maximum notch depth of the filter, maximum feedback is required. A schematic of a notch filter with maximum notch depth is illustrated in figure 6.5(b):

$$f_{\text{notch}} = 1/(2\pi R_1 C_1), \tag{6.3}$$

$$R_1 = R_2 = 2R_3, \tag{6.4}$$

$$C_1 = C_2 = C_3/2, \tag{6.5}$$

$$Q = (1 + R_5/R_4)/4. \tag{6.6}$$

In [48], a twin-T notch filter is applied to remove line frequency (50 Hz). C_1 and C_2 are chosen as 10 nF in the work. Therefore, R_1 and R_2 are selected as 318 kΩ. The value of R_3 is one-half of R_1, which is 159 kΩ, and the capacitance of C_3 is 20 nF. Equation (6.7) calculates the theoretical notch frequency, f_{notch}, based on the components employed:

$$f_{\text{notch}} = 1/(2\pi \times 318k \times 10n) = 50.05 \quad \text{Hz}. \tag{6.7}$$

6.4 Design examples of analog front-end

The work in [51] presents a signal-processing method to analyze the periodic electrical signals generated from a sweep-type triboelectric linear motion sensor, as shown in figure 6.6. The raw signal was buffered by a preamplifier (AD620) and then sampled by a data-acquisition unit (NI PCI-6259). The preamplifier helps to convert the high-impedance signal to a low-impedance signal and amplifies the raw signal output. An external load of 10 MΩ was picked to prevent the output signal of the sensor from leading to saturation distortions of the amplifier. After the data-acquisition system sampled the signal, the signal was transmitted to a computer for further processing. A digital filter algorithm was developed and processed by

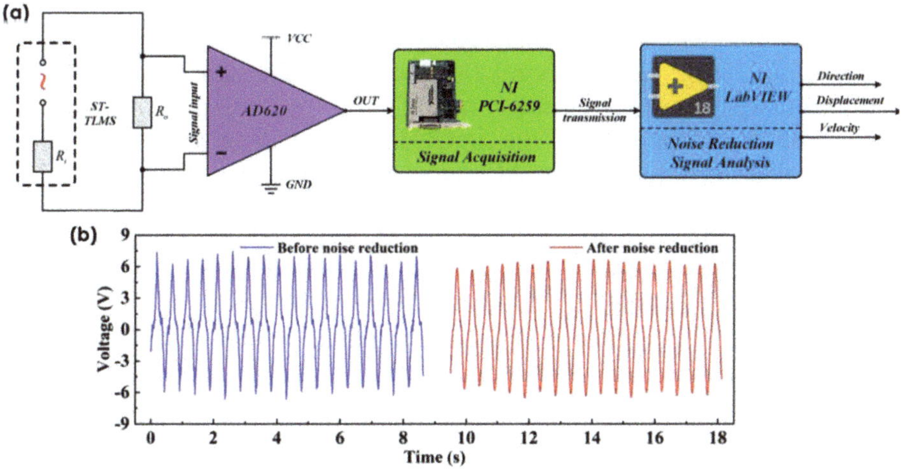

Figure 6.6. A signal-processing method for processing periodic triboelectric signals. (a) Block schematic of the signal-processing circuit; (b) the processed signal. Reproduced with permission from [51]. Copyright 2020, Elsevier.

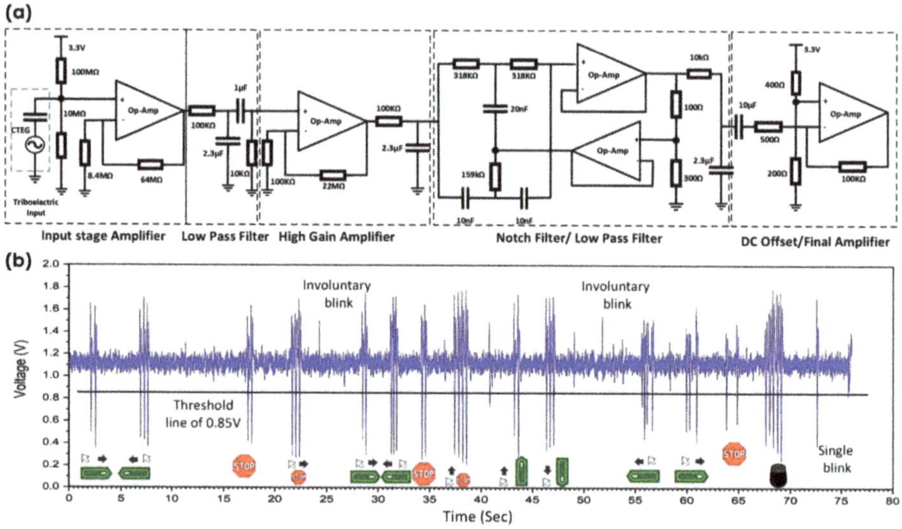

Figure 6.7. Conditioning circuit for a non-contact, eye-motion-triggered triboelectric-based HMI. (a) Circuit diagram; (b) a sample of the signal for triggering control commands. Reproduced with permission from [48]. Copyright 2020, Elsevier.

LabVIEW and Matlab software, realizing real-time data analysis and visualization. The processed signal waveform is depicted in figure 6.6(b).

A conditioning circuit was developed to detect the electrical signal generated by a TENG-based eye-motion HMI, and then applied to a hand-free computer control system by Anaya *et al* [48]. As depicted in figure 6.7(a), the circuit consists of a high-input-impedance preamplifier, a high-gain amplifier to amplify the low-amplitude eye-motion signal, a notch filter to reduce 50 Hz line interference and an amplifier with DC offset to obtain an AC output signal. Because the eye-motion displacement is small, the amplitude of the original sensor signal is relatively weak. An amplifier with a gain of 220 was used to boost the strength of the signal and to obtain the desired sensitivity. Besides, each stage was followed by a low-pass filter to suppress the high-frequency component of the signal. In the last stage, a DC bias voltage of 1 V was applied, resulting in the AC signal swinging between 0 and 2 V at the output, which was within the sensing range of the built-in ADC of the MCU (MSP430G2553). Figure 6.7(b) shows a sample of the output of the system. Based on the number of eye blinks, a corresponding instruction can be activated to move the cursor of a computer on the screen.

Charoonsuk *et al* proposed a triboelectric tactile sensor that can be used to control smart appliances [52]. The system was developed by integrating the TENG tactile sensor with the signal-processing circuit, as presented in figure 6.8. The signal-processing circuit is mainly comprised of an inverting amplifier, a comparator and an MCU with a WiFi transmitter. The processed signal was exerted to trigger the corresponding control command to be sent in the wireless communication system to remotely turn on/off the smart appliances (e.g., lamps and electric fans). The sensor

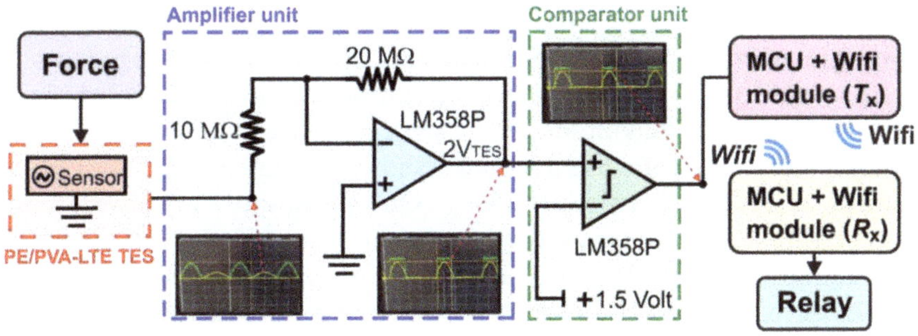

Figure 6.8. Block diagram representing the sensing system of a tactile sensor for wireless controlling. Reproduced with permission from [52]. Copyright 2021, Elsevier.

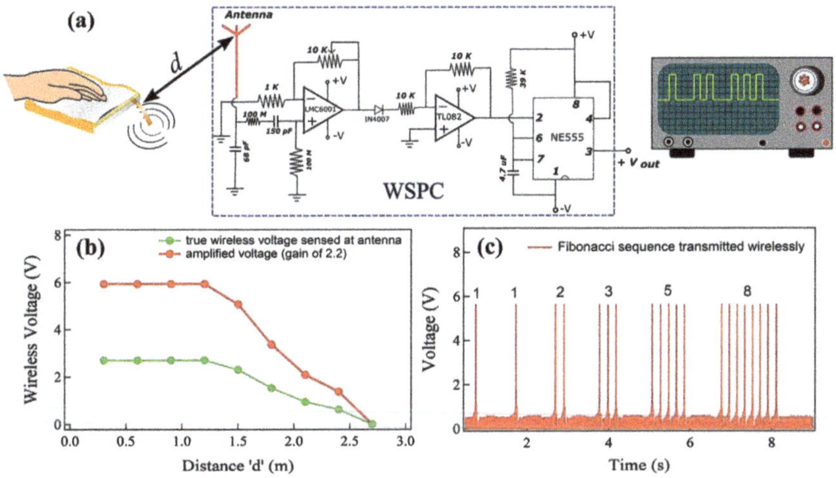

Figure 6.9. A wireless TENG system. (a) Signal-processing circuit for detecting the electric field generated by the TENG; (b) the amplitude of the signal sensed by the conditioning circuit as a function of the distance between the TENG and the receiver; (c) a Fibonacci sequence is represented by the number of pulses transmitted. Reproduced with permission from [53]. Copyright 2017, Wiley.

output was connected to an inverting amplifier to intensify the incoming sensor signal. The high-value resistors (i.e., 10 and 20 MΩ) utilized in the circuit offered the amplifier a gain of 2 and provided the high input impedance. It should be noted that the system used a single power supply; thus, the negative output pulses were clipped. A comparator unit with a threshold voltage of 1.5 V was employed to convert the signal pluses to square-shaped pulses and filter out any interference signal lower than the threshold voltage.

A 'wireless' TENG is demonstrated in [53]. The key to this TENG is that it can generate a very high output voltage >2 kV and create a strong electric field around it. The electric field was able to be sensed at a distance of about 3 m, which enabled wireless control of electronic devices. A signal-processing circuit was designed to effectively detect the electric field generated by the TENG, as illustrated in figure 6.9(a). The circuit mainly includes a preamplifier (LMC6001), an inverting amplifier (TL082) and a pulse-shaping integrated circuit (IC; NE555) to output square pulses. The preamplifier was configured with a gain of 2.2 to raise the amplitude of the detected signal. The relationship between the magnitude of the sensed voltages and the distance between the sensor and the sensing circuit's antenna is shown in figure 6.9(b). The output of the preamplifier was rectified by a diode (IN4007). Only the positive pulses were reserved for further processing due to the positive pulse revealing a higher signal-to-noise ratio. An inverting amplifier with unity gain was employed to convert the positive signal pulses to negative pulses in order to be compatible with the pulse-shaping IC's trigger input (active low). When a negative-going pulse was applied to the trigger input, the pulse-shaping IC would produce a square pulse with a 0.2 s duration in the output. The pulse-shaping IC reshaped the signal pulse and eliminated any input pulse bouncing. Finally, the number of generated pulses can be used to represent a user-defined passcode, such as the Fibonacci sequence (figure 6.9(c)).

Most studies have demonstrated TENGs either as active sensors or energy harvesters, and cannot realize both functions concurrently. One of the difficulties in achieving simultaneous energy harvesting and sensing in a TENG is that gaining one will inevitably affect the other. For example, the amplitude of the electrical output signal for sensing purposes will drop since the energy-harvesting unit could acquire the most energy. However, it has been observed that the frequency of the signal is scarcely affected by the energy-harvesting process. Based on the frequency-sensing approach, Lu *et al* proposed an appropriate strategy to extract the signal and the energy from a TENG, which achieved signal sensing and energy harvesting from a TENG simultaneously [54]. The circuit structure diagram is shown in figure 6.10(a), which consists of a signal-processing unit, a power management unit and an MCU with a wireless transmitter. The signal-processing unit involves a low-power, zero-crossing comparator based on an operational amplifier to transform the TENG output signals to a square waveform for frequency detection (figure 6.10(b)). Due to the system having a single power supply (3.3 V), the signal input terminals were biased by a DC offset of 1.5 V to detect the full AC signal. Besides, 22 MΩ resistors were utilized to build the voltage divider to establish a high input impedance. Figure 6.10(c) illustrates the voltage at different points of the rectification circuit. It should be noted that the sensor output (V_{ac}) was dropped since the capacitor had received most of the energy. However, the frequency of the sensor output is retained, which can be extracted for sensing purposes. Figure 6.10(d) demonstrates the energy-harvesting capability of the system. The device was able to provide a stable 3.3 V output to drive the system. Figures 6.10(e) and (f) depict the experimental result of the signal-processing circuit. The pulse signal created by the triboelectric sensor was captured by the signal-acquisition circuit successfully with excellent quality. Finally, a self-powered system that can collect wind energy to power the sensing circuit, in order to detect rotation rate and wind speed concurrently, was successfully established.

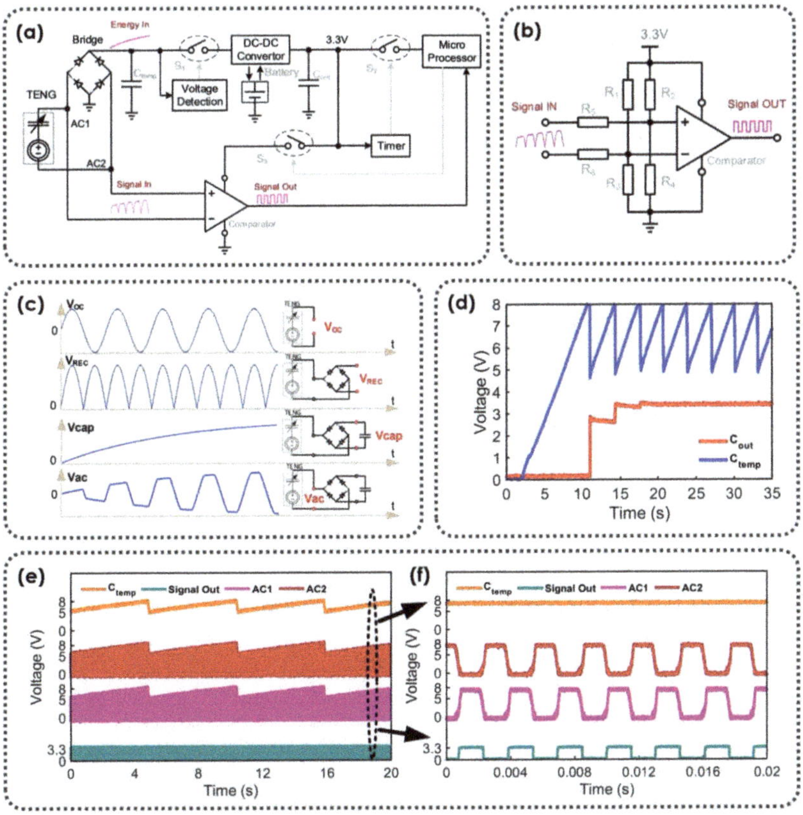

Figure 6.10. Schematic diagram of a concurrent sensing and energy-harvesting system. (a) Circuit diagram of the entire system; (b) schematic of the signal-processing circuit; (c) voltage at each port of the full-wave rectifier; (d) experimental result of the energy-harvesting circuit; (e) experimental result of the signal-conditioning circuit. Reproduced with permission from [54]. Copyright 2020, Elsevier.

6.5 Case study: circuitry for a grating-patterned triboelectric human–machine interface

Recently, more creative and intelligent HMIs based on TENGs have been extensively developed. A TENG can act as a self-powered sensor to detect human interaction and provide useful information for controlling external objects by converting a mechanical triggering into a self-generated electrical signal. This section presents a portable signal-processing circuit designed to process the AC signal generated by a grating-structured, TENG-based HMI, realizing real-time control in practical applications [55]. Based on the grating structure design principle, a triboelectric HMI with only one sensing electrode has been developed for controlling remote robotic devices. The developed control interface uses only one sensing electrode to establish direction sensing and trigger control instructions. Based on a peak-detection algorithm, different control instructions will be triggered

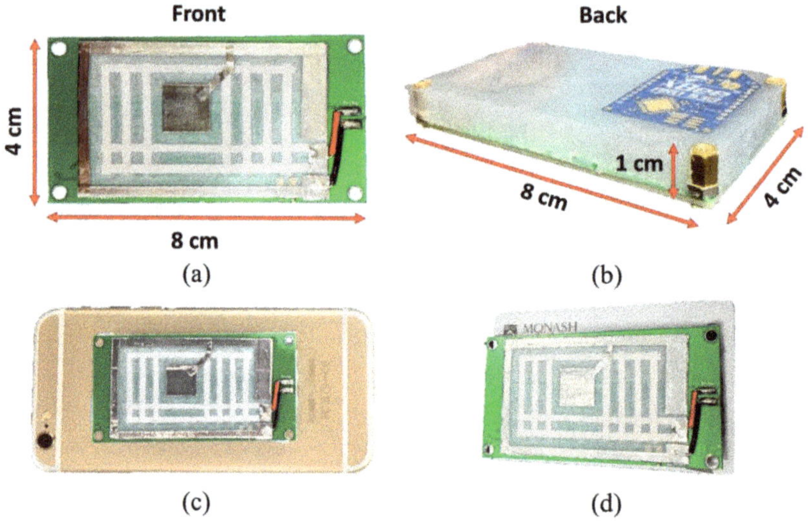

Figure 6.11. A fabricated device. (a) A TENG sensor at the front side; (b) the signal-processing circuit is enclosed by a silicone rubber substrate; (c) the device is attached to a mobile phone; (d) the device is attached to an ID card. Reproduced under the terms of the CC BY 4.0 license from [55].

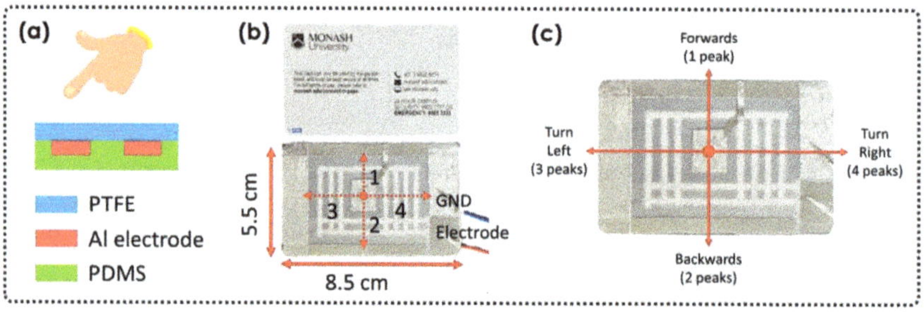

Figure 6.12. Design of a triboelectric-based sensor with a single-output-electrode and grounded outer shielding. (a) Structure layout of the triboelectric-based sensor; (b) an ID-card-sized sensor based on a peak-detection mechanism; (c) instructions are formed to control the movement of robots. Reproduced under the terms of the CC BY 4.0 license from [55].

according to the number of output peaks generated by the sensor. Figure 6.11 shows a prototype device by integrating the TENG sensor with the signal-processing circuit. A remote mobile robot control system is designed and constructed using the fabricated device to evaluate the feasibility and performance of the system, demonstrating its applicability, reliability and real-time control capability.

6.5.1 Design and working mechanism of a triboelectric sensor

Figure 6.12 shows the design of a triboelectric-based sensor. The triboelectric-based sensor consists of three layers, as illustrated in figure 6.12(a). The top layer is a

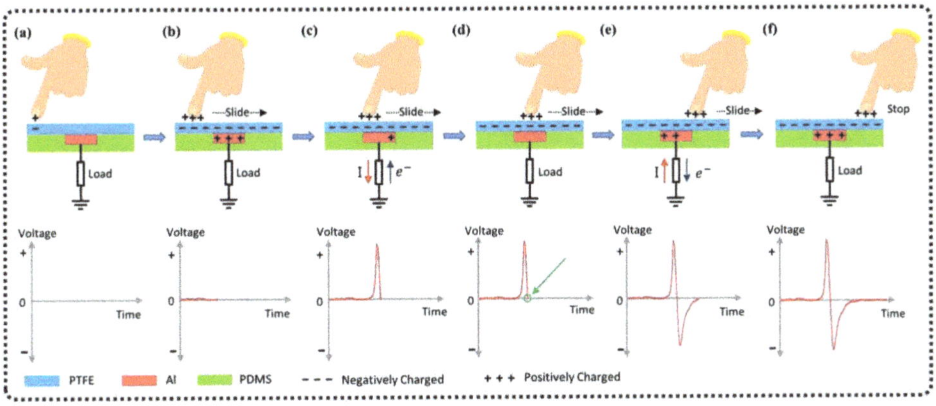

Figure 6.13. Working mechanism of the triboelectric-based, self-powered sensor. (a) The initial contact position of the sensor; (b)–(f) when the finger slides across the electrode, a current is generated. The states in (b)–(f) form a sliding operation cycle. Reproduced under the terms of the CC BY 4.0 license from [55].

polytetrafluoroethylene (PTFE) film with a thickness of 0.1 mm, which serves as the negative triboelectric material. The middle layer is the electrode layer, which is made of aluminum (Al) foil strips (0.09 mm in thickness). The bottom layer is a polydimethylsiloxane (PDMS) substrate. PDMS is used as the substrate because it is a flexible insulator. Due to its flexibility, the sensor can be easily attached to different devices, such as mobile phones, tablets and laptops. An ID-card-sized sensor is fabricated, as depicted in figure 6.12(b). The sensor has a single-signal-output-electrode and is surrounded by grounded Al foil in order to minimize the influence of unwanted electrostatic induction from the ambient environment. The width of the Al electrode strip is 2 mm with 3 mm spacing between two adjacent electrodes. The sensor is based on a peak-detection sensing algorithm. As the finger slides from the middle Al pad and across different numbers of electrodes in different directions, it will result in different numbers of output peaks corresponding to the number of electrodes. Therefore, direction sensing can be realized accordingly. When different numbers of peaks are detected, corresponding instructions will be sent to control the action of a mobile robot according to the instructions given in figure 6.12(c).

Figure 6.13 illustrates the working mechanism of the triboelectric-based sensor. The output signal is generated based on the contact electrification and electrostatic induction effects [56]. Initially, both the triboelectric sensor and the finger are uncharged. When the finger contacts and slides on the PTFE surface, as depicted in figure 6.13(a), because of the large difference in electron-attracting abilities, the PTFE film attracts the electrons from the finger, resulting in the net negatively charged PTFE surface and net positively charged finger skin. The negative charges on the PTFE surface will induce positive charges on the Al electrode due to the electrostatic induction effect, as shown in figure 6.13(b). When the finger approaches the electrode, positive charges on the finger will attract electrons flowing from the ground to the Al electrode in order to balance out the excessive positive charges on

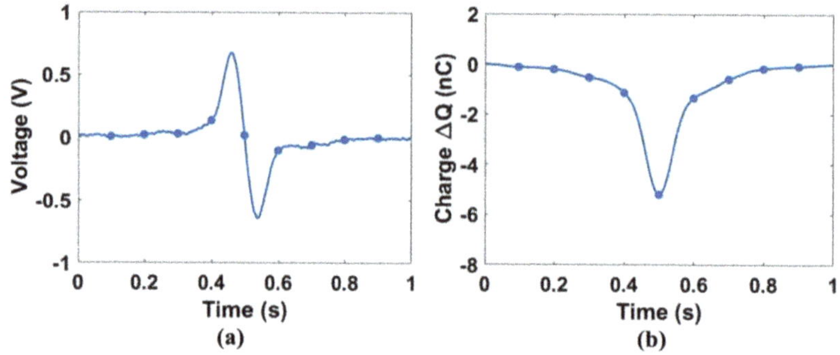

Figure 6.14. (a) The real-time output voltage and (b) charge transferred across the load when the finger is sliding across an Al electrode with a width of 4 mm. Reproduced under the terms of the CC BY 4.0 license from [55].

the electrode. Thus, a positive voltage will be generated on the load, as depicted in figure 6.13(c). The positive voltage signal is induced until all the positive charges on the finger are balanced out by the induced electrons, as shown in figure 6.13(d). Accordingly, when the finger slides away from the strip electrode, a negative voltage is generated on the load when referenced to the ground (figure 6.13(e)). Finally, as presented in figure 6.13(f), after the finger leaves the electrode, a new electrostatic equilibrium between the positive and negative charges will be reached. The real-time output voltage and the quantity of charge transferred across the load when the finger is sliding across an Al electrode with a width of 4 mm are depicted in figure 6.14. When the positive voltage peak is generated, electrons flow from the ground to the electrode. Therefore, negative charges are transferred across the load. While in the negative part of the output voltage, electrons flow back from the electrode to the ground. As a result, the charge transferred across the load returns to zero. The result is consistent with the aforementioned working mechanism in figure 6.13. In this way, an AC signal is generated without any external power supply. If this signal can be detected effectively, it can be used to control robotic devices and electronic systems for IoT applications.

6.5.2 Signal-conditioning circuit

Figure 6.15 illustrates a circuit diagram of a signal-processing circuit. There are three major stages in the signal-processing circuit, comprising an input stage, an amplifier stage and a comparator stage.

6.5.2.1 Input stage

The signal generated from the triboelectric-based sensor is an AC signal. However, a single power supply is usually used in the circuit design due to its simplicity. In the single-supply system, the negative half part of the AC signal will be clipped. In order to obtain both the positive and the negative parts of the AC signal, a DC offset needs to be applied to the AC signal. The DC offset is selected as one-half of the power

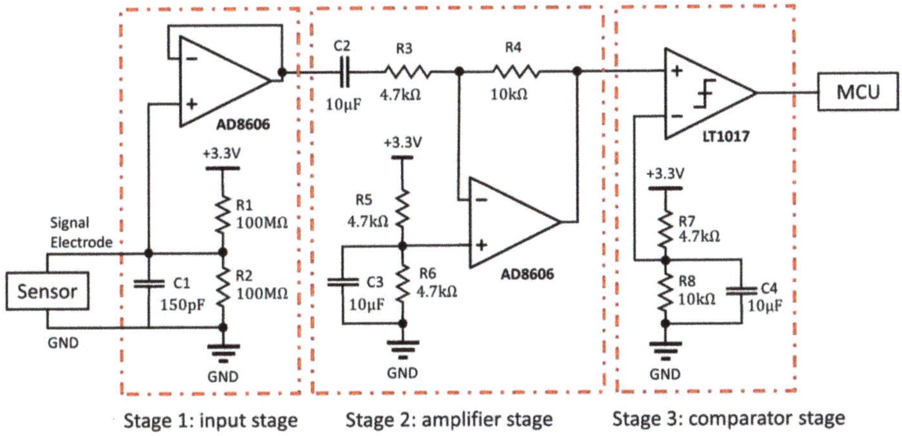

Figure 6.15. Circuit diagram of a signal-processing circuit for processing the voltage signal generated by triboelectric-based sensors. Reproduced under the terms of the CC BY 4.0 license from [55].

supply voltage to allow for maximum AC signal scaling. The power supply voltage in this system is 3.3 V. Thus, the AC signal is shifted and centered at 1.65 V.

A voltage divider composed of two resistors is used to provide the DC offset voltage. Because triboelectric-based sensors have high output impedance, resistors with high resistance values are employed in the voltage divider in order to provide a high input impedance of the signal-processing circuit. Therefore, two 100 MΩ resistors are chosen. The biased AC signal is then buffered by a voltage follower because it has a very high input impedance and will allow the stage 2 to receive sufficient voltage.

6.5.2.2 Amplifier stage

An AC coupled inverting amplifier is implemented to amplify the AC signal received from stage 1. The gain of the inverting amplifier is 2.13, which is set by the resistors R_3 and R_4, assuming the AC coupling capacitor C_2 is shorted in AC signal. The frequency response of the amplifier is simulated by the LTspice® SPICE simulation software (figure 6.16). The lower cut-off frequency (f_L) of the amplifier is 3.39 Hz. The simulation result is well matched with the measured result, and the bandwidth is sufficient for the signal-sensing application in this work.

6.5.2.3 Comparator stage

A non-inverting comparator is designed to compare the amplified signal with a threshold voltage. The comparator is used to identify a peak in the signal and filter out the noise in the signal. If the signal is higher than the threshold voltage ($V_+ > V_-$), the output will saturate toward the power supply voltage; otherwise, the output will saturate toward the ground. When the finger slides across the sensor, voltage peaks will be generated. The peak voltage will be higher than the threshold voltage. Thus, the output of the comparator will be 3.3 V. When the signal is below the threshold, the output will be 0 V.

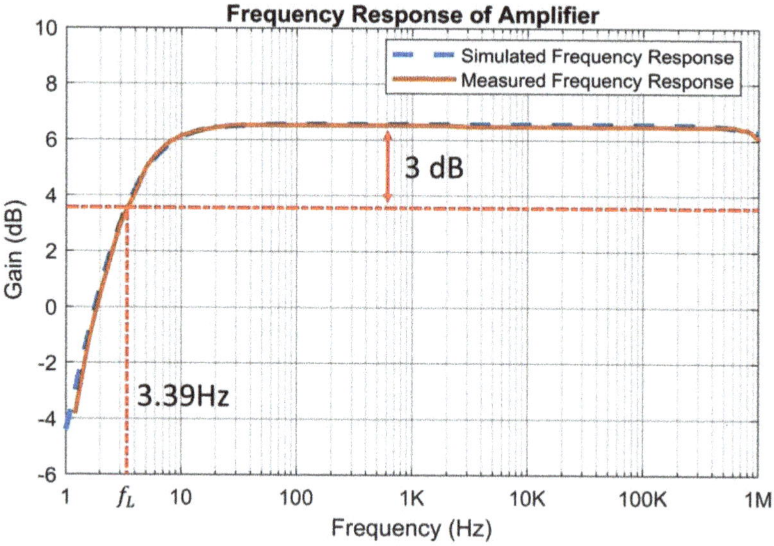

Figure 6.16. Simulated and measured frequency response of the amplifier. Reproduced under the terms of the CC BY 4.0 license from [55].

The average signal output from stage 2 is the DC offset of 1.65 V, which is one-half of the power supply voltage. Therefore, a voltage higher than 1.65 V needs to be chosen as the threshold voltage. Two-thirds of the power supply voltage (2.2 V) is chosen in this work. The output of the comparator is then sampled by the ADC embedded on the MCU at a frequency of 500 Hz for data processing.

6.5.3 Design of the printed circuit board

A printed circuit board (PCB) of the signal-processing circuit has been designed and fabricated, as shown in figure 6.17. The PCB is 4 cm wide and 8 cm long, which makes it suitable to be attached to mobile phones and ID cards. The MCU used in the controlling circuit is an ATmega328P (from Atmel®). It is a high-performance and low-power MCU. It has a built-in 8-channel, 10-bit ADC, which can be used to sample the output from the comparator. The MCU operates at 3.3 V with a clock speed of 8 MHz. After the signal processing, the MCU will send corresponding instructions to control the mobile robot wirelessly through the XBee-PRO® S3B module. The module operates at 900 MHz and provides long-range wireless communication between devices. This is sufficient to cover the indoor and outdoor control of the mobile robots. A rechargeable battery (LIR2450, from Multicomp®) is employed to power the MCU and the signal-processing circuit to process the signal generated from the triboelectric-based sensors.

6.5.4 Experimental results

To avoid sweaty hands affecting the performance of the sensor, and to overcome the limitation that the fingertip is larger than the spacing between the strip electrodes, a

Figure 6.17. The PCB design of the signal-processing circuit to process the signal generated from the triboelectric-based sensor. Reproduced under the terms of the CC BY 4.0 license from [55].

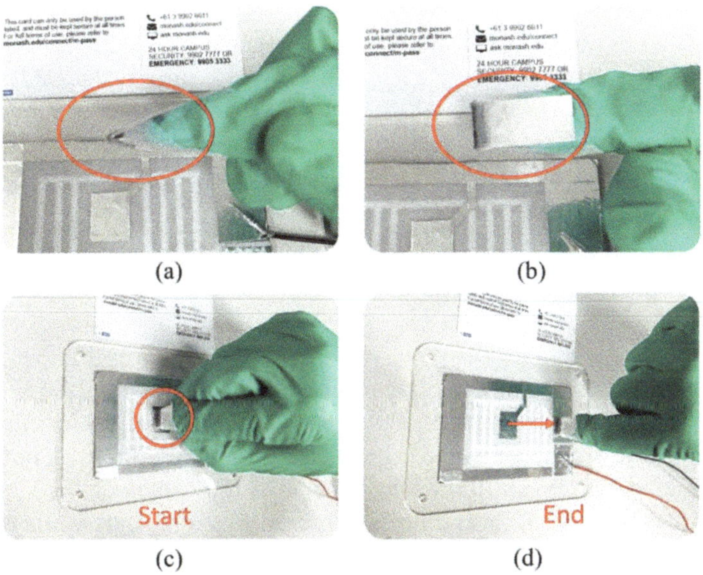

Figure 6.18. The ID-card-sized sensor system. (a) and (b) are the side view of finger cot; (c) a finger slide starts from the middle Al pad; (d) sliding ends at the outer Al shield. Reproduced under the terms of the CC BY 4.0 license from [55].

glove with a finger cot is deployed, as shown in figures 6.18(a) and (b). A textile coated with poly(3,4-ethylenedioxythiophene):poly(styrene sulfonate) is attached to the tip of the finger cot to increase the triboelectric output [57]. Figures 6.18(c) and (d) illustrate a sliding gesture starting at the center grounded Al pad and ending at the outer grounded shield.

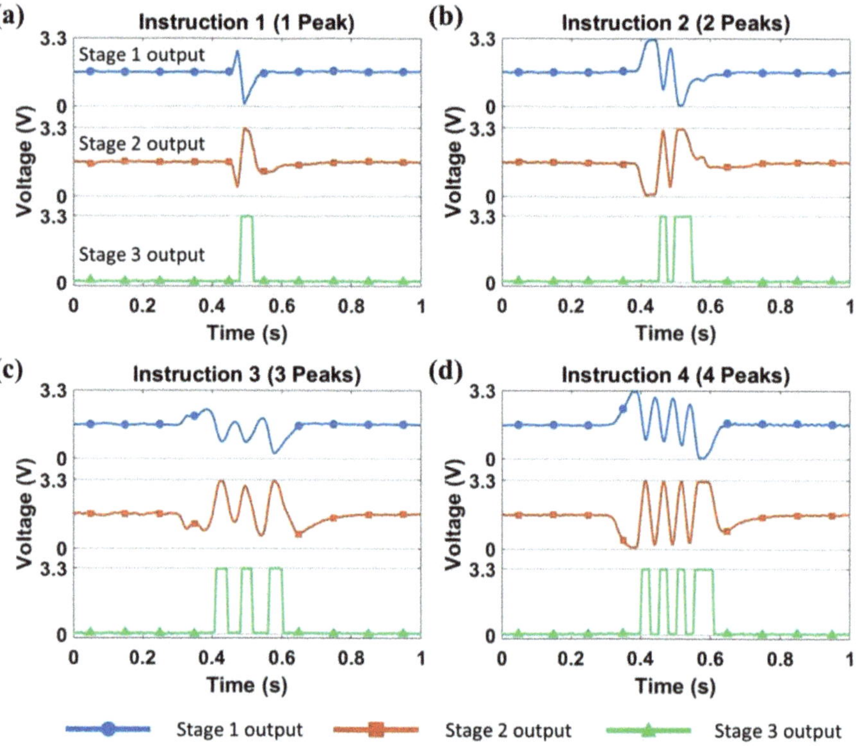

Figure 6.19. The output signal at different stages of the signal-processing circuit under normal sliding speed (75 mm s^{-1}): (a) one peak, (b) two peaks, (c) three peaks, (d) four peaks. Reproduced under the terms of the CC BY 4.0 license from [55].

Figure 6.19 shows the output voltage signals of the ID-card-sized sensor at three different stages of the signal-processing circuit under normal (~75 mm s^{-1}) sliding speed. The output signal of stage 1, which is the AC signal generated by the triboelectric-based sensor, shifted up by 1.65 V to the center of the supply voltage. The output of the inverting amplifier is centered at 1.65 V, as well, while the AC voltage has been amplified by the gain of 2.13. As can be seen from the plot, some peaks are clipped. However, clipping is not an issue in this work because the peaks are transformed to square pulses by the comparator in stage 3, and the number of peaks can be easily identified at this stage. The system is capable of detecting peaks under various sliding speeds, as shown in figure 6.20. With higher sliding speed, the amplitude of the output voltage will increase, and the width of the output voltage pulse will be narrower due to the shorter time of pulse generation. To evaluate the accuracy of the system, experiments have been conducted on seven subjects to validate the accuracy of the system. In the experiment, the subject wears the glove and triggers each instruction 100 times. The recognition accuracy equals the number of correct recognitions divided by the total number of instructions triggered. Table 6.3 is the recognition accuracy of the ID-card-sized sensor. The result suggests

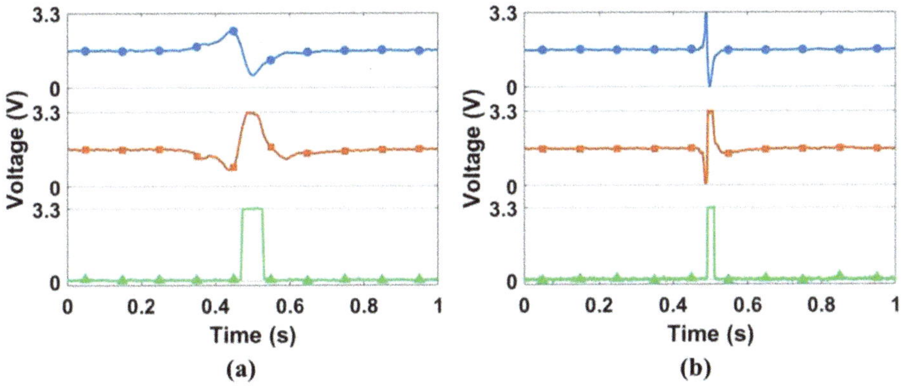

Figure 6.20. The generated output signal at (a) slow (~45 mm s^{-1}) and (b) fast (~200 mm s^{-1}) sliding speeds. Reproduced under the terms of the CC BY 4.0 license from [55].

Table 6.3. Recognition accuracy of the ID-card-sized sensor (out of 100 measurements). Reproduced under the terms of the CC BY 4.0 license from [55].

Subject	Instruction 1 (One Peak)	Instruction 2 (Two Peaks)	Instruction 3 (Three Peaks)	Instruction 4 (Four Peaks)
Subject 1	100%	100%	100%	100%
Subject 2	100%	98%	100%	100%
Subject 3	100%	100%	100%	100%
Subject 4	100%	100%	96%	100%
Subject 5	100%	100%	100%	100%
Subject 6	99%	100%	99%	100%
Subject 7	100%	98%	99%	99%
Average	99.86%	99.43%	99.14%	99.86%

that a high degree of recognition accuracy has been achieved, which is above 99% on average.

6.6 Summary

The emergence of TENGs has initiated a significant breakthrough in self-powered sensing. When considering the applications of triboelectric-based sensors in practical environments, analog front-end circuitry must be combined with these sensors to process the generated electrical signals in real time. The conditioning circuit design should take into account the characteristics of the sensors as well as their application fields. This chapter provides an overview of signal-conditioning circuits for triboelectric-based sensors along with several design examples.

A signal-conditioning circuit that can process the real-time signal generated by a triboelectric-based, grating-structured control interface is presented in a case study.

The system is able to generate control instructions when a finger slides on the control interface in different directions. According to the experimental results, the device has a high recognition accuracy of more than 99%, demonstrating the reliability and applicability of the proposed signal-conditioning circuit.

References

[1] Cheng T, Gao Q and Wang Z L 2019 The current development and future outlook of triboelectric nanogenerators: a survey of literature *Adv. Mater. Technol.* **4** 1800588

[2] Chen B D *et al* 2018 Ultrafine capillary-tube triboelectric nanogenerator as active sensor for microliquid biological and chemical sensing *Adv. Mater. Technol.* **3** 1700229

[3] Wang Z L and Wang A C 2019 On the origin of contact-electrification *Mater. Today* **30** 34–51

[4] Chen B, Yang Y and Wang Z L 2018 Scavenging wind energy by triboelectric nanogenerators *Adv. Energy Mater.* **8** 1702649

[5] Niu S and Wang Z L 2015 Theoretical systems of triboelectric nanogenerators *Nano Energy* **14** 161–92

[6] Wu C, Wang A C, Ding W, Guo H and Wang Z L 2019 Triboelectric nanogenerator: a foundation of the energy for the new era *Adv. Energy Mater.* **9** 1802906

[7] Pu X, An S, Tang Q, Guo H and Hu C 2021 Wearable triboelectric sensors for biomedical monitoring and human-machine interface *iScience* **24** 102027

[8] Zhou L, Liu D, Li S, Zhao Z, Zhang C, Yin X, Liu L, Cui S, Wang Z L and Wang J 2020 Rationally designed dual-mode triboelectric nanogenerator for harvesting mechanical energy by both electrostatic induction and dielectric breakdown effects *Adv. Energy Mater.* **10** 2000965

[9] Qiu Y, Yang D, Li B, Shao S and Hu L 2018 Wearable triboelectric nanogenerators based on hybridized triboelectric modes for harvesting mechanical energy *RSC Adv.* **8** 26243–50

[10] Liu L, Shi Q, Ho J S and Lee C 2019 Study of thin film blue energy harvester based on triboelectric nanogenerator and seashore IoT applications *Nano Energy* **66** 104167

[11] Luo J *et al* 2019 Flexible and durable wood-based triboelectric nanogenerators for self-powered sensing in athletic big data analytics *Nat. Commun.* **10** 5147

[12] Wu C, Ding W, Liu R, Wang J, Wang A C, Wang J, Li S, Zi Y and Wang Z L 2018 Keystroke dynamics enabled authentication and identification using triboelectric nanogenerator array *Mater. Today* **21** 216–22

[13] Pu X, Guo H, Tang Q, Chen J, Feng L, Liu G, Wang X, Xi Y and Hu C 2018; Wang Z L Rotation sensing and gesture control of a robot joint via triboelectric quantization sensor *Nano Energy* **54** 453–60

[14] Wang X, Que M, Chen M, Han X, Li X, Pan C and Wang Z L 2017 Full dynamic-range pressure sensor matrix based on optical and electrical dual-mode sensing *Adv. Mater.* **29** 1605817

[15] Yan Z *et al* 2021 Flexible high-resolution triboelectric sensor array based on patterned laser-induced graphene for self-powered real-time tactile sensing *Adv. Funct. Mater.* **31** 2100709

[16] Liu H, Dong W, Li Y, Li F, Geng J, Zhu M, Chen T, Zhang H, Sun L and Lee C 2020 An epidermal semg tattoo-like patch as a new human-machine interface for patients with loss of voice *Microsyst. Nanoeng.* **6** 16

[17] He Q, Wu Y, Feng Z, Sun C, Fan W, Zhou Z, Meng K, Fan E and Yang J 2019 Triboelectric vibration sensor for a human-machine interface built on ubiquitous surfaces *Nano Energy* **59** 689–96
[18] Tang Q *et al* 2018 Whirligig-inspired triboelectric nanogenerator with ultrahigh specific output as reliable portable instant power supply for personal health monitoring devices *Nano Energy* **47** 74–80
[19] Cho S, Yun Y, Jang S, Ra Y, Choi J H, Hwang H J, Choi D and Choi D 2020 Universal biomechanical energy harvesting from joint movements using a direction-switchable triboelectric nanogenerator *Nano Energy* **71** 104584
[20] Chen L, Shi Q, Sun Y, Nguyen T, Lee C and Soh S 2018 Controlling surface charge generated by contact electrification: strategies and applications *Adv. Mater.* **30** 1802405
[21] Li S *et al* 2020 A fully self-powered vibration monitoring system driven by dual-mode triboelectric nanogenerators *ACS Nano* **14** 2475–82
[22] Garcia C, Trendafilova I, Guzman de Villoria R and Sanchez del Rio J 2018 Self-powered pressure sensor based on the triboelectric effect and its analysis using dynamic mechanical analysis *Nano Energy* **50** 401–9
[23] Su Y *et al* 2020 Alveolus-inspired active membrane sensors for self-powered wearable chemical sensing and breath analysis *ACS Nano* **14** 6067–75
[24] Zhang Z, Zhang J, Zhang H, Wang H, Hu Z, Xuan W, Dong S and Luo J 2019 A portable triboelectric nanogenerator for real-time respiration monitoring *Nanoscale Res. Lett.* **14** 354
[25] Liu H, Wang H, Lyu Y, He C and Liu Z 2020 A novel triboelectric nanogenerator based on carbon fiber reinforced composite lamina and as a self-powered displacement sensor *Microelectron. Eng.* **224** 111231
[26] Niu S, Wang S, Lin L, Liu Y, Zhou Y S, Hu Y and Wang Z L 2013 Theoretical study of contact-mode triboelectric nanogenerators as an effective power source *Energy Environ. Sci.* **6** 3576–83
[27] Liu C *et al* 2020 A self-powered and high sensitivity acceleration sensor with V-Q-a model based on triboelectric nanogenerators (TENGs) *Nano Energy* **67** 104228
[28] Kim J-N, Lee J, Go T W, Rajabi-Abhari A, Mahato M, Park J Y, Lee H and Oh I-K 2020 Skin-attachable and biofriendly chitosan-diatom triboelectric nanogenerator *Nano Energy* **75** 104904
[29] Wang S, Lin L, Xie Y, Jing Q, Niu S and Wang Z L 2013 Sliding-triboelectric nanogenerators based on in-plane charge-separation mechanism *Nano Lett.* **13** 2226–33
[30] Lin L, Wang S, Xie Y, Jing Q, Niu S, Hu Y and Wang Z L 2013 Segmentally structured disk triboelectric nanogenerator for harvesting rotational mechanical energy *Nano Lett.* **13** 2916–23
[31] Pu X, Liu M, Chen X, Sun J, Du C, Zhang Y, Zhai J, Hu W and Wang Z L 2017 Ultrastretchable, transparent triboelectric nanogenerator as electronic skin for biomechanical energy harvesting and tactile sensing *Sci. Adv.* **3**
[32] Shi Q, Qiu C, He T, Wu F, Zhu M, Dziuban J A, Walczak R, Yuce M R and Lee C 2019 Triboelectric single-electrode-output control interface using patterned grid electrode *Nano Energy* **60** 545–56
[33] Lin Z, Chen J, Li X, Zhou Z, Meng K, Wei W, Yang J and Wang Z L 2017 Triboelectric nanogenerator enabled body sensor network for self-powered human heart-rate monitoring *ACS Nano* **11** 8830–7

[34] Xi Y, Guo H, Zi Y, Li X, Wang J, Deng J, Li S, Hu C, Cao X and Wang Z L 2017 Multifunctional TENG for blue energy scavenging and self-powered wind-speed sensor *Adv. Energy Mater.* **7** 1602397

[35] Pan S and Zhang Z 2019 Fundamental theories and basic principles of triboelectric effect: a review *Friction* **7** 2–17

[36] Dharmasena R and Silva S 2019 Towards optimized triboelectric nanogenerators *Nano Energy* **62** 530–49

[37] Fang C, Tong T, Bu T, Cao Y, Xu S, Qi Y and Zhang C 2020 Overview of power management for triboelectric nanogenerators *Adv. Intell. Syst.* **2** 1900129

[38] Xu S, Ding W, Guo H, Wang X and Wang Z L 2019 Boost the performance of triboelectric nanogenerators through circuit oscillation *Adv. Energy Mater.* **9** 1900772

[39] Qin H, Gu G, Shang W, Luo H, Zhang W, Cui P, Zhang B, Guo J, Cheng G and Du Z 2020 A universal and passive power management circuit with high efficiency for pulsed triboelectric nanogenerator *Nano Energy* **68** 104372

[40] Hu Y, Yue Q, Lu S, Yang D, Shi S, Zhang X and Yu H 2018 An adaptable interface conditioning circuit based on triboelectric nanogenerators for self-powered sensors *Micromachines* **9** 105

[41] Yang J, Yang F, Zhao L, Shang W, Qin H, Wang S, Jiang X, Cheng G and Du Z 2018 Managing and optimizing the output performances of a triboelectric nanogenerator by a self-powered electrostatic vibrator switch *Nano Energy* **46** 220–8

[42] Starecki T 2014 Analog front-end circuitry in piezoelectric and microphone detection of photoacoustic signals *Int. J. Thermophys.* **35** 2124–39

[43] Yañez Y, Garcia-Hernandez M, Salazar J, Turo A and Chavez J 2005 Designing amplifiers with very low output noise for high impedance piezoelectric transducers *NDT E Int.* **38** 491–6

[44] Mallineni S S K, Behlow H, Podila R and Rao A M 2018 A low-cost approach for measuring electrical load currents in triboelectric nanogenerators *Nanotechnol. Rev.* **7** 149–56

[45] Williams A B and Taylor F J 2006 *Electronic Filter Design Handbook* 4th edn (New York: McGraw-Hill Education)

[46] Karki J 2000 *Active low-pass filter design* SLOA049B Texas Instruments Application Report

[47] Pu X, Guo H, Chen J, Wang X, Xi Y, Hu C and Wang Z L 2017 Eye motion triggered self-powered mechnosensational communication system using triboelectric nanogenerator *Sci. Adv.* **8** e1700694

[48] Vera Anaya D, He T, Lee C and Yuce M R 2020 Self-powered eye motion sensor based on triboelectric interaction and near-field electrostatic induction for wearable assistive technologies *Nano Energy* **72** 104675

[49] Winder S 2002 Bandstop filters *Analog and Digital Filter Design (Second Edition)* ed S Winder 2nd edn (Woburn: Newnes) ch 7 pp 199–222

[50] Zumbahlen H 2008 Analog filters, *Linear Circuit Design Handbook* ed H Zumbahlen (Burlington: Newnes) ch 8 pp 581–679

[51] Xie Z, Dong J, Yang F, Xu R, Gao Q, Cheng T and Wang Z L 2020 Sweep-type triboelectric linear motion sensor with staggered electrode *Ext. Mech. Lett.* **37** 100713

[52] Charoonsuk T, Muanghlua R, Sriphan S, Pongampai S and Vittayakorn N 2021 Utilization of commodity thermoplastic polyethylene (PE) by enhanced sensing performance with liquid

phase electrolyte for a flexible and transparent triboelectric tactile sensor *Sustain. Mater. Technol.* **27** e00239

[53] Mallineni S S K, Dong Y, Behlow H, Rao A M and Podila R 2018 A wireless triboelectric nanogenerator *Adv. Energy Mater.* **8** 1702736

[54] Lu S, Gao L, Chen X, Tong D, Lei W, Yuan P, Mu X and Yu H 2020 Simultaneous energy harvesting and signal sensing from a single triboelectric nanogenerator for intelligent self-powered wireless sensing systems *Nano Energy* **75** 104813

[55] Qiu C, Wu F, Shi Q, Lee C and Yuce M R 2019 Sensors and control interface methods based on triboelectric nanogenerator in IoT applications *IEEE Access* **7** 92745–57

[56] Niu S, Wang S, Liu Y, Zhou Y S, Lin L, Hu Y, Pradel K C and Wang Z L 2014 A theoretical study of grating structured triboelectric nanogenerators *Energy Environ. Sci.* **7** 2339–49

[57] He T, Shi Q, Wang H, Wen F, Chen T, Ouyang J and Lee C 2019 Beyond energy harvesting —multi-functional triboelectric nanosensors on a textile *Nano Energy* **57** 338–52

IOP Publishing

Systems for Printed Flexible Sensors
Design and implementation
Tarikul Islam, Subhas Chandra Mukhopadhyay and Boby George

Chapter 7

Flexible interfacing circuits for wearable sensors and wireless communication

Moupali Chakraborty, Oliver Ozioko and Ravinder Dahiya

7.1 Introduction

The interest in wearable systems for various applications, including monitoring of body signals (e.g. for health monitoring) [1–5], robotics [6–9], prosthetics [10–13], rehabilitation [14–17] and human–machine interfaces [14, 18], is rapidly increasing. Flexible and wearable sensors are the fundamental elements of these systems as they serve as the primary source of data acquisition, and are utilized for sensing external stimuli such as pressure, temperature, humidity, orientation [19–23], etc. To be able to acquire these signals from the sensors and utilize them for real-time applications, they are interfaced to the systems using various circuits [24] depending on the operating mechanism of the sensor [25, 26]. Sensors used in flexible systems generally convert external mechanical stimuli into an electrical signal; different mechanisms, such as piezoresistive, capacitive, piezoelectric and triboelectric effects, have been explored. For proper interaction processes, wearability and efficient extraction of data [24, 25, 27, 28], these sensors and their accompanying interfacing circuits need to be flexible. As a result, there is a significant interest in making the electronics which drive these sensors not only efficient but flexible and wearable, as the conventional rigid electronics are bulky and not suitable for conforming to the human skin. Therefore, mechanical flexibility and stretchability are becoming important considerations for the development of interfacing circuits for wearable applications and wireless communications [29].

In recent years, interfacing circuits for wearable sensors have substantially advanced from the usual rigid types to more advanced types which are bendable, flexible, soft and also stretchable. The components and substrates of the new generation of electronics are becoming lightweight and thinner to allow them to conform to the curvy surfaces of the human body. A typical wearable system

doi:10.1088/978-0-7503-3935-3ch7

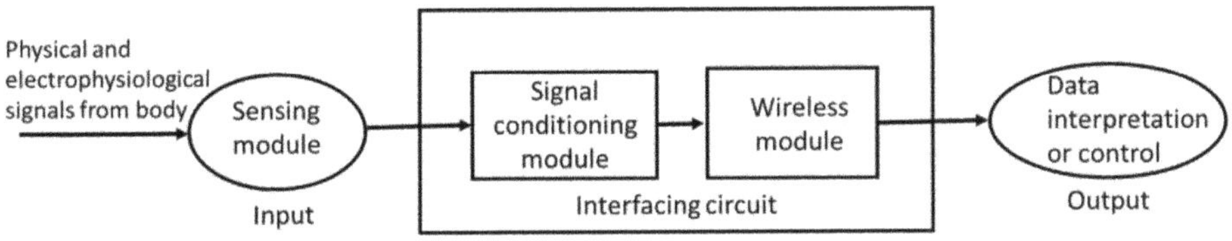

Figure 7.1. Four different modules of a typical wearable system.

comprises four basic modules, as shown in figure 7.1: (1) an input, sensing module, which senses the physical and electrophysiological signals from the wearer or its surroundings; (2) a signal-processing module, which processes the signal from the sensors; (3) a wireless module, which enables the processed signals to be transmitted wirelessly; and (4) an output, data interpretation/control, where the data from the sensors are interpreted and used to effect a system control. The interfacing circuit acts as the connector between the sensing module of the input (1) to the output (4) [30].

This chapter mainly focuses on modules (2) and (3), i.e. it summarizes the different sensor mechanisms and accompanying interfacing circuits utilized for wearable systems as well as wireless applications. This includes non-self-powered systems, such as piezoresistive and capacitive, and self-powered types, such as piezoelectric sensors and the rapidly growing area of triboelectric nanogenerators (TENGs), both of which require no power source to be able to obtain their output signal.

7.2 Wearable sensors and interfacing circuits

Flexible wearable sensors are the fundamental input of the interfacing circuits which monitor external stimuli. These sensors generally convert the external mechanical stimuli into electrical signals, which are read by an accompanying and suitable interfacing circuit. In this section, we present different sensor mechanisms and accompanying interfacing circuits for non-self-powered sensors (e.g. piezoresistive and capacitive) and self-powered sensors (e.g. piezoelectric sensors and TENGs).

7.2.1 Non-self-powered sensors and interfacing circuits

This section focuses on non-self-powered sensors, which require an external power source for their operation. The most used sensors for wearable applications are piezoresistive and capacitive types, which are widely used due to their simple fabrication techniques and simple interfacing circuits. In this section we will discuss the operating principles of different non-self-powered sensors and their interfacing circuits:

7.2.1.1 Piezoresistive sensors and interfacing circuit

Piezoresistive sensors are frequently used in flexible and wearable systems for applications such as health monitoring [31], electronic skins [26, 32], robotics [33], etc, because of their low power consumption, direct interfacing electronics, wide detection range and extensive fabrication processes [34]. The piezoresistive effect involves a change in the resistance, R, of an electric conductor ($R = \rho l/A$) due to externally applied forces, where ρ denotes the material resistivity (Ω·cm), l is the length of the conductor and A is the transverse section area. Wearable applications which currently use this type of sensor include pressure sensors, strain sensors, temperature sensors and multifunctional sensors [35]. In flexible piezoresistive pressure sensors, for instance, mechanical stress on the sensors causes a change in the resistance of the piezoresistive material used. Various types of materials have been utilized, including carbonized nanofibers, graphene and carbon nanotubes; some are realized by embedding these piezoresistive materials into an elastomeric polymer [15, 36]. Subsequently, substrate materials are an essential part of flexible electronics. The most frequently used substrate materials are polydimethylsiloxane (PDMS), elastomers, self-healing materials, hydrogels and ultraviolet (UV) cross-linkable materials [37]. Among these, PDMS and elastomers are the commonly used substrate material for stretchable electronic devices because of their flexible mechanical properties, transparency and stability [17, 37].

Piezoresistive sensors are commonly termed as non-self-powered sensors as they need power supply circuits for their operation [38]. The interfacing circuit for the piezoresistive sensor depends upon the number of measurands and the sensing elements. As the sensor is resistive, the electrical voltage output can be easily measured using a power supply. Figure 7.2(a)(i) shows a single piezoresistive sensor whose resistance changes with the measurand and, accordingly, the output electronic signal changes, which can be calibrated with the measurand. This single resistive sensor can be modeled as

$$R_x = R + \Delta R, \tag{7.1}$$

where R is the resistance when no measurand is present, and ΔR is the increment of the resistance due to the change in the measurand. The basic interface needs to be amplified by a low-noise circuit. An operational amplifier (op-amp) is an important stage as it prevents the loading effect of the piezoresistive sensor when it is connected with other electronic circuits for further communication. Figure 7.2(a)(ii) shows a measurement configuration with an op-amp. A general signal-conditioning circuit consists of a Wheatstone bridge through the generation of a non-zero voltage, V, in response to the bridge imbalance generated by one or more of the resistive elements. Basically, three arms of the Wheatstone bridge circuit are made of dummy sensors to compensate the environmental effects, and a single arm consists of the piezoresistive sensor [39]. Figure 7.2(a)(iii) shows an interfacing circuit using a full Wheatstone bridge.

A novel tactile sensing and actuation device (SensAct) that utilizes a soft piezoresistive pressure sensor to control an integrated actuator for application in prosthetics, robotics, etc, is presented in [17]. Figure 7.2(b) represents an interfacing circuit used to wirelessly create vibrotactile feedback depending on how much force is applied on the soft piezoresistive sensor. This involves three main modules: (1) a sensing module, (2) an actuation module, and (3) a Bluetooth communication module. The entire platform is controlled by a 32-bit Arm Cortex®-M3 microcontroller

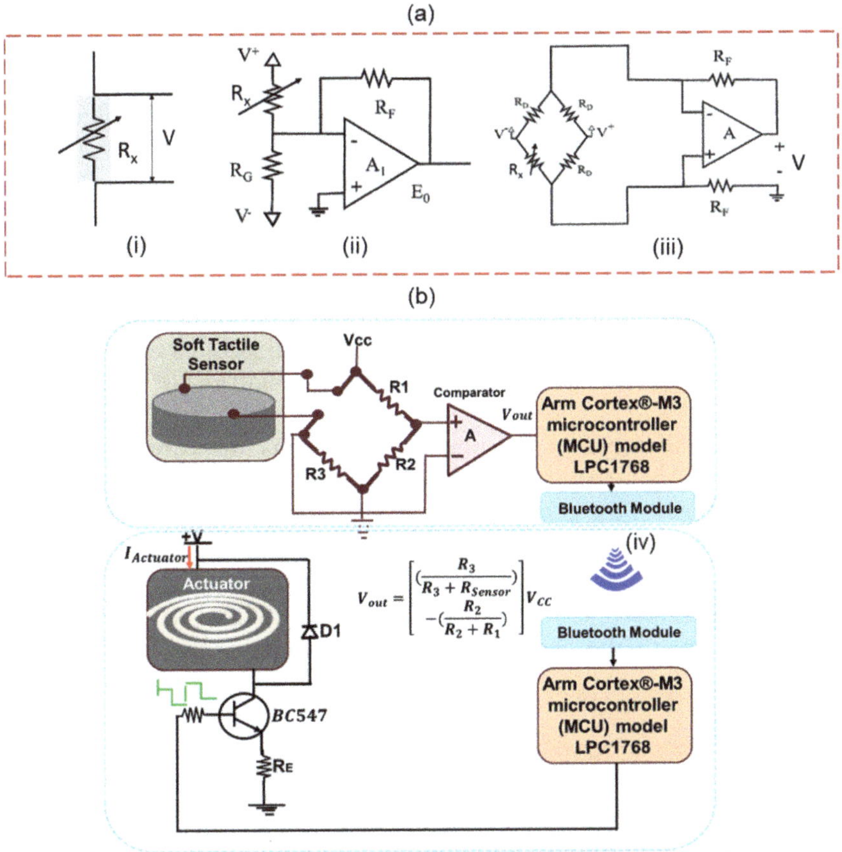

Figure 7.2. (a) Interfacing circuit for piezoresistive sensors: (i) single element measurement, (ii) differential measurement, and (iii) Wheatstone bridge method. (b) Interfacing circuit for wirelessly controlling an actuator using its integrated piezoresistive soft sensor [17]. John Wiley & Sons. Reproduced under the terms of creative common license (CC-BY 4.0), Copyright 2021.

(MCU), model LPC1768 (NXP Semiconductors, Eindhoven, The Netherlands), which can operate at CPU frequencies of up to 100 MHz. The sensing module contains the sensing element along with the resistance required for the Wheatstone bridge circuit interface that measures the resistance of the sensing module. This circuit can be implemented on a flexible substrate and used for wearable applications, including prostheses, for remote mapping of pressure sensation via vibrotactile feedback. Similarly, in [36], a polylactic-acid-film-based wearable pressure sensor is reported to predict the early signs of Parkinson's disease in patients.

7.2.1.2 Capacitive sensors and interfacing circuits

Another widely explored type of wearable sensors are capacitive sensors. These sensors are comparable to piezoresistive sensors, as both require an external power supply to obtain their output signal [38]. However, they generally have faster response times and lower power consumption. The capacitance (C) of a parallel plate capacitor can be calculated by the formula $C = \varepsilon_r \varepsilon_o A/d$, where d, A, ε_r and ε_o represent the distance between the two electrodes, the effective cross-sectional area, the relative permittivity of the dielectric medium and the permittivity of the free space, respectively. The sensors usually consist of two parallel flexible conducting electrodes separated by a dielectric layer [16, 40]. For instance, a capacitive touch-sensing module was fabricated in [16], using a facile planar capacitive structure realized by bonding two layers of single-sided flexible printed circuit board (FPCB) with ~35 μm of copper on a polyimide (PI) substrate. A polyvinyl chloride film, which serves as a dielectric layer, was sandwiched between the non-conducting surface of one FPCB with the conducting copper surface of the other. Similarly, another type of capacitive sensor is realized by PDMS gold electrodes with surface convex microarrays in [40], where polyvinylidene fluoride (PVDF) acts as the dielectric layer.

Like piezoresistive sensors, the interfacing circuits for capacitive sensors are simple and depend upon the number of measurands and the sensing elements. In a single capacitive sensor, the capacitance changes with the measurand are defined as

$$C_x = C + \Delta C, \tag{7.2}$$

where C is the capacitance when no measurand is present and ΔC is the increment of the capacitance due to the change in the measurand (figure 7.3(a)(i)). Electrical

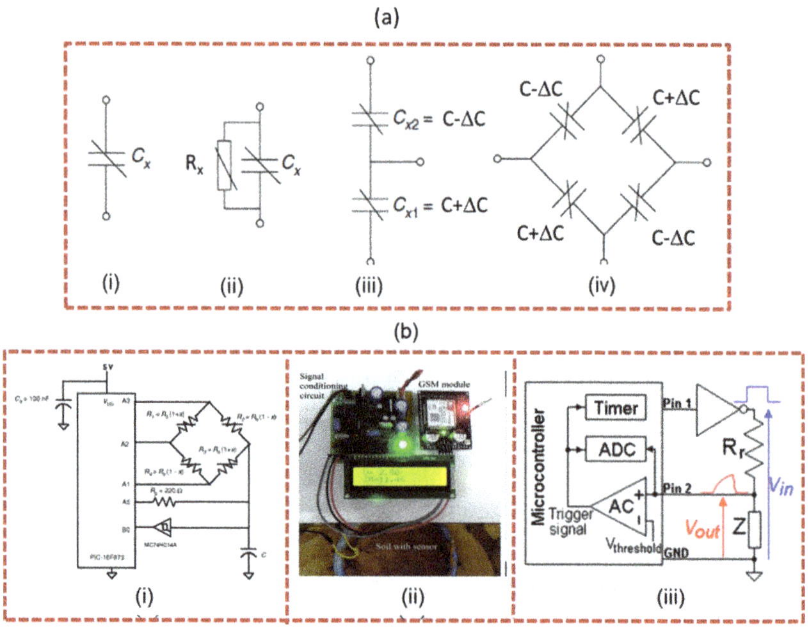

Figure 7.3. (a) Interfacing circuit for capacitive sensors: (i) single element measurement, (ii) lossy single element with measurement, (iii) differential measurement, and (iv) Wheatstone bridge method. (b) Interfacing circuit including microcontroller: (i) piezoresistive sensors (reprinted from [41], © 2006, with permission from Elsevier); (ii) a lossy capacitive sensor (© 2019 IEEE, reprinted, with permission, from [42]); and (iii) integrated resistive and capacitive sensors (reprinted from [43], © 2018, with permission from Elsevier).

signal is then measured across this capacitance. Apart from the ideal capacitor, the real capacitance is sometimes modeled by connecting a parallel resistance with the capacitance, where the resistance also changes with the measurand. This lossy capacitance gives the accurate behavior of the practical sensor (figure 7.3(a)(ii)). Figure 7.3(a)(iii) shows the differential mode of measurement of the capacitive sensors where two measurands are connected in a push–pull combination. This can be measured by C_{x1} and C_{x2} in the equations:

$$C_{x1} = C + \Delta C,$$

$$C_{x2} = C - \Delta C. \tag{7.3}$$

Like piezoresistive sensors, capacitive sensors can also be connected in a Wheatstone bridge circuit, as shown in figure 7.3(a)(iv).

In wearable applications, usually the sensors are powered by battery for a complete standalone system. Therefore, there is a need for data-acquisition systems, which should be energy efficient as well as small and cheap. These requirements are satisfied by a low-power on-chip MCU, which is widely used as the central controller. Moreover, the greatest advantage of MCUs is the conversion of the analog signals from the sensors to suitable digital signals by using an analog-to-digital converter (ADC) module. This digital form of the data can then be used further for processing, transmitting and decision making.

An example of an interfacing circuit is shown in [41], where piezoresistive pressure sensors are mounted in a Wheatstone bridge configuration and finally processed by the MCU (figure 7.3(b)(i)). Similarly, in [42, 44], lossy capacitive sensors are processed by an MCU-based interfacing circuit (figure 7.3(b)(ii)). Normally, an MCU can operate more than one input signal. Such types of application are shown in [43], where both resistive and capacitive sensors are integrated with the MCU circuit (figure 7.3(b)(iii)).

7.2.2 Chemical sensors and interfacing circuits

This section summarizes the basic principles of chemical sensors and the accompanying interfacing circuits for non-self-powered types. Chemical sensing is part of an information-acquisition process in which a sensor serves as an analyzer for a chemical system in real time [45–47]. The chemical sensor provides this information by generating measurable electrical signals as a result of the interaction (physical or chemical) between the sensor and the target analyte [46, 47] within the chosen environment and condition. Generally, a chemical sensor has two primary elements: (1) a recognition element, which is responsible for interacting with the analyte under test, and (2) a transducer, which transforms the sensor's interaction with the analyte into a measurable quantity suitable for the required analysis. Chemical sensors carry out their sensing functions either by measuring the physical interaction of the sensor and the analyte or by distinguishing the chemical changes that occur as a result of this interaction. Chemical sensors come in various types depending on their working principle; these include optical, electrochemical, mass-sensitive and paramagnetic

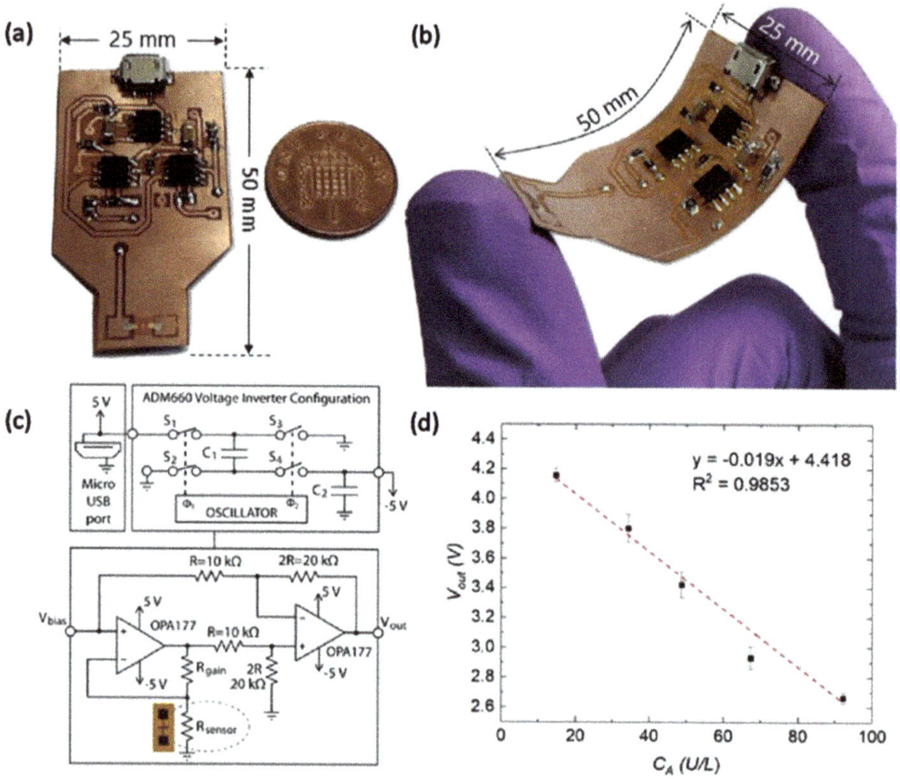

Figure 7.4. (a) and (b) Image of the flexible printed circuit board for a chemical sensor which is able to measure the concentration of α-amylase in serum. (c) Functional circuit diagram of the interfacing circuit. (d) Calibration curve of the output voltage of the readout circuit as a function of α-amylase concentration. Reprinted from [50], Copyright (2020), with permission from Elsevier.

chemical sensors [19, 46–49]. Depending on the working principle, interfacing circuits are designed to measure the corresponding electrical quantities such as resistance, capacitance, piezoelectricity, etc, which change during the physical/chemical interaction.

In [50], and as shown in figure 7.4, a flexible patch sensor with the capability to measure the concentration of α-amylase in serum was presented. The detection was achieved through the generation of mixing vortices (~12 rpm) inside the micro-droplet of aqueous starch-$FeSO_4$ solution. The large (~80%) variation in the electrical resistance across the droplet was correlated to detect the level of the α-amylase in the analyte.

Figures 7.4(a) and (b) show the sensor and the integrated flexible interfacing circuit. The chemical sensor is resistive and so the signal-conditioning circuit employed a voltage divider, a voltage inverter based on the ADM660 and a dual-supply op-amp (OPA177; Texas Instruments, USA) (figure 7.4(c)). This was designed to accurately measure the change in resistance that occurs as a result of the analyte with minimal interference due to noise from the electrolyte. The op-amp requires a dual power supply (±5 V) and so the ADM660 inverter was utilized to obtain −5 V from the supply. The positive supply of +5 V was obtained through a micro-USB connector (figure 7.4(a)), and the bias voltages were obtained from the same source using a resistor-based voltage divider. Figure 7.4(d) shows the calibration curve for the sensor. A similar approach has also been adopted for pH measurements in buffer solution (6–8 pH values) with a printed thick film potentiometric pH sensor with a platinum counter electrode [51]. Chemical sensors which not only measure pH but are also integrated to a wireless communication system have been explored [49].

Although this section has focused on non-self-powered chemical sensors, self-powered chemical sensors have also been reported; a comprehensive review on this topic can be found in [46]. A general overview of self-powered sensors and their interfacing circuits are discussed in the following section of this chapter.

7.2.3 Self-powered systems and interfacing circuits

Nowadays, self-powered wearable systems are essential requirements for the minimization of the overall power consumption and to move towards more compact and battery-free applications. This has led to a huge interest in the adoption of self-powered sensors for the realization of wearable devices. The most commonly used self-powered sensors are piezoelectric and triboelectric sensors, which have diverse applications due to their self-powering capacity. In this section we will discuss piezoelectric and triboelectric sensors with their interfacing circuits.

7.2.3.1 Piezoelectric sensors and interfacing circuits

Piezoelectric sensors are one of the types of self-powered sensors and therefore require no external power to obtain an output signal [52–58]. These types of sensors leverage electrical charges that are generated on piezoelectric materials, such as PVDF, zirconate titanate (PZT), lithium niobate, glycine, etc, to measure an external stimulus [27, 54, 55, 59, 60]. For piezoelectric pressure sensors, for instance, a piezoelectric material is sandwiched between a positive and negative electrode, and when pressure is applied on the material electrical charges are generated, which are then measured as an output voltage [60]. This voltage is proportional to the amount of pressure applied [61, 62]. Many piezoelectric sensors have been reported for different wearable applications including wearable health monitoring [63, 64], etc. Most commonly, the generated voltage is measured using a charge amplifier, which is a current integrator whose output voltage is proportional to the integrated value of the input current or the total injected electrical charge [52]. The charge amplifier amplifies the charge from the piezoelectric sensor and generates a voltage signal at the output [64]. Figure 7.5(a) shows a conventional charge amplifier circuit, where V_{in}, R_{in} and C_{in} represents the model of a piezoelectric sensor at frequencies below the resonance frequency of the piezoelectric sensor and R_f is the feedback resistor [65, 66]. R_f is essential as it helps to create a DC path for the input current of the op-amp's inverting input. For the

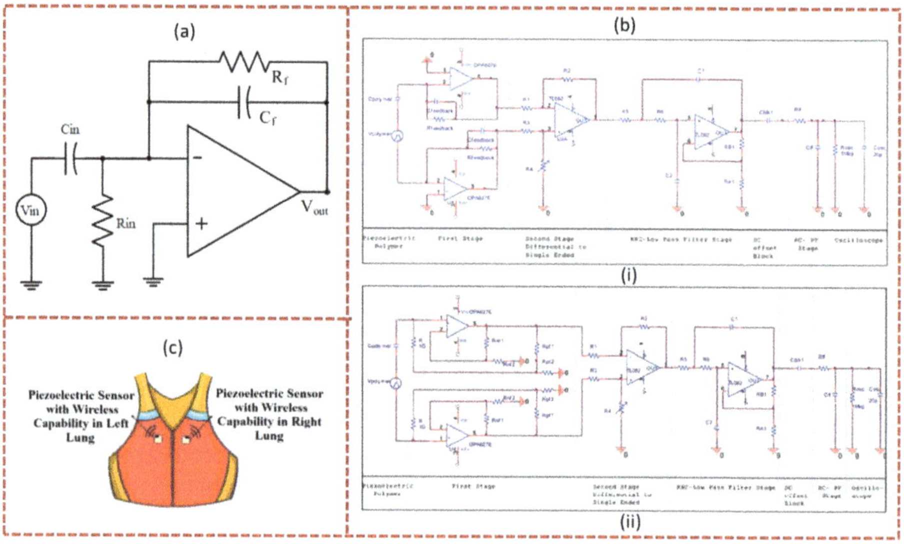

Figure 7.5. Interfacing circuits for piezoelectric sensors: (a) conventional charge amplifier circuit; (b) (i) schematic of three stage charge amplifier [25], (ii) schematic of three stage voltage amplifier [25], reproduced with permission, copyright 2013, Springer Science + Business Media, Dordrecht; (c) schematic of the sensor location in jacket. (c) Reprinted from [68], Copyright (2020), with permission from Elsevier.

realization of low cut-off frequencies for some quasi-static applications, the value of R_f is usually chosen to be very high (typically in the GΩ and TΩ range). The designs of three-stage charge and voltage amplifiers, to read the taxels on a piezoelectric polymer-micoelectrode-based tactile sensing chip [25], is shown in figure 7.5(b), where figure 7.5(b)(i) shows a schematic of a charge amplifier. The independence of the amplifier output from the cable capacitances and the minimization of the charge leakage through the stray capacitance around the sensor make the charge amplifier a preferred choice, especially when long connecting cables are present [58]. A schematic of a voltage amplifier is shown in figure 7.5(b)(ii). These are preferred if the ambient temperature variation is high, since they exhibit less temperature dependence and hence are useful for measuring the piezoelectric polymer film response.

Some existing wearable sensor prototypes have utilized similar interfacing circuits to demonstrate the application of their sensor for real-time applications. In [67], a piezoelectric sensor for breath monitoring was presented. It comprises a PDMS polymer-based design to create the main structure for the proposed sensor patch, a PVDF polymer sensing film layer, and a Mylar layer. During breathing, there is a mechanical stimulation of the piezoelectric sensor as a result of the pulsatile vibrations and periodical deformations on the chest wall of the human body. The signal from the sensor is amplified and, following this, the processed analog signals are sent to an MCU, which converts the analog signals into digital signals. Following this conversion is the wireless transmission of the digital signal via Zigbee standard to an RF receiver, which was linked to a computer with a RS232 serial port. Similarly, in [68], two piezoelectric electrodes were utilized on the chest to detect chest vibrations. Figure 7.5(c) schematically shows how the sensors could be located on a jacket. The output charge from the sensor is converted to voltage, amplified by a charge amplifier and converted to digital signal using an ADC.

In [63], self-powered, real-time arterial pulse monitoring using ultra-thin epidermal piezoelectric sensors was presented. The piezoelectric sensor is based on PZT thin film realized by transferring a piezoelectric thin film onto an ultra-thin plastic (thickness 4.8 μm) using inorganic-based laser lift-off. They demonstrated a self-powered piezoelectric pulse sensor by interfacing the sensor using a signal-processing circuit composed of amplifiers, comparators and output modules to identify the tiny arterial pulse. The pulse signal was wirelessly transmitted to a smartphone for a real-time monitoring system using a MCU and a Bluetooth transmitter [63]. The interfacing circuit, used by the authors to validate their device, comprises an LM358 op-amp, passive band-pass filters (BPFs), a red light-emitting diode (LED) and a speaker. The BPFs filter the unwanted DC noise and high-frequency noise components. A voltage threshold was set and a comparator employed to generate a constant voltage output (−5 or +5) depending on the output from the second non-inverting amplifier. When the threshold is exceeded, the LED blinks and the speaker turns on. The LED and speaker were synchronously operated corresponding to the radial artery pulse waveform. Although the circuit was not realized on a flexible substrate, it demonstrates a good concept for realizing an interfacing circuit for such applications of a piezoelectric sensor. A similar design approach could be adopted with the components now realized on a flexible substrate, as the authors acknowledged.

7.2.3.2 Triboelectric sensors and interfacing circuits

TENGs are newly emerging renewable energy harvesters that convert mechanical energy to electrical energy using the triboelectric effect [69–74]. When two contact surfaces with different electron affinities repeatedly touch, they cause repeated cycles of charge transfer, and repeated generation of an electric field which generates alternating current. TENG devices usually deliver high voltage but low current output, and therefore a power-management circuit is often required to enable them to be used in practical applications. In [69], a TENG-based body sensor network system for measuring human heart rate was presented. It comprises a downy-structure-based triboelectric nanogenerator (D-TENG), a power-management circuit, a heart-rate sensor, a signal-processing unit and a Bluetooth module for wireless data transmission. The D-TENG delivered a high voltage (up to 540 V) but relatively low current output (~20 μA), resulting in a large output impedance and thus affecting its applicability as a power source. To remedy this, the authors utilized a power-management circuit (figures 7.6(a) and (b)).

Figure 7.6(a) depicts the complete power-supplying system, which includes a TENG, a power-management circuit and the commercial electronics. A more detailed description of the power-management circuit is shown in figure 7.6(b). It provides a matching condition between the TENG output and the power requirements of the other electronic components of the system. It consists of a transformer, a rectifier and a low-leakage energy storage capacitor.

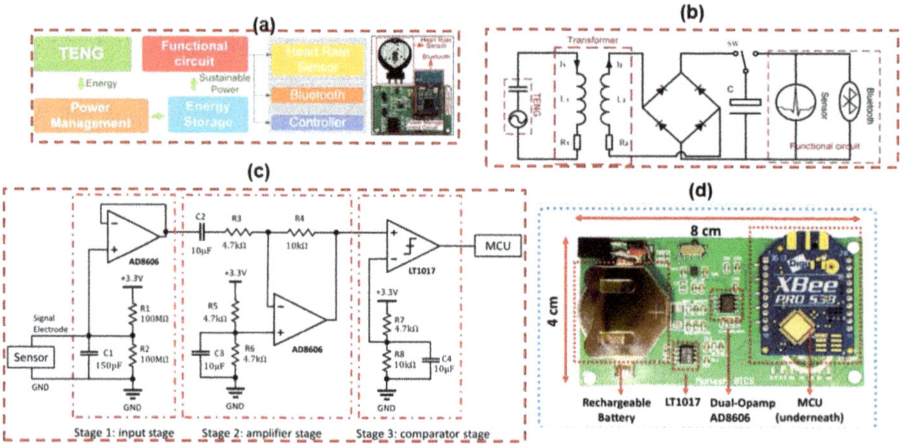

Figure 7.6. Interfacing circuits to utilize TENG for self-powering applications. (a) Power-supplying system. (b) Power-management circuit. (a)–(b) Reprinted with permission from [69], © 2017, American Chemical Society. (c) Signal-conditioning circuit for use in the Internet-of-Things. (d) PCB implementation to control a mobile robot in real time. (c)–(d) © 2019 IEEE. Reprinted, with permission, from [75].

A TENG-based sensor capable of being used for the Internet-of-Things (IoT) has also been explored [75]. A signal-processing circuit was designed to process the output of the sensor, which is used to control a mobile robot in real time. The circuit has three main signal-processing stages, including an input stage, a namplifier stage and a comparator stage (figure 7.6(c)). Since the input from the TENG is AC, and a single power supply is required, the negative half part of the AC signal was clipped and a DC offset applied to the AC signal to obtain both the positive and the negative parts of the AC signal. In this case, the power supply voltage in this system is 3.3 V and the AC signal is shifted and centered at 1.65 V. For impedance matching, a voltage divider, employing two 100 MΩ resistors, was used to provide the DC offset voltage. To amplify the signal, an AC coupled inverting amplifier with a gain of −2.13 and lower cut-off frequency (f_L) of 3.39 Hz was utilized. As a final stage, a non-inverting comparator was designed to compare the amplified signal with a threshold voltage. The output of the comparator is then sampled by the ADC embedded on the ATmega328P at a frequency of 500 Hz for data processing and controlling. Control signals were then sent from the MCU via a XBee-PRO® S3B module (figure 7.6(d)) to control the mobile robot. This demonstrates a typical practical example of integrating TENGs to real applications and a similar approach could be followed to realize flexible wearable applications based on TENGs.

7.3 Interfacing circuits for wireless communication

The interfacing circuit for both wireless communication and charging depends on the printed antennas. Electromagnetic fields are generated by the radio integrated circuit (IC) at a frequency range of 6000 MHz and delivered to the antenna for communicating over the air. Alternatively, for wireless charging, the antenna is used for inductive coupling at lower frequencies of 10–50 MHz. Wireless technologies such near-field communication (NFC) and radio frequency identification (RFID) allow the same antenna and system for both commanication and charging [76]. In recent years, sensing tags based on RFID technology have become a growing interest, as RFID-based smart labels can be placed on products to sense the product's quality and safety. Typically, the interfacing circuit of such tags contains an antenna, silicon IC and sensors [77, 78].

For example, a RFID printed tag was placed on a cardboard box to check the conditions of the packaging in [79]. The tag has a commercial force sensor and a printed opening detector, where the mass applied to the packaging can be measured as well as the opening of the box detected. These two sensors provide relevant information during all stages of the supply chain. The ultra-high frequency of the SL900A RFID chip, with a resonating antenna of 868 MHz, is used, which can be read by an additional RFID reader. The packaging and the circuitry are shown in figure 7.7. Similarly, a fully printed, multisensing, semi-passive high-frequency RFID smart label is reported in [78] by utilizing inkjet printing techniques. An onboard battery power source is used to read the humidity and ammonia sensor and also for recording the data.

Nowadays, NFC, a subset of RFID technology in the high-frequency band, is growing rapidly for short-range wireless systems where the main application is system identification. Usually, for ultra-high-frequency RFID systems, some specialized equipment and reader antennas are required. However, in NFC-based tags, interaction between the reader or NFC-enabled device, such as a smartphone, occurs in the proximity of 1–3 cm to transmit the data safely, which gives NFC a benefit over RFID. For this reason, various physical and chemical sensors have been integrated with NFC to provide a safe communication for applications such as smart food packaging and wearable health monitoring to the end user.

A PDMS microchannel with a poly(3,4-ethylenedioxythiophene) polystyrene sulfonate (PEDOT:PSS) conductive polymer-based strain sensor is integrated with an NFC tag to detect strain in packaging [80]. The strain sensors transfer the mechanical deformation into an electrical resistance change. A custom NFC tag

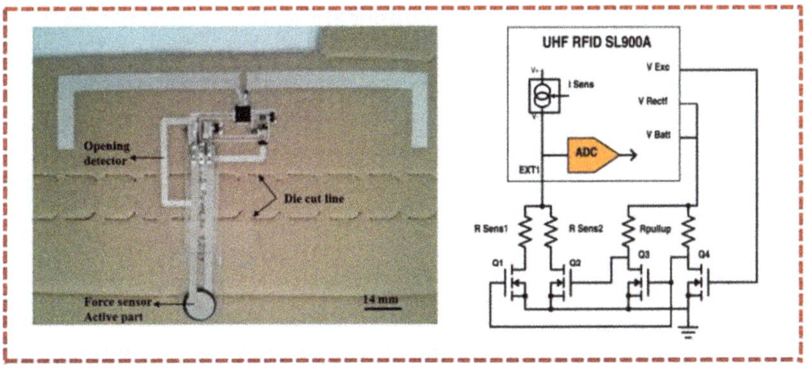

Figure 7.7. RFID tag with commercial force sensor. Reproduced with permission from [79].

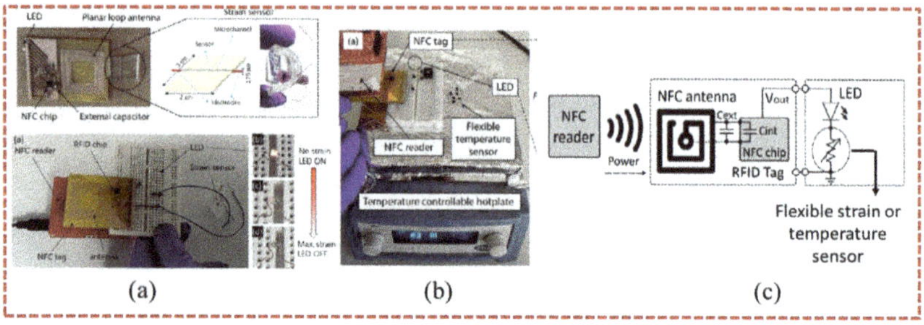

Figure 7.8. A custom NFC tag with (a) a custom strain sensor (© 2020 IEEE, reprinted, with permission, from [80]), (b) a custom temperature sensor; and (c) the interfacing circuit designed in the rigid tag for both sensors. ((b)–(c) © 2020 IEEE, reprinted, with permission, from [81]).

system with an LED was developed, as shown in figure 7.8(a), where the applied strain modulates the light intensity of the LED. The tag antenna has a printed circuit board (PCB) planar coil, which is designed in such a way that it matches the internal capacitor of the NFC transponder chip, RF430FRL154H, from Texas Instruments. The resonating frequency of the tag is designed at 13.56 MHz for the ease of compatibility with NFC-enabled devices. A similar concept is utilized in [81], where a PEDOT:PSS-based temperature sensor in the range of 25 °C to 90 °C is integrated with an LED-connected NFC tag system for intelligent packaging, as shown in figure 7.8(b). Here, the light intensity changes with the temperature. In both these types of application, the tags are fabricated on a rigid FR4 substrate, and the NFC-based passive antenna is read by a commercial reader form Texas Instruments (figure 7.8(c)).

A flexible version of the same NFC tag system has been developed by demonstrating its application in health monitoring as a smart bandage for wound dressing [22]. The flexible PCB is fabricated by UV etching of a flexible PI substrate with copper on one side. The custom-designed strain and temperature sensors, as described in [80, 81], are integrated in an NFC transponder chip, as shown in figure 7.9. In this literature, an NFC-enabled smartphone is used to read the flexible tag for the real-time recording of data from sensors. Data are processed by a user-friendly, custom-developed smartphone application, the SenseAble app.

Apart from health monitoring, a flexible version of the same NFC tag was also developed for better integration with the packaging and to demonstrate its application in food spoilage detection systems [82], as shown in figure 7.10. The tag is encapsulated in PDMS to ensure protection against moisture and liquid, which shows that the consumer can easily detect food spoilage at an early stage by using a smartphone.

7.4 Conclusion and future perspective

In this chapter we have presented some common functional mechanisms of flexible and wearable sensors including their interfacing circuits and flexible circuits for wireless communication. These comprise non-self-powered sensors (piezoresistive

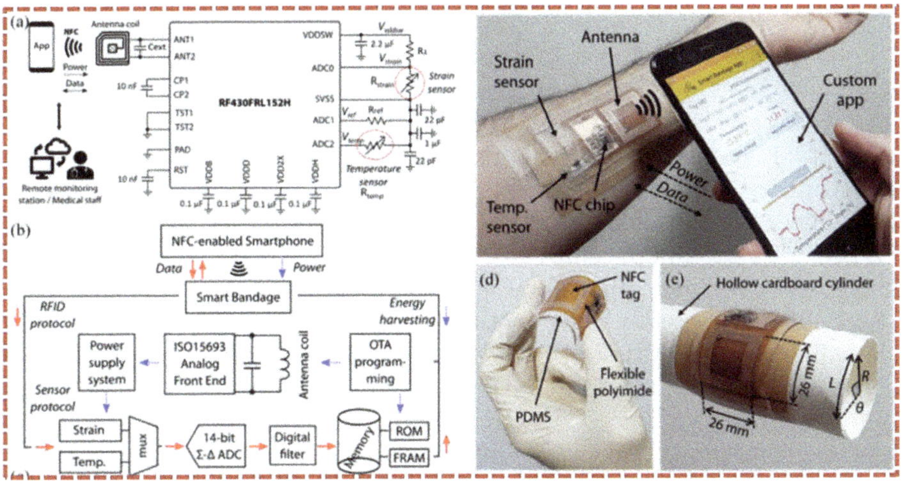

Figure 7.9. The flexible NFC tag system with its application in health monitoring as a smart bandage for wound dressing. © 2020 IEEE. Reprinted, with permission, from [22].

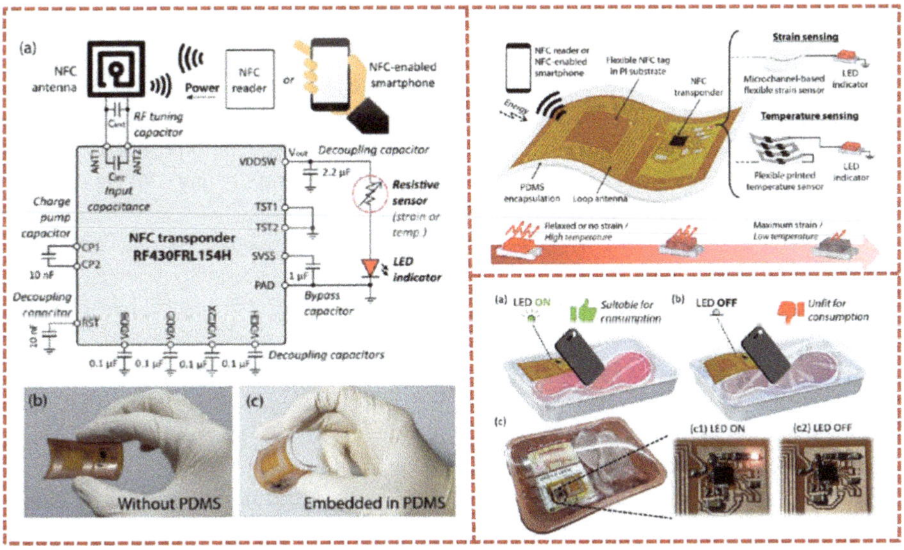

Figure 7.10. The flexible NFC tag system with its application in a food spoilage detection system. © 2021 IEEE. Reprinted, with permission, from [82].

and capacitive sensors) and self-powered sensors (piezoelectric sensors and TENGs). Following the advances in flexible and printed electronics, interfacing circuits for wearable applications are gradually shifting from conventional rigid versions to more flexible versions, which will lead to the realization of very comfortable and highly conformal wearable systems [71]. In recent years, interest in the realization of circuits with printing technologies such as screen printing, contact printing, etc, has been rapidly increasing, and many researchers are now also exploring the use of biodegradable materials including paper [83, 84] for the realization of disposable circuits. Rapid advances in 3D/4D printing technologies [85–88] are also revolutionizing wearable systems, and interest in 3D printed circuits [89, 90] is also increasing.

However, despite the huge interest in flexible electronics, many challenges still exist which need to be tackled in order to fully implement them in real applications. This is because conventional chips are still largely fabricated using rigid silicon, and the integration of these rigid chips on flexible substrates is still a big hurdle. Although there are recently reported techniques for thinning these chips to achieve flexibility [91], it is highly desirable to realize fully flexible electronic circuits wherein the components are all seamlessly integrated onto a flexible substrate. Performance and reliability are also other issues with flexible circuits, as some of the materials used for realizing the electrodes are still not as reliable as the application demands. However, this chapter will help anyone who is planning to design an interfacing circuit for wearable sensors to have an overview of what is possible and the techniques for reading these sensors, including knowledge of some existing materials that have been utilized.

References

[1] Hosseini E *et al* 2021 Healing and monitoring of chronic wounds: advances in wearable technologies *From A to Z: Wearables in Modern Medicine* (Amsterdam: Elsevier)

[2] Dahiya R S 2015 Epidermial electronics: flexible electronics for biomedical applications *Handbook of Bioelectronics—Directly interfacing Electronics and Biological Systems* (Cambridge: Cambridge University Press)

[3] Manjakkal L *et al* 2021 Energy autonomous sweat based wearable systems *Adv. Mater.* **33** 2100899

[4] Bhattacharjee M, Nikbakhtnasrabadi F and Dahiya R 2021 Printed chipless antenna as flexible temperature sensor *IEEE Internet Things J.* **8** 5101–10

[5] Nikbakhtnasrabadi F *et al* 2021 Textile-based stretchable microstrip antenna with intrinsic strain sensing *ACS Appl. Electron. Mater.* **3** 2233–46

[6] Mukherjee R, Ganguly P and Dahiya R 2021 Bioinspired distributed energy in robotics and enabling technologies *Adv. Intel. Syst.* 2100036

[7] Dahiya R, Akinwande D and Chang J S 2019 Flexible electronic skin: from humanoids to humans [scanning the issue] *Proc. IEEE.* **107** 2011–5

[8] Dahiya R 2019 E-skin: from humanoids to humans [point of view] *Proc. IEEE.* **107** 247–52

[9] Escobedo P *et al* 2020 Energy generating electronic skin with intrinsic tactile sensing without touch sensors *IEEE Trans. Rob.* **37** 683–90

[10] Dahiya R *et al* 2019 Large-area soft e-skin: the challenges beyond sensor designs *Proc. IEEE.* **107** 2016–33

[11] Chirila R, Ntagios M and Dahiya R 2020 3D printed wearable exoskeleton human-machine interfacing device *2020 IEEE Sensors.* (Piscataway, NJ: IEEE)

[12] Ntagios M *et al* 2020 Robotic hands with intrinsic tactile sensing via 3D printed soft pressure sensors *Adv. Intell. Syst.* **2** 1900080

[13] Soni M and Dahiya R 2020 Soft eSkin: distributed touch sensing with harmonized energy and computing *Philos. Trans. R. Soc. A.* **378** 20190156

[14] Ozioko O and Dahiya R 2021 Smart tactile gloves for haptic interaction, communication and rehabilitation *Adv. Intell. Syst.* **4** 2100091

[15] Ozioko O *et al* 2020 Wearable assistive tactile communication interface based on integrated touch sensors and actuators *IEEE Trans. Neural Syst. Rehab. Eng.* **28** 1344–52

[16] Ozioko O *et al* 2020 Tacsac: a wearable haptic device with capacitive touch-sensing capability for tactile display *Sensors.* **20** 4780

[17] Ozioko O *et al* 2021 SensAct: the soft and squishy tactile sensor with integrated flexible actuator *Adv. Intell. Syst.* **3** 1900145

[18] Christou A *et al* 2020 3D touch surface for interactive pseudo-holographic displays *Adv. Intell. Syst.* **4** 2000126

[19] Manjakkal L, Dervin S and Dahiya R 2020 Flexible potentiometric pH sensors for wearable systems *RSC Adv.* **10** 8594–617

[20] Bhattacharjee M *et al* 2020 PEDOT: PSS microchannel-based highly sensitive stretchable strain sensor *Adv. Electron. Mater.* **6** 2000445

[21] Yogeswaran N, Hosseini E S and Dahiya R 2020 Graphene based low voltage field effect transistor coupled with biodegradable piezoelectric material based dynamic pressure sensor *ACS Appl. Mater. Interfaces.* **12** 54035–40

[22] Escobedo P *et al* 2020 Smart bandage with wireless strain and temperature sensors and battery-less NFC tag *IEEE Internet Things J.* **8** 5093–100

[23] Kumaresan Y, Ozioko O and Dahiya R 2021 Multifunctional electronic skin with a stack of temperature and pressure sensor arrays *IEEE Sens. J.* **21** 26243–51

[24] Ren Y, Wang X and Liu J 2019 Fabrication of high-resolution flexible circuits and sensors based on liquid metal inks by spraying and wiping processing *IEEE Trans. Biomed, Circuits Syst.* **13** 1545–51

[25] Dahiya R S and Valle M 2012 *Robotic Tactile Sensing: Technologies and System.* (Berlin: Springer Science & Business Media)

[26] Dahiya R S *et al* 2009 Tactile sensing—from humans to humanoids *IEEE Trans. Rob.* **26** 1–20

[27] Hosseini E S *et al* 2020 Biodegradable materials for sustainable health monitoring devices *ACS Appl. Bio Mater.* **4** 163–94

[28] Dahiya R S *et al* 2013 Directions toward effective utilization of tactile skin: a review *IEEE Sens. J.* **13** 4121–38

[29] Khan Y *et al* 2020 A new frontier of printed electronics: flexible hybrid electronics *Adv. Mater.* **32** 1905279

[30] Yin R *et al* 2021 Wearable sensors-enabled human–machine interaction systems: from design to application *Adv. Funct. Mater.* **31** 2008936

[31] Trung T Q and Lee N E 2016 Flexible and stretchable physical sensor integrated platforms for wearable human-activity monitoringand personal healthcare *Adv. Mater.* **28** 4338–72

[32] Sundaram S *et al* 2019 Learning the signatures of the human grasp using a scalable tactile glove *Nature.* **569** 698–702

[33] Osborn L E *et al* 2018 Prosthesis with neuromorphic multilayered e-dermis perceives touch and pain *Sci. Robot.* **3**

[34] Wang Z *et al* 2019 Full 3D printing of stretchable piezoresistive sensor with hierarchical porosity and multimodulus architecture *Adv. Funct. Mater.* **29** 1807569

[35] Zhang S *et al* 2020 A review of electronic skin: soft electronics and sensors for human health *J. Mater. Chem. B.* **8** 852–62
[36] Guo Y *et al* 2019 A wearable transient pressure sensor made with MXene nanosheets for sensitive broad-range human–machine interfacing *Nano Lett.* **19** 1143–50
[37] He J *et al* 2020 Recent advances of wearable and flexible piezoresistivity pressure sensor devices and its future prospects *J. Materiomics.* **6** 86–101
[38] Carr J J 1993 *Sensors and Circuits* (Englewood Cliffs, NJ: PTR Prentice-Hall)
[39] Farshidi E 2011 A low-voltage current-mode Wheatstone bridge using CMOS transistors *Int. J. Elect. Comp. Sci.* **5** 1368–72
[40] Xiong Y *et al* 2020 A flexible, ultra-highly sensitive and stable capacitive pressure sensor with convex microarrays for motion and health monitoring *Nano Energy.* **70** 104436
[41] Jordana J and Pallas Areny R 2006 A simple, efficient interface circuit for piezoresistive pressure sensors *Sens. Actuators, A.* **127** 69–73
[42] Chakraborty M, Kalita A and Biswas K 2018 PMMA-coated capacitive type soil moisture sensor: design, fabrication, and testing *IEEE Trans. Instrum. Meas.* **68** 189–96
[43] Czaja Z 2018 Time-domain measurement methods for R, L and C sensors based on a versatile direct sensor-to-microcontroller interface circuit *Sens. Actuators, A.* **274** 199–210
[44] Reverter F and Casas Ò 2010 A microcontroller-based interface circuit for lossy capacitive sensors *Meas. Sci. Technol.* **21** 065203
[45] Bhattacharjee M, Vilouras A and Dahiya R 2019 Microdroplet based organic vapour sensor on a disposable GO-Chitosan flexible substrate *2019 IEEE Int. Conf. on Flexible and Printable Sensors and Systems (FLEPS)* (Piscataway, NJ: IEEE)
[46] Sahoo S *et al* 2021 Recent developments in self-powered smart chemical sensors for wearable electronics *Nano Res.* **14** 3669–89
[47] Gründler P 2007 *Chemical Sensors: An Introduction for Scientists and Engineers.* (Berlin: Springer Science & Business Media)
[48] Israr-Qadir M *et al* 2017 Zinc oxide-based self-powered potentiometric chemical sensors for biomolecules and metal ions *Sensors* **17** 1645
[49] Marsh P *et al* 2020 Flexible iridium oxide based pH sensor integrated with inductively coupled wireless transmission system for wearable applications *IEEE Sens. J.* **20** 5130–8
[50] Bhattacharjee M *et al* 2020 Microdroplet based disposable sensor patch for detection of α-amylase in human blood serum *Biosens. Bioelectron.* **165** 112333
[51] Escobedo P *et al* 2020 Flexible potentiostat readout circuit patch for electrochemical and biosensor applications *2020 IEEE Int. Conf. on Flexible and Printable Sensors and Systems (FLEPS)* (Piscataway, NJ: IEEE)
[52] Alnasser E 2020 A novel low output offset voltage charge amplifier for piezoelectric sensors *EEE Sensors J.* **20** 5360–67
[53] Dai Y and Gao S 2021 A flexible multi-functional smart skin for force, touch position, proximity, and humidity sensing for humanoid robots *IEEE Sens. J.* **21** 26355-63
[54] Steinem C and Janshoff A 2007 *Piezoelectric Sensors.* vol 5 (Berlin: Springer Science & Business Media)
[55] Rupitsch S J 2018 *Piezoelectric Sensors and Actuators: Fundamentals and Applications.* (Berlin: Springer)
[56] Natta L *et al* 2021 Conformable AlN piezoelectric sensors as a non-invasive approach for swallowing disorder assessment *ACS Sensors* **6** 1761–9

[57] Kim K *et al* 2021 Static force measurement using piezoelectric sensors *J. Sensors* **2021** 6664200
[58] Dahiya R 1999 Piezoelectric tactile sensors *Wiley Encyclopedia of Electrical and Electronics Engineering* (New York: Wiley) pp 1–15
[59] Zhang S and Yu F 2011 Piezoelectric materials for high temperature sensors *J. Am. Ceram. Soc.* **94** 3153–70
[60] Hosseini E S *et al* 2020 Glycine–chitosan-based flexible biodegradable piezoelectric pressure sensor *ACS Appl. Mater. Interfaces* **12** 9008–16
[61] Dahiya R S, Valle M and Lorenzelli L 2009 SPICE model for lossy piezoelectric polymers *IEEE Trans. Ultrason. Ferroelectr. Freq. Control* **56** 387–95
[62] Dahiya R S *et al* 2014 Tactile sensing chips with POSFET array and integrated interface electronics *IEEE Sens. J.* **14** 3448–57
[63] Park D Y *et al* 2017 Self-powered real-time arterial pulse monitoring using ultrathin epidermal piezoelectric sensors *Adv. Mater.* **29** 1702308
[64] Panahi A, Hassanzadeh A and Moulavi A 2020 Design of a low cost, double triangle, piezoelectric sensor for respiratory monitoring applications *Sens. Bio-Sens. Res.* **30** 100378
[65] Webster J G and Eren H 2018 *Measurement, Instrumentation, and Sensors Handbook: Two-Volume Set.* (Boca Raton, FL: CRC Press)
[66] Giannelli P *et al* 2018 A buffered single-supply charge amplifier for high-impedance piezoelectric sensors *IEEE Trans. Instrum. Meas.* **68** 368–76
[67] Chiu Y-Y *et al* 2013 Development of a piezoelectric polyvinylidene fluoride polymer-based sensor patch for simultaneous heartbeat and respiration monitoring *The 8th Annual IEEE Int. Conf. on Nano/Micro Engineered and Molecular Systems* (Piscataway, NJ: IEEE)
[68] Mahbub I *et al* 2017 A low-power wireless piezoelectric sensor-based respiration monitoring system realized in CMOS process *IEEE Sens. J.* **17** 1858–64
[69] Lin Z *et al* 2017 Triboelectric nanogenerator enabled body sensor network for self-powered human heart-rate monitoring *ACS Nano* **11** 8830–7
[70] Wang Z L 2020 Triboelectric nanogenerator (TENG)—sparking an energy and sensor revolution *Adv. Energy Mater.* **10** 2000137
[71] Xiao X *et al* 2021 Wearable triboelectric nanogenerators for therapeutics *Trends Chem.* **3** 279–90
[72] Min G *et al* 2021 Origin of the contact force-dependent response of triboelectric nanogenerators *Nano Energy* **83** 105829
[73] Xu Y *et al* 2020 A unified contact force-dependent model for triboelectric nanogenerators accounting for surface roughness *Nano Energy* **76** 105067
[74] Min G *et al* 2019 Triboelectric nanogenerator with enhanced performance via an optimized low permittivity substrate *IEEE Sens. J.* **20** 6856–62
[75] Qiu C *et al* 2019 Sensors and control interface methods based on triboelectric nanogenerator in IoT applications *IEEE Access* **7** 92745–57
[76] Finkenzeller K 2010 *RFID Handbook: Fundamentals and Applications in Contactless Smart Cards, Radio Frequency Identification and Near-Field Communication* (New York: Wiley)
[77] Virtanen J *et al* 2011 Inkjet-printed humidity sensor for passive UHF RFID systems *IEEE Trans. Instrum. Meas.* **60** 2768–77
[78] Quintero A V *et al* 2016 Smart RFID label with a printed multisensor platform for environmental monitoring *Flex. Print. Electron.* **1** 025003

[79] Fernández-Salmerón J *et al* 2015 Passive UHF RFID tag with multiple sensing capabilities *Sensors* **15** 26769–82

[80] Escobedo P *et al* 2020 Flexible strain sensor with NFC tag for food packaging *2020 IEEE Int. Conf. on Flexible and Printable Sensors and Systems (FLEPS)* (Piscataway, NJ: IEEE)

[81] Bhattacharjee M *et al* 2020 Printed flexible temperature sensor with NFC interface *2020 IEEE Int. Conf. on Flexible and Printable Sensors and Systems (FLEPS)* (Piscataway, NJ: IEEE)

[82] Escobedo P *et al* 2021 Flexible strain and temperature sensing NFC tag for smart food packaging applications *IEEE Sens. J.* **21** 26406–14

[83] Conti S *et al* 2020 Low-voltage 2D materials-based printed field-effect transistors for integrated digital and analog electronics on paper *Nat. Commun.* **11** 1–9

[84] Ha D, Fang Z and Zhitenev N B 2018 Paper in electronic and optoelectronic devices *Adv. Electron. Mater.* **4** 1700593

[85] Zhang Y *et al* 2021 4D printing of magnetoactive soft materials for on-demand magnetic actuation transformation *ACS Appl. Mater. Interfaces.* **13** 4174–84

[86] Wang Y and Li X 2021 4D printing reversible actuator with strain self-sensing function via structural design *Compos. Part B: Eng.* **211** 108644

[87] Tang Y *et al* 2021 Recent advances of 4D printing technologies toward soft tactile sensors *Front. Mater.* **8** 110

[88] Lee J *et al* 2017 A review on 3D printed smart devices for 4D printing *Int. J. Precis. Eng.* **4** 373–83

[89] Hong F, Myant C and Boyle D E 2021 Thermoformed circuit boards: fabrication of highly conductive freeform 3D printed circuit boards with heat bending *Proc. of the 2021 CHI Conf. on Human Factors in Computing Systems*

[90] Kwok S W *et al* 2017 Electrically conductive filament for 3D-printed circuits and sensors *Appl. Mater. Today* **9** 167–75

[91] Kumaresan Y, Ma S and Dahiya R 2021 PMMA sacrificial layer based reliable debonding of ultra-thin chips after lapping *Microelectron. Eng.* **247** 111588

Chapter 8

Compact and efficient wireless power and information transfer systems for IoT sensors and implants

Ramesh K Pokharel and Adel Barakat

In this chapter, we discuss simultaneous wireless power and information transfer (WPIT) systems for Internet-of-Things (IoT) sensors and implants. The necessity of simultaneous WPIT is explained and examples from the literature are given. Then, the reasons behind selecting a two-channel system over single-channel and dual-band systems are detailed. After that, figure-8 inductance is deliberated in terms of self-inductance, losses, mutual coupling profile with identical figure-8 inductance, and cross-coupling with defected ground structure (DGS) inductance. Finally, a simultaneous WPIT system using two coupled sets of figure-8 and DGS inductances is designed, fabricated and characterized. The system performance is discussed in terms of transfer efficiency in both channels, and isolation between the power and information channel.

8.1 Introduction

An indispensable necessity for IoT sensors as well as implant applications is that of simultaneous wireless power and information transfer (WPIT) [1–17]. Supplying the required energy wirelessly can enable a reduced-size battery or even eliminate the need for an attached battery depending on the application needs. The utilization of batteries has several drawbacks, including bulkiness and toxicity in the case of biomedical implants. These associated disadvantages of batteries encourage the use of wireless power transfer technology to avoid such drawbacks.

Generally, IoT sensors and implants are expected to be used in harsh and sensitive environments such as human body tissue. Therefore, resonant inductive coupling with low-frequency carriers has been the preferred method to transport the wireless

doi:10.1088/978-0-7503-3935-3ch8

power for IoT sensors and implants [3–15]. Such a strategy has been favored in order to avoid the low wireless power transfer efficiency that results from absorption by the receiver's surrounding environment, such as biological tissue, which is also a necessity for tissue safety [5–9, 18]. In addition, the resonant nature enables the power to be focused on a narrow band such that a high wireless power transfer efficiency can be achieved [5–9].

Alternatively, in the case of wireless information transfer, a channel with a comparatively broader bandwidth than that of the wireless power transfer is vital to empower a smooth information transfer of a few mega-bits-per-second (Mbps) [5–9]. A high enough frequency carrier for transferring the information is one likely solution to guarantee the need for this broader bandwidth. Since information is transmitted with a relatively low power level, no restriction is put in the frequency selection as it may encounter low absorption by the receiver's surrounding medium. In summary, simultaneous WPIT to IoT sensors and implants have distinct requirements as regards the power and information transfer channels, which need to be carefully considered during the design phase of a system that has a simultaneous operation.

8.2 Examples from the literature

During the early stages of the development of WPIT systems, the same frequency carrier for simultaneous power and information transfer was used [3, 4]. However, the final performance of these WPIT systems was limited due to the distinct bandwidth and efficiency requirements described above. The use of a low-frequency carrier may limit the information transfer rate due to the narrow bandwidth, while the use of a high-frequency carrier may result in a degraded efficiency due to the receiver's surrounding medium's absorption [5]. Another alternative is the use of a two-band WPIT system, which uses two coexisting inductive coupling modules that operate at two different frequency carriers to overcome the limitations of single-frequency carrier systems [5–9]. Such a strategy would guarantee isolation between the power and information channels of these two-band WPIT systems [5–9], which is essential for two key reasons: (1) the avoidance of degradation of the efficiency by any undesired cross-coupling; (2) the reduction of the burden and complexity associated with the information filter in the receiving side.

In two-band WPIT systems, several approaches have been proposed. The placement of the resonant inductors in a coplanar configuration was presented in [6, 7] to support two-band WPIT for simultaneous operation. However, the strong cross-coupling resulted in degraded isolation performances. In [8, 9], another approach was investigated by using a perpendicular arrangement of these inductors to decrease the cross-coupling between the two modules. Still, the profile of this system was moderately large, and the fabricated designs have some non-idealities, which affected the targeted isolation between the two channels. Another approach was the utilization of mathematical analysis [5] to increase isolation. However, the employment of this mathematical analysis resulted in small inductors for the information channel to achieve high isolation, which caused efficiency degradation in the information link. Another alternative is to design the two-band WPIT system using shared input/output ports [10–15]. Nevertheless, diplexers may be required to separate the information and power signals, which can cause an increase in the overall system cost and size.

In this chapter, we explain the utilization of a DGS resonator in coaxial configuration with a figure-8 inductor [16] in a simultaneous WPIT system.

8.3 Figure-8 inductor

8.3.1 Structure of a figure-8 inductor

Figure 8.1(a) shows a three-dimensional (3D) electromagnetic (EM) model of coupled figure-8 inductors. Each figure-8 inductor consists of two identical current loops. These two current loops are connected in series and routed such that the magnetic fluxes generated by each of them are equal in magnitude and 180° out of phase in the axial direction with that of the other current loop. This phase difference

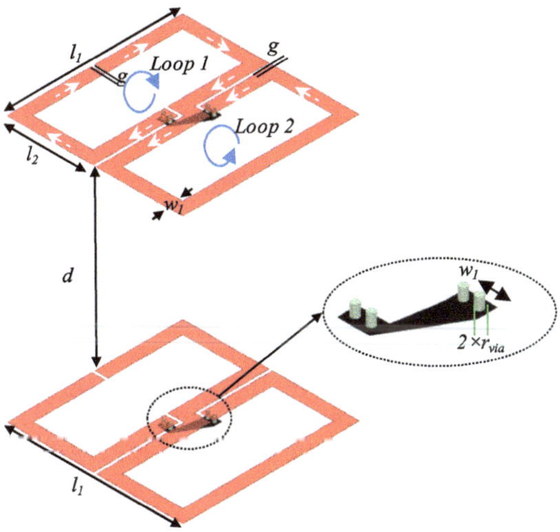

(a)

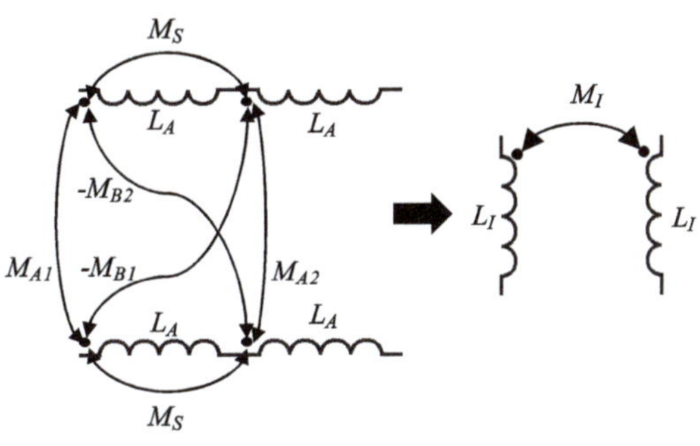

(b)

Figure 8.1. Coupled figure-8 inductors. (a) EM model. (b) Equivalent and simplified circuits.

can be understood clearly from figure 8.1(a) as the current rotates in 'loop 1' in a clockwise direction and rotates in 'loop 2' in an anti-clockwise direction.

The out-of-phase manner is required to cancel the mutual inductance when the figure-8 loop inductor is placed in the same substrate as a conventional spiral inductor, as explained later in the following sections.

8.3.2 Self-inductance of the figure-8 inductor

The equivalent circuit of the coupled printed figure-8 inductors is shown in figure 8.1(b). Each loop of the figure-8 inductor has a self-inductance, L_A, and an intra-mutual inductance, M_S, that exists between them. So, the equivalent inductance of the figure-8 inductor, L_I, can be calculated as $L_I = 2 \times (L_A + M_S)$. Wheeler's formula [19] and Neumann's formula [20] are used to estimate the self and mutual inductances, respectively, for given dimensions of coupled current loops.

8.3.3 Losses of the figure-8 inductor

Two types of losses are associated with an inductive loop: conductive losses [21] and radiation losses [22]. These losses are modeled as resistances, R_c and R_r, which represent the conduction and radiation losses, respectively. Each current loop has a total loss as the sum of these two resistances. Then, the overall loss resistance, R_I, of a figure-8 inductor can be calculated as $R_I = 2 \times (R_c + R_r)$. This R_I is double the sum of R_c and R_r since the two loops of the figure-8 inductor are connected in series.

8.3.4 Mutual inductance between two figure-8 inductors

On the one hand, each of the two coaxially aligned current loops of the figure-8 inductor has additive mutual inductances, M_{A1} and M_{A2}, as they have the same phase angle as the magnetic flux, as can be interpreted from figure 8.1(a). On the other hand, the non-aligned loops have opposing mutual inductances, M_{B1} and M_{B2}, due to being 180° out of phase in their magnetic fluxes. In return, they encounter different signs in the overall mutual inductance, M_I, as can be interpreted from the equivalent circuit in figure 8.1(b), where $M_I = M_{A1} + M_{A2} - M_{B1} - M_{B2}$.

Example dimensions and their corresponding equivalent circuit parameters of two coupled figure-8 inductors at perfect alignment are listed in table 8.1. These dimensions are selected to maximize the obtainable efficiency of a WPIT system that consists of these two coupled figure-8 inductors, as we detail in the following sections. Furthermore, we show the mutual inductance profile of the coupled figure-8 inductors during lateral misalignment in figure 8.2. First, the EM simulation model used in the extraction of the mutual coupling is shown in figure 8.2(a). Each inductive loop is excited by a lumped port.

The four-port Z-parameters are calculated at each misalignment position. Then, the mutual inductances are extracted from the matrix of these four-port Z-parameters. On the one hand, in figure 8.2(b), each of M_{A1}, M_{A2}, M_{B1} and M_{B2} has a decreasing relationship with $\pm\Delta x$ as the coupled loops move away. Hence, M_I

Table 8.1. Dimensions of the coupled figure-8 inductors and their corresponding EM-extracted equivalent circuit parameters at 100 MHz.

Symbol	Value	Unit
l_1	27	mm
l_2	13.25	mm
w_1	2.1	mm
g	0.5	mm
d	14	mm
r_{via}	0.25	mm
L_A	34	nH
M_S	4	nH
L_I	76	nH
M_{A1}, M_{A2}	2.1	nH
M_{B1}, M_{B2}	0.5	nH
M_I	3.2	nH
R_I	0.16	Ω

does not have nulls, as can be interpreted from figure 8.2(b). On the one hand, in figure 8.2(c), each of M_{A1} and M_{A2} has a decreasing trend with $\pm\Delta y$ as the coupled loops move away.

Nonetheless, M_{B1} has an increasing trend when moving with $-\Delta y$ as the loops become closer and a decreasing trend when moving with $+\Delta y$ as the loops become more separated. Also, M_{B2} has an increasing trend when moving with $+\Delta y$ and a decreasing trend when moving with $-\Delta y$ due to the same reasons as M_{B1}. Hence, M_D encounters two nulls, as can be interpreted from figure 8.2(c). The mutual inductance profile of M_I is required to understand the lateral misalignment performance of a WPIT system based on these coupled figure-8 inductors, as detailed in later sections.

8.3.5 Cross-coupling with DGS inductor

Minimal cross-coupling is the most important feature of the figure-8 inductor when placed in the same substrate as a DGS inductor, as shown in the EM model in figure 8.3(a). Each inductive loop is excited by a lumped port. Then, the three-port Z-parameters are calculated and the cross-mutual inductances (M_{C1} and M_{C2}) are extracted from the matrix of these three-port Z-parameters, as shown in figure 8.3(b).

M_{C1} and M_{C2} have the same magnitude and are with opposite signs. Hence, their summation, $M_{C1} - M_{C2}$, is less than 0.4 nH. In return, high isolation is expected between a WPIT system utilizing a coupled version of this configuration.

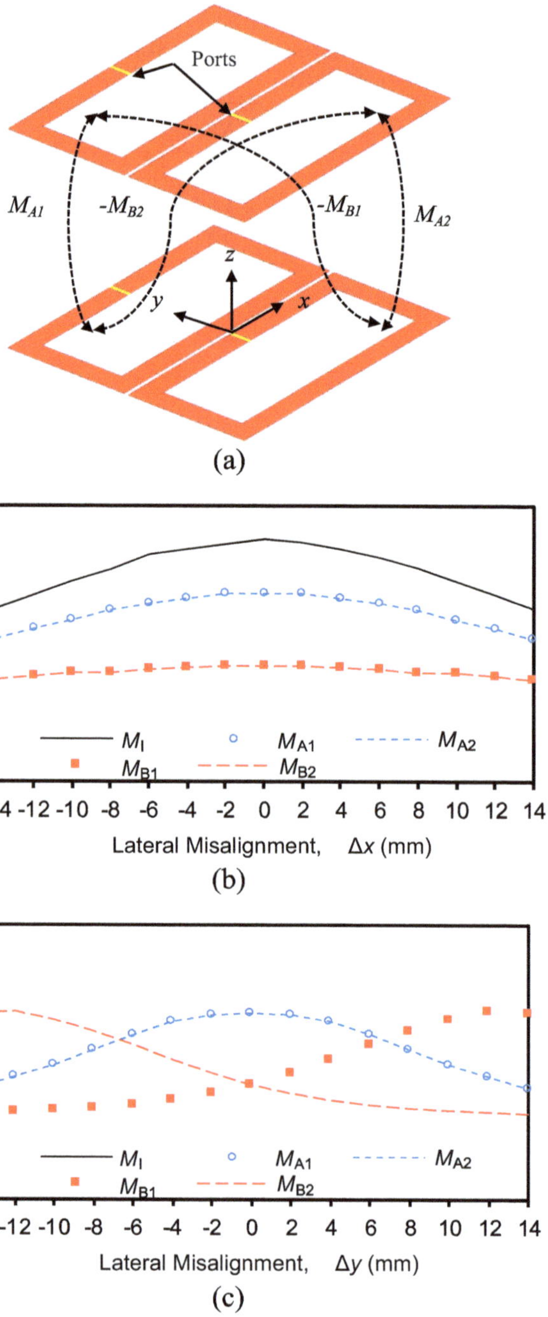

Figure 8.2. Mutual inductance profile of the coupled figure-8 inductors at 100 MHz and d = 14 mm. (a) 3D layout of the EM setup. (b) Lateral misalignment (x-axis). (c) Lateral misalignment (y-axis).

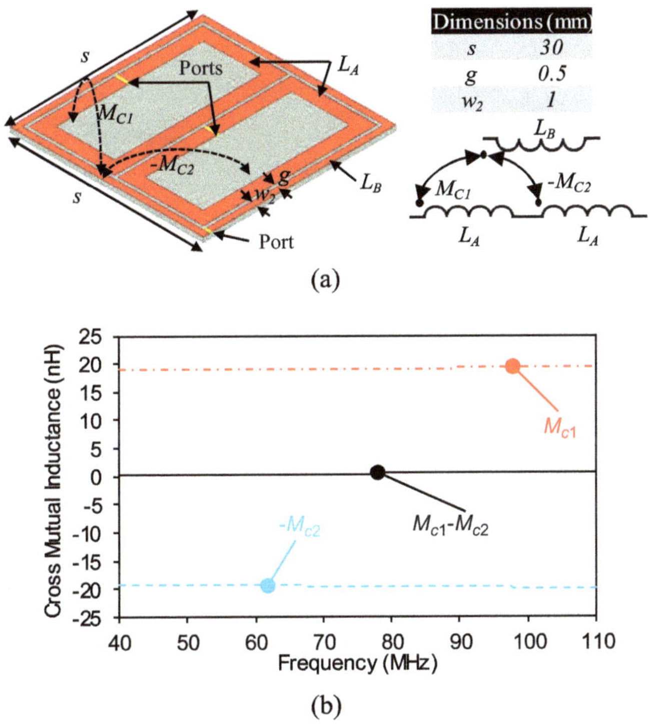

Figure 8.3. Cross-coupling between figure-8 and DGS inductors. (a) EM model and equivalent circuit. (b) Extracted mutual inductance.

8.4 System design and discussions

8.4.1 System layout

Figure 8.4(a) shows a simultaneous WPIT system based on figure-8 and DGS inductors. The coupled outer DGSs are used for implementing the power channel at 50 MHz and the coupled inner figure-8 inductors for the information channel at 100 MHz. The dimensions are listed in table 8.1, figure 8.3(a) and figure 8.4(a). These dimensions are selected to maximize the coupled quality (Q) factor [23], $kQ = \omega M / R$, for each type of coupled inductors, as shown in figure 8.4(b), where M is the mutual coupling between the inductors and R is the resistive losses of each inductor. The corresponding maximum obtainable efficiency, $\eta_{\max}$, is 81.2% at 50 MHz and 82.5% at 100 MHz. $\eta_{\max}$ is calculated using (8.1) [24]:

$$\eta_{\max} = \left(kQ / \left[1 + \sqrt{1 + (\ kQ)^2}\right]\right)^2. \tag{8.1}$$

An equivalent circuit of the simultaneous WPIT system is shown in figure 8.5. In this system, each of the information and power channels can be designed independently due to the minimal cross-coupling between the inductors of each subsystem. We have utilized the design method in [22] to estimate the capacitor values, which are listed in table 8.2. These capacitors were implemented using surface mount

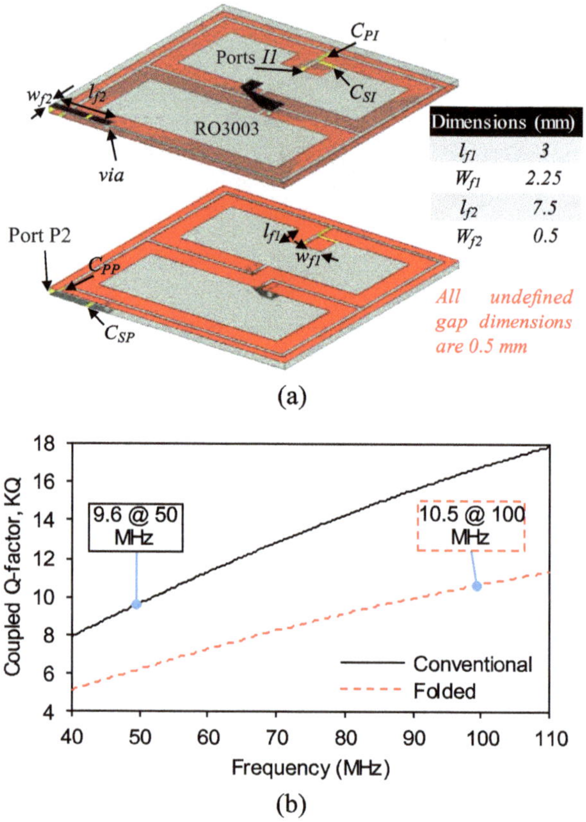

Figure 8.4. Simultaneous WPIT system based on coupled figure-8 and DGS inductors. (a) EM model. (b) Coupled quality factor (KQ). Reproduced from [16].

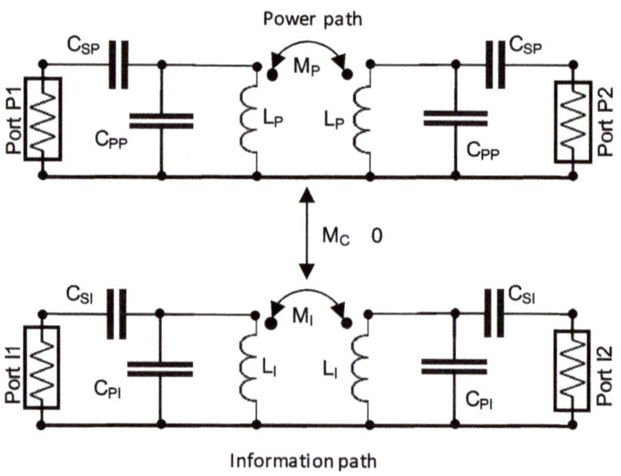

Figure 8.5. Equivalent circuit of the simultaneous WPIT system in figure 8.4(a). Reproduced from [16].

Table 8.2 Summary of the SMCs used in the fabrication. Reproduced from [16].

Resonator (frequency)	Capacitor	Value (pF)	SMC part number
Inner (100 MHz)	C_{PD}	$12 \times 2 = 24$	GJM1555C1H120FB01D
	C_{SD}	$3.3 \times 2 = 6.6$	GJM1555C1H3R3BB01D
Outer (50 MHz)	C_{PP}	$47 \times 2 = 94$	GRM1552C1H470JA01D
	C_{SP}	33	GRM1551X1H330JAG1D

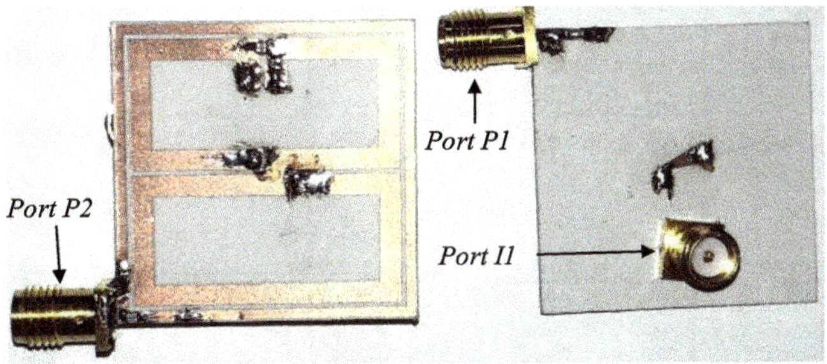

Figure 8.6. Fabricated simultaneous WPIT system. (a) Bottom view. (b) Top view. Reproduced from [16].

capacitors (SMCs). For the power channel, L_P and M_P are 82.6 and 9 nH, respectively. For the information channel, L_D and M_D are given in table 8.1. All ports in figure 8.5 are 50 Ω ports to enable characterization with a vector network analyzer and other RF measurement tools.

8.4.2 Fabrication

The fabricated simultaneous WPIT system is shown in figure 8.6. This simultaneous WPIT system was fabricated using the MITS FP-21T Precision 50 μm line/space prototyping machine. RO3003 substrate ($\varepsilon_r = 3$, $\tan \delta = 0.0013$ and $h = 0.76$ mm and $t = 17$ μm) was used for this fabrication as well as during the EM simulations, which was performed using the ANSYS High-Frequency Structure Simulator (HFSS). The part numbers of the capacitors used in the fabrication are summarized in table 8.2. These capacitor values are selected to achieve resonance at 50 MHz and 100 MHz for the power and information channels, respectively, with the necessary admittance inversion at every frequency.

8.4.3 Measurement category 1: S-parameters

In the S-parameters measurement category, we characterized the transmission, reflection and isolation characteristics of the fabricated system employing the measured S-parameters. These S-parameters measurements were done using a Keysight vector network analyzer (PNA series part number: N5222A), which allows two-port measurements only. Four cases of S-parameters measurements were performed to enable the full characterization of the four-port system as the fabricated simultaneous WPIT system is symmetric. Block diagrams of these four measurement cases are shown in figure 8.7(a). Two 50 Ω SMA terminators

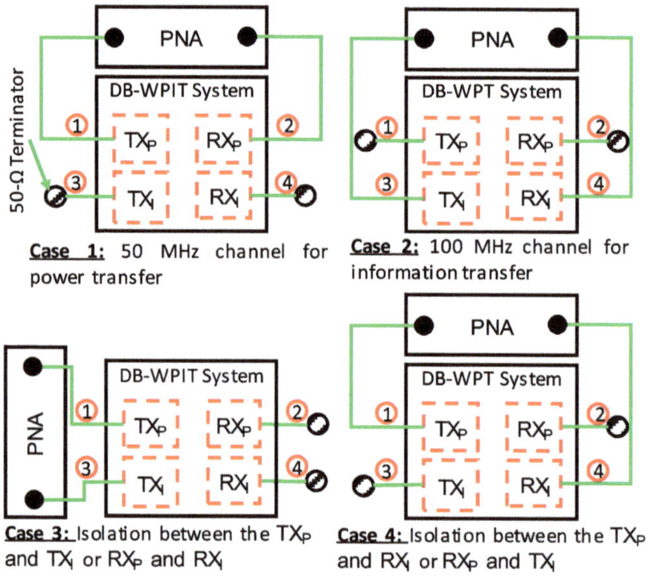

(a)

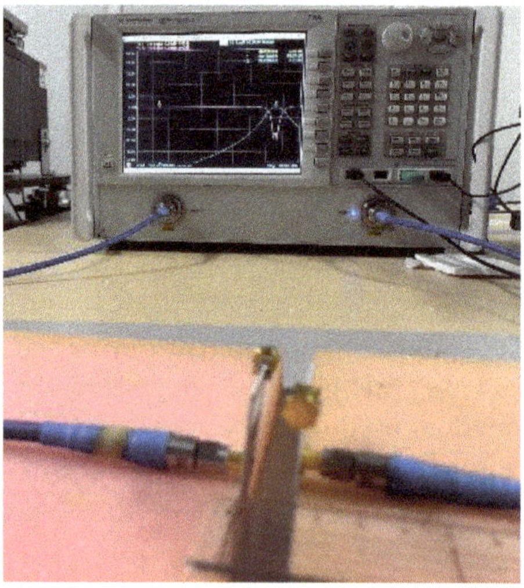

(b)

Figure 8.7. Measurement setup for two-port S-parameters. (a) Block diagram illustrating the four cases of the two-port S-parameters measurements. (b) Photo of the setup during the 100 MHz system characterization. Reproduced from [16].

(Amphenol RF part number: 132360) are used to terminate the unconnected ports in each set of these measurements, where TX_P, TX_I, RX_P and RX_I are the transmitters and receivers of the power and information carriers, respectively. A photo of the measurement setup of the S-parameters is shown in figure 8.7(b).

The EM-simulated and measured S-parameters are in good agreement, as shown in figure 8.8, which considers the perfectly aligned case. The measured transmission coefficients are −1.15 and −1.2 dB at 50 and 100 MHz, respectively, as shown in figure 8.8(a). These measured values are about 0.2 and 0.25 dB, respectively, lower than those of the corresponding simulated ones. The isolation between in-plane and off-plane modules is compared with the EM-simulated ones in figures 8.8(b) and (c), respectively, and a fair agreement exists. The measured isolation is more than 34.5 dB and 34 dB from 40 MHz to 110 MHz for the in-plane and off-plane modules, respectively.

8.4.4 Measurement category 2: misalignment

Misalignment causes changes in the mutual inductance that results in a degradation in the WPIT system efficiency [25–27]. A study of the efficiency of the simultaneous WPIT system during misalignment in the three principal axes was conducted, and the measured performances are shown in figure 8.9. The efficiency, η_{measured}, is calculated from the S-parameters as $\eta_{\text{measured}} = |S_{21}|^2$. When the separation, d, varies between 10–14 mm, the η_{measured} variations are within 65%–78% and 48%–76% at 50 MHz and 100 MHz, respectively, as shown in figure 8.9(a).

For the lateral misalignment, the 50 MHz frequency carrier shows an identical performance when misalignment occurs in the x-axis or y-axis. η_{measured} varies within 56.2%–78% for $\Delta x = \pm 14$ mm or $\Delta y = \pm 14$ mm as shown in figures 8.9(b) and (c) for Δx and Δy, respectively. This performance is due to the symmetry of the loop used in the implementation of this 50 MHz simultaneous WPIT subsystem.

On the contrary, the 100 MHz frequency carrier has distinct performance when misalignment occurs in the x-axis and y-axis. On the one hand, η_{measured} variations are within 48.3%–76% when the increment, Δx, varies between ±14 mm, as shown in figure 8.9(b).

On the other hand, η_{measured} variations are within 50%–76% when the increment, Δy, varies between ±5 mm, as shown in figure 8.9(c).

Otherwise, η_{measured} is lower than 50% and even nulls occur. The reason for the limited η_{measured} when misalignment occurs in the y-axis and the existence of nulls for the 100 MHz frequency carrier is explained through the mutual inductance profile in figure 8.3(c).

8.4.5 Measurement category 3: simultaneous wireless power and information transfer

In figure 8.10(a), a block diagram of the test setup for the simultaneous WPIT system is shown. In this setup, a Keysight signal generator (part number: E8257D) was connected to port 1 (power TX) with a frequency of 50 MHz, and an Anritsu signal generator (part number: MG3710A) was connected to port 3 (information TX) with a frequency of 100 MHz. The output power of both of these signal

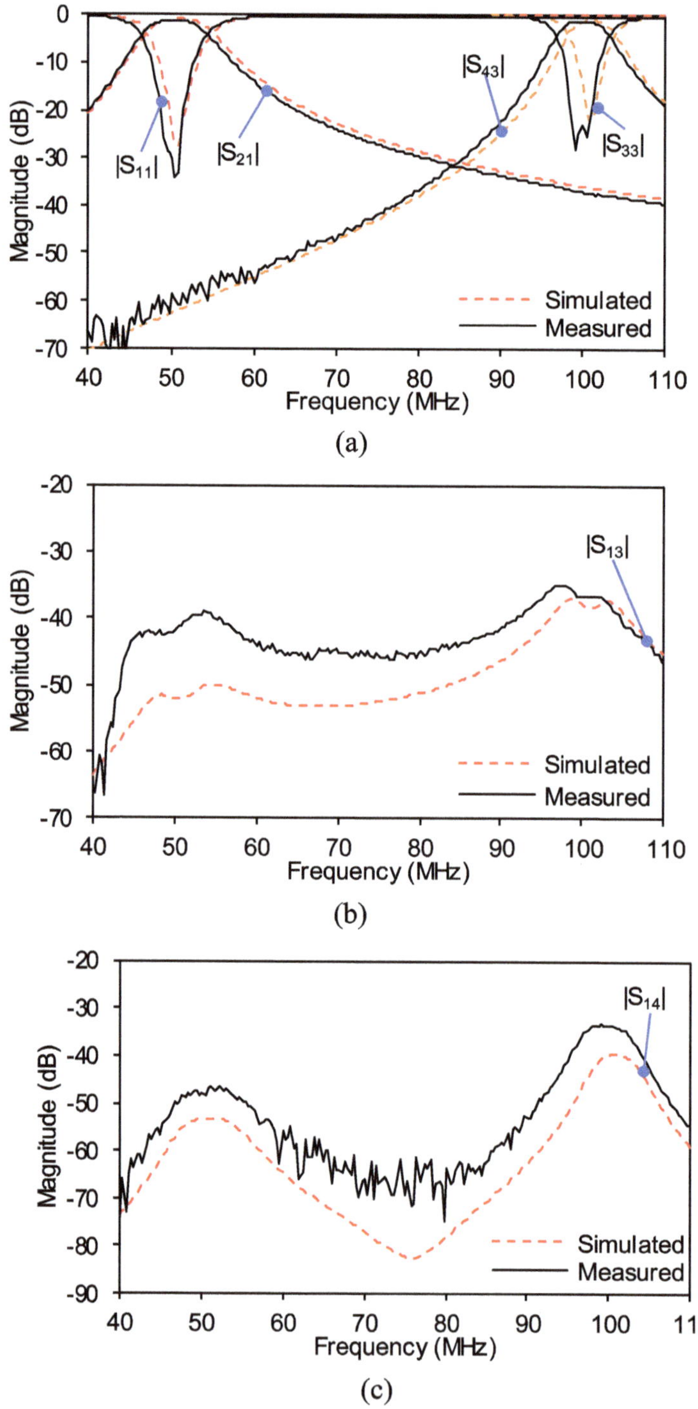

Figure 8.8. Comparison between measured and EM-simulated S-parameters. (a) Reflection ($|S_{11}|$ and $|S_{33}|$) and transmission ($|S_{21}|$ and $|S_{43}|$). (b) In-plane isolation ($|S_{13}|$). (c) Off-plane isolation ($|S_{14}|$). Reproduced from [16].

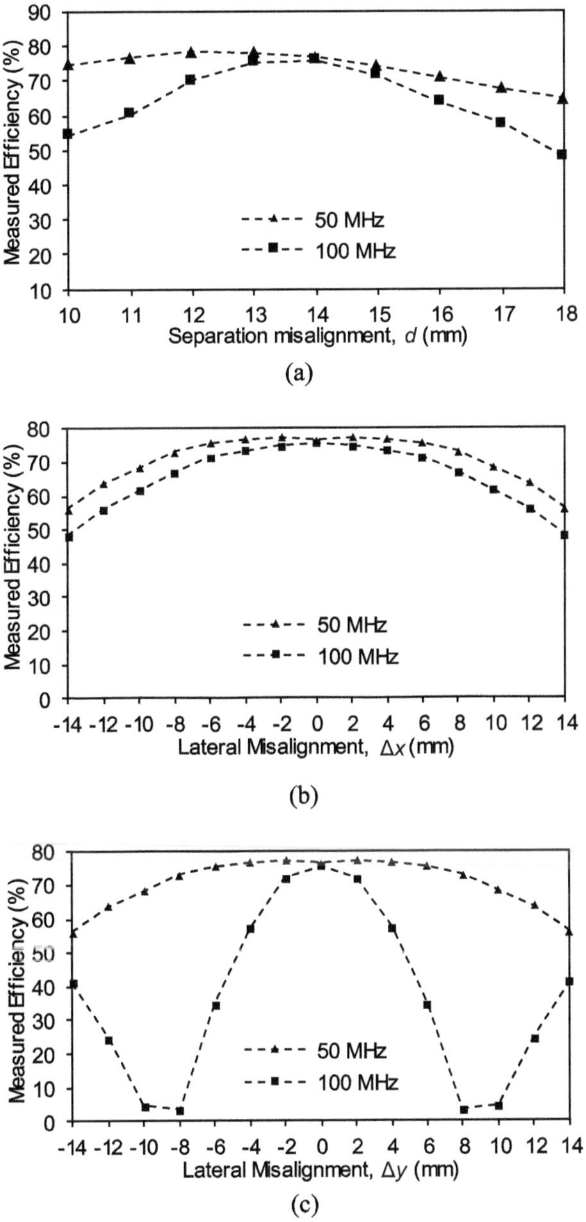

Figure 8.9. Measured performance of the simultaneous WPIT system during misalignment in the three principal planes. (a) Separation misalignment (z-axis). (b) Lateral misalignment (x-axis). (c) Lateral misalignment (y-axis).

generators and their output powers were varied between ±10 dBm during the measurements. This setup has two cases. In case 1, the 50 Ω SMA terminator (Amphenol RF part number: 132360) was connected to port 2, and the Keysight signal analyzer (part number: N9010A) was connected to port 4 to enable efficiency

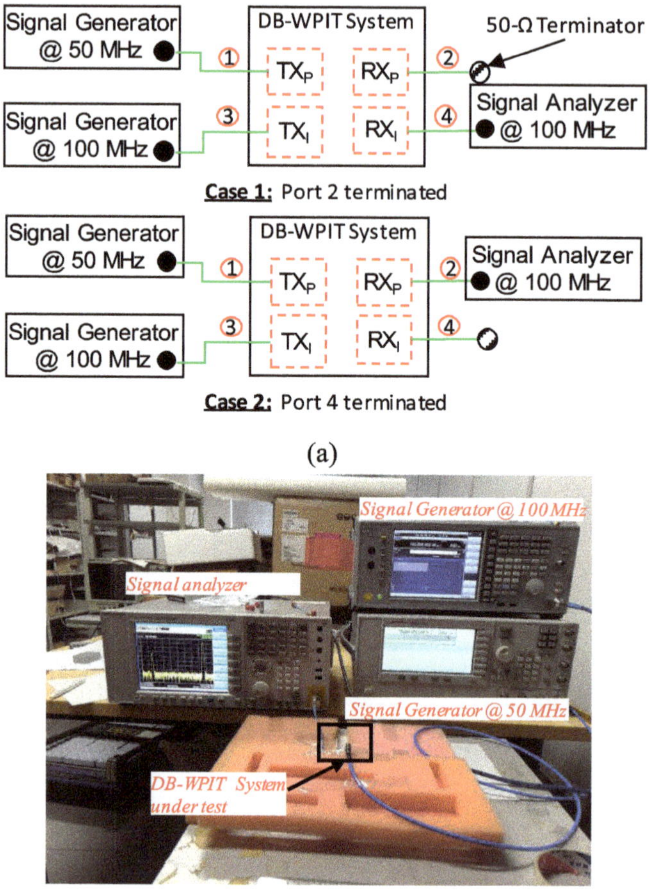

(a)

(b)

Figure 8.10. Simultaneous WPIT test setup. (a) Block diagram. (b) Photo of the test setup. (c) Measured power level at port 4 'information RX' resulting from input power at port 1 'power TX' at 50 MHz.

measurements of the 100 MHz channel. In case 2, the 50 Ω SMA terminator was connected to port 4, and the Keysight signal analyzer was connected to port 2 to enable the efficiency measurement of the 50 MHz channel.

Isolation between the two channels is calculated in both cases. A photo of the system during the experiment is shown in figure 8.10(b). Then, the measured output power at port 4 resulting from the 50 MHz frequency carrier is shown in figure 8.11(a). According to figure 8.11(a), a linear relationship exists between the input and output powers, and the isolation is 42 dB at 50 MHz.

The measured output power at port 2 was linear with the input power at port 1, as shown in figure 8.11(b), with an insertion loss of 1.15 dB. Similarly, the measured output power at port 4 is linear with the input power at port 3, as shown in figure 8.11(c), with an insertion loss of 1.2 dB. The resulting efficiencies are calculated using (8.2) [25] and are 78% and 76% at the 50 MHz and 100 MHz

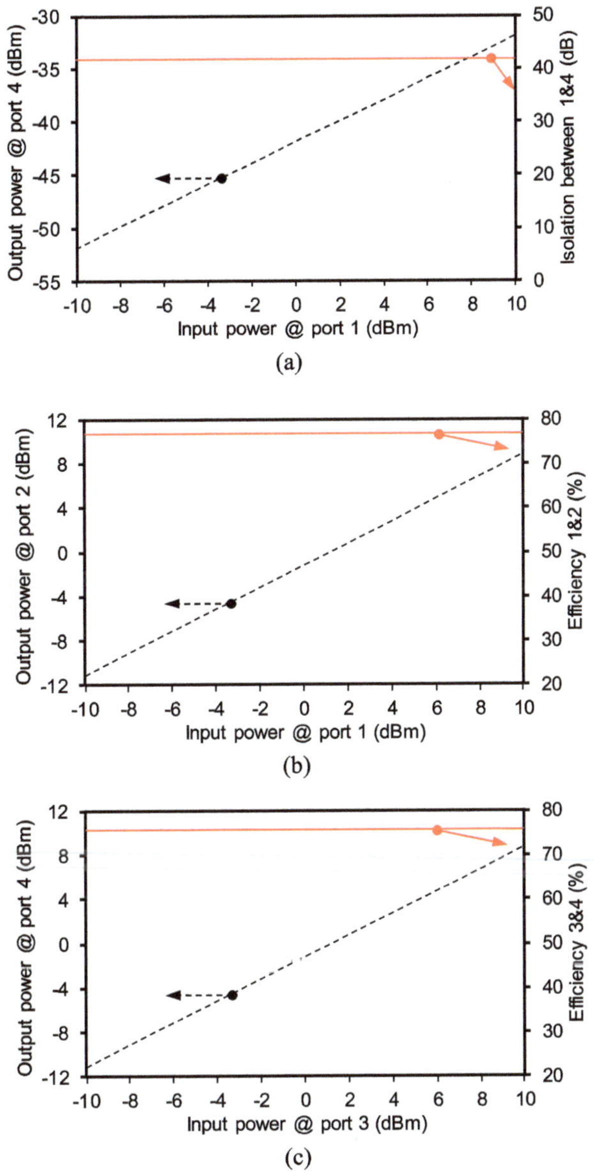

Figure 8.11. Measured isolation and efficiency of the simultaneous WPIT. (a) Isolation. (b) Efficiency of the 50 MHz frequency carrier. (c) Efficiency of the 100 MHz frequency carrier.

channels, respectively; the measured efficiencies are independent of the input powers as shown in figures 8.11(b) and (c), respectively:

$$\eta_{\text{measured}} = 10^{0.1\times(P_{\text{out}}-P_{\text{in}})} \times 100\%, \tag{8.2}$$

where P_{out} and P_{in} are the output and input powers in dBm.

We are working on two issues to realize an actual prototype. First, we are replacing the substrate used with a flexible copper cladding substrate, which is expected to be more applicable for implant application. Second, signal-processing modules are being developed in complementary metal-oxide-semiconductor technology and, together with sensing devices, will be attached within the same area of this simultaneous WPIT such that the full system has a compact size.

8.5 Conclusion

In this chapter figure-8 inductance characteristics have been discussed. A figure-8 inductance has two loops, which are routed such that the generated magnetic flux densities at their axes are equal in magnitude and 180° out of phase. The implementation of a figure-8 inductance in the same substrate as a DGS inductance showed a minimal cross-coupling between them. Then, a simultaneous, highly isolated WPIT system was realized by utilizing two coupled sets of this implementation. The operating frequencies of this simultaneous WPIT system were designed independently, thanks to the minimized cross-coupling. This simultaneous WPIT system was fabricated for operation at 50 MHz for the power channel and 100 MHz for the information channel. A wide set of measurements were performed for the characterization of this simultaneous WPIT system. The EM-simulated and measured S-parameters showed good agreement. The in-plane and off-plane isolations between the power and information channels were more than 34.5 dB and 34 dB, respectively, from 40 MHz to 110 MHz based on measurements of the S-parameters. The measured efficiencies for the perfectly aligned simultaneous WPIT system are 78% and 76% at 50 MHz and 100 MHz, respectively. Furthermore, a simultaneous transmission test using two signal generators was conducted, which confirmed the efficiency of the power and information channels and the isolation between them. Future directions have been explained using a flexible copper cladding substrate.

References

[1] Xue R, Cheng K and Je M 2013 High-efficiency wireless power transfer for biomedical implants by optimal resonant load transformation *IEEE Trans. Circuits Syst.* **60** 867–74

[2] Wang G, Liu W, Sivaprakasam M and Kendir G A 2005 Design and analysis of an adaptive transcutaneous power telemetry for biomedical implants *IEEE Trans. Circuits Syst.* I **52** 2109–17

[3] Troyk P R and Schwan M A K 1992 Closed-loop class E transcutaneous power and data link for microImplants *IEEE Trans. Biomed. Eng.* **39** 589–99

[4] Liu W, Vichienchom K, Clements M, DeMarco S C, Hughes C, McGucken E, Humayun M S, De Juan E, Weiland J D and Greenberg R 2000 A neuro-stimulus chip with telemetry unit for retinal prosthetic device *IEEE J. Solid-State Circuits* **35** 1487–97

[5] Wang G, Wang P, Tang Y and Liu W 2012 Analysis of dual band power and data telemetry for biomedical implants *IEEE Trans. Biomed. Circuits Syst.* **6** 208–15

[6] Wang G, Liu W, Sivaprakasam M, Zhou M, Weiland J D and Humayun M S 2006 A dual band wireless power and data telemetry for retinal prosthesis *Proc. of IEEE 26th Engineering in Medicine and Biology Society Conf.* pp 4392–5

[7] Theogarajan L S and Low-Power A 2008 Fully implantable 15-channel retinal stimulator chip *IEEE J. Solid-State Circuits* **43** 2322–37

[8] Ghovanloo M and Atluri S 2007 A wide-band power-efficient inductive wireless link for implantable microelectronic devices using multiple carriers *IEEE Trans. Circuits Syst.* I **54** 2211–21

[9] Jow U and Ghovanloo M 2010 Optimization of data coils in a multiband wireless link for neuroprosthetic implantable devices *IEEE Trans. Biomed. Circuits Syst.* **4** 301–10

[10] Kim J, Wei G, Kim M, Ryo H, Ri P and Zhu C 2018 A splitting frequencies-based wireless power and information simultaneous transfer method *IEEE Trans. Circuits Syst.* I **65** 4434–45

[11] Barakat A, Yoshitomi K and Pokharel R K 2019 Design and implementation of dual-mode inductors for dual-band wireless power transfer systems *IEEE Trans. Circuits Syst. Express Briefs* **66** 1287–91

[12] Sharaf R, Abdel-Rahman A B, Abd El-Hameed A S, Barakat A, Hekal S and Allam A 2019 A new compact dual-band wireless power transfer system using interlaced resonators *IEEE Microwave Wirel. Compon. Lett.* **29** 498–500

[13] Kung M-L and Lin K-H 2018 Dual-band coil module with repeaters for diverse wireless power transfer applications *IEEE Trans. Microwave Theory Tech.* **66** 332–45

[14] Tahar F, Barakat A, Saad R, Yoshitomi K and Pokharel R K 2017 Dual-band defected ground structures wireless power transfer system with independent external and inter-resonator coupling *IEEE Trans. Circuits Syst. Express Briefs* **64** 1372–6

[15] Kung M-L and Lin K-H 2015 Enhanced analysis and design method of dual-band coil module for near-field wireless power transfer systems *IEEE Trans. Microwave Theory Tech.* **63** 821–32

[16] Barakat A, Pokharel R K, Alshhawy S, Yoshitomi K and Kawasaki S 2020 High isolation simultaneous wireless power and information transfer system using coexisting DGS resonators and figure-8 inductors *Digest of IEEE Int. Microwave Symp.* pp 1172–5

[17] Barakat A, Alshhawy S, Yoshitomi K and Pokharel R K 2021 Simultaneous wireless power and information transfer using coupled co-existing defected ground structure resonators *IEEE Trans. Circuits Syst. Express Briefs* **68** 632–6

[18] FCC 2009 *Code of federal regulations* FCC 15 47

[19] Mohan S S, Hershenson M M, Boyd S P and Lee T H 1999 Simple Accurate Expressions for planar spiral inductors *IEEE J. Solid-State Circuits* **34** 1419–24

[20] Raju S, Wu R, Chan M and Yue C P 2014 Modeling of mutual coupling between planar inductors in wireless power applications *IEEE Trans. Power Electron.* **29** 481–90

[21] Kuhn W B and Ibrahim N M 2001 Analysis of current crowding effects in multiturn spiral inductors *IEEE Trans. Microwave Theory Tech.* **49** 31–8

[22] Hui D, Yisheng Z and Baishan Z 2005 Research on the electromagnetic radiation of a PCB planar inductor *Proc. of Asia-Pacific Microwave Conf.* pp 1–3

[23] Ohira T 2017 The kQ product as viewed by an analog circuit engineer *IEEE Circuits Syst. Mag.* **17** 27–32

[24] Imura T and Hori Y 2011 Maximizing air gap and efficiency of magnetic resonant coupling for wireless power transfer using equivalent circuit and neumann formula *IEEE Trans. Ind. Electron.* **58** 4746–52

[25] Barakat A, Yoshitomi K and Pokharel R K 2018 Design approach for efficient wireless power transfer systems during lateral misalignment *IEEE Trans. Microwave Theory Tech.* **66** 4170–7
[26] Alshhawy S, Barakat A, Yoshitomi K and Pokharel R K 2019 Separation misalignment insensitive WPT system using two-plane printed inductors *IEEE Microwave Wirel. Compon. Lett.* **29** 683–6
[27] Lee J, Lim Y S, Yang W J and Lim S O 2014 Wireless power transfer system adaptive to change in coil separation *IEEE Trans. Antennas Propag.* **62** 889–97

IOP Publishing

Systems for Printed Flexible Sensors
Design and implementation
Tarikul Islam, Subhas Chandra Mukhopadhyay and Boby George

Chapter 9

Analysis and design considerations of relaxation-oscillator-based flexible sensor circuits

Shahid Malik and Laxmeesha Somappa

9.1 Introduction

In this chapter, relaxation-oscillator-based sensor interface circuits for resistive/capacitive and leaky resistive/capacitive sensors will be discussed. A brief introduction and working of a conventional relaxation-oscillator-based interface circuit is presented, followed by appropriate pointers to advanced topologies which are referenced for prospective readers. Finally, a detailed design guide along with supported modeling is provided so that one can devise relaxation oscillator designs to required design criteria. The design methodology can be used as a reference for the optimal choice of oscillator components for efficient design.

Sensors represent the first element in the development of measurement and instrumentation systems. Sensors are used in virtually all aspects of our day-to-day life, including monitoring, security, surveillance and awareness in general. In addition, sensors are also widely used in industrial and scientific applications. Many significant developments are made each day in the field of sensors with new fabrication methods such as micro-electromechanical systems and microfluidic systems [1]. Capacitive and resistive sensors are widely used in industrial, scientific and biomedical applications for measurement of displacement, force, pressure, temperature, humidity, liquid level, solid contaminants in gas pipelines and biomedical implants [2–4].

Signal-conditioning circuits are used to convert the sensors' information into a suitable form for further processing [5, 6]. Depending on the output of the sensor, the signal-conditioning circuits may involve one or more operations such as amplification, filtering, linearization, non-ideality compensation, and so on. Signal-conditioning circuits for resistive and capacitive sensors are generally based on bridge-based circuits, relaxation-oscillator-based resistance/capacitance-to-frequency converters and direct digital converters [7–9].

Most signal-conditioning circuits for resistive and capacitive (*R*-*C*) sensors are designed based on the assumption that the sensor can be represented in an electrical

doi:10.1088/978-0-7503-3935-3ch9

equivalent form by either a capacitor in the case of the capacitive sensor or a resistor in the case of the resistive sensors [10]. However, the assumption is not valid for many sensing applications such as water level sensors, humidity sensors, fingerprint sensing and soil moisture sensors. The practical electrical equivalent model of sensors consists of a leakage shunt resistor and/or parasitic capacitor with the sensor element [11]. Therefore, the signal-conditioning circuit needs to be designed such that the information about the measurand is not affected by the shunt resistor and/or parasitic capacitor. In addition, in the case of impedance sensors where both R and C carry the measurand information, both parameters' simultaneous measurement is needed. Moreover, almost all the sensors are affected by the parasitic capacitors. Therefore, signal-conditioning circuits should also provide a provision to compensate for such parasitic capacitors' effects on the sensor parameters' measurement.

Flexible capacitive and resistive sensors are widely used in the design and development of wearable, portable and low-power sensing applications such as blood pressure monitoring, respiration monitoring and electronic skins for flexible and hybrid electronics [12, 13]. The signal-conditioning circuits for such sensing applications should be miniaturized, low power and highly sensitive. Conventional voltage-mode-based circuits are not preferable because of their complex architecture. Relaxation-oscillator-based circuits are preferred for such applications because of their simple architecture, low power consumption, high sensitivity and quasi-digital output [14, 15].

9.2 Relaxation-oscillator-based circuits

Oscillator-based signal-conditioning circuits are preferred as sensor interface circuits because of their inherent advantages in terms of high noise immunity, direct-to-digital conversion using a pulse counter, high resolution and high sensitivity [3, 16, 17]. Different oscillator-based signal-conditioning circuits have been developed based on the type of sensor in use. Relaxation-oscillator-based interface circuits have been developed for various sensors configuration. Conventional relaxation oscillator circuits, based on an integrator and scheme trigger, are utilized for lossless resistive and capacitive sensors. Conventional relaxation oscillator circuits have been extended for bridge-based resistive sensors, resistive sensors with parasitic capacitance, leaky capacitive sensors and differential capacitive sensors.

Relaxation-oscillator-based circuits for bridge-based resistive sensors are reported in [18, 19]. The circuits provide a linear change in the frequency of the relaxation oscillator with respect to the change in the sensor resistance. The circuit in [18] is suitable for strain-gauge-based full-bridge sensors. This is extended using active-bridge circuits, as reported in [19].

A relaxation-oscillator-based circuit for a leaky capacitive sensor is reported in [11]. The circuit provides a wide range of compensation for the shunt conductance, but the measurement range for incremental change in the sensor capacitance is very small. A relaxation oscillator interface for capacitive sensor capable of compensating the shunt conductance is reported in [20]. Though the circuit in [20] can measure a very low value of the sensor capacitance (0–1.2 pF), this solution is not suitable for a shunt conductance higher than 2.3 MΩ. Any higher value of shunt conductance significantly affects the response of capacitive sensors and needs effective compensation.

9.2.1 Conventional relaxation-oscillator-based interface

Relaxation-oscillator-based signal-conditioning circuits are used to convert the sensor resistance into a change in the time/frequency of the output signal. The most common topology of an relaxation oscillator circuit for sensing applications is shown in figure 9.1. A conventional relaxation oscillator circuit consists of an integrator and a Schmitt trigger. The integrator is implemented using an operational amplifier (op-amp), OA_1, a resistor, R_x, and a capacitor, C_i. The resistance R_x represents the sensor for the resistive interface. Circuits based on this topology are presented in [21–26] (figure 9.2).

9.2.2 Analysis for time period T

The R_x and C_x in figure 9.1 can work as a sensor. For example, in the case of a capacitive sensor, the sensor will be connected as C_x and R_x will be a reference element. On the contrary, R_x will act as the sensor element in the case of a resistive sensor and C_x as a reference element. The timing waveform is shown in figure 9.2. The expression for the voltage, $V_R(t)$, can be derived as follows:

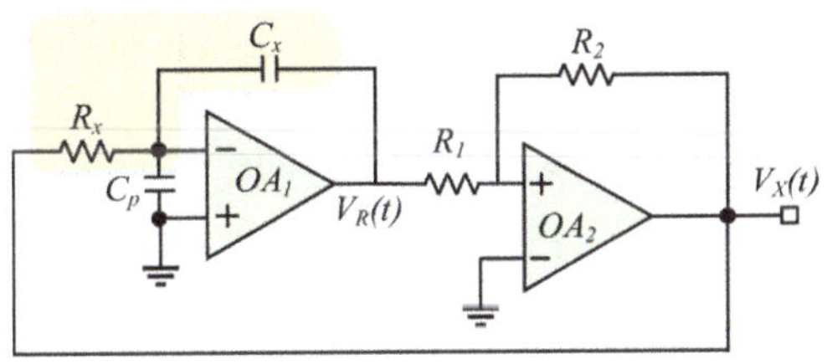

Figure 9.1. Schematic diagram of the conventional relaxation oscillator interface for resistive and capacitive sensors.

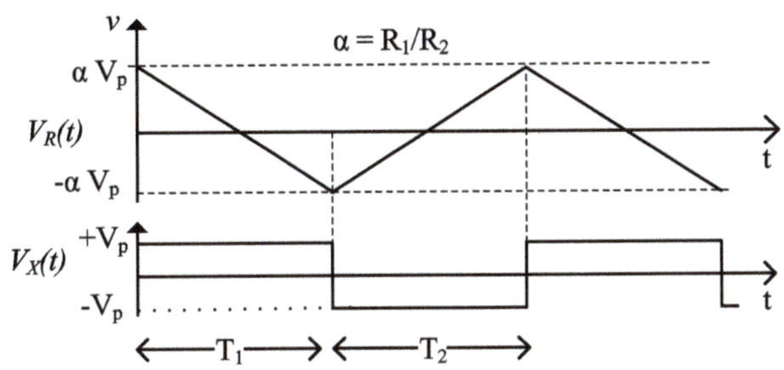

Figure 9.2. Waveforms at key test points in the relaxation-oscillator-based interface for resistive and capacitive sensors.

$$V_R(t) = -\frac{1}{R_x C_x} \int_0^t V_p \, dt. \tag{9.1}$$

The voltage $V_R(t)$ is compared with the threshold voltage of the Schmitt trigger. The threshold voltage of the Schmitt trigger can be given as $\pm V_p \times (R_2/R_1)$. At $t = T/2$, the voltage $V_R(t)$ can be written as follows:

$$V_R(t) = -\frac{T \; V_p}{2 \; R_x \; C_x} = -2 \; V_p \frac{R_2}{R_1}. \tag{9.2}$$

Therefore, the expression of the time period T of the square-wave output $V_X(t)$ can be derived as follows:

$$T = 4 \; R_x \; C_x \frac{R_2}{R_1}. \tag{9.3}$$

9.2.3 Conventional relaxation-oscillator-based interface for bridge sensors

A relaxation-oscillator-based circuit for a full Wheatstone bridge strain sensor is reported in [18, 19, 27–29]. The circuits provide a linear variation in the frequency of the oscillator with respect to variation in the ΔR of the sensor.

The relaxation oscillator circuit for bridge-based resistive sensors, reported in [18, 19], is shown in figure 9.3. The output of the bridge circuit is connected at the input of the oscillator. When the bridge is at a balanced condition, no voltage is generated at the INA output and the current through resistor R_n is zero. The relaxation oscillator oscillates at a frequency, f, when the bridge is balanced. Any variation in ΔR linearly varies the frequency of the oscillator. The expression of the frequency f of the oscillator can be written as follows:

$$f = f_o\left(1 + \frac{\Delta R}{R_o} \frac{G \; R_i}{R_n}\right), \tag{9.4}$$

where f_o is the free-running frequency of the oscillator when ΔR is zero (i.e. the bridge is balanced). Any variation in ΔR varies the frequency of the oscillator, which can be measured to monitor the incremental change in the sensor resistance.

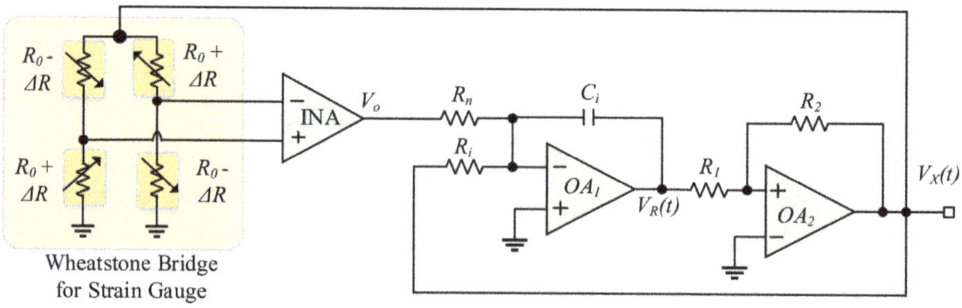

Figure 9.3. Schematic diagram of a relaxation oscillator circuit with active bridge topology to linearize the output voltage of the conventional Wheatstone bridge (reported in [18]).

The circuit shown in figure 9.3 is not suitable for single-element resistive sensors in a bridge configuration. An active-bridge-based resistance measurement circuit for single-element resistive sensors is shown in figure 9.4. The circuit utilizes an op-amp in an inverting configuration to form the active bridge. A detailed expression and analysis of the active-bridge circuit is reported in [30].

9.2.4 Relaxation oscillator circuits for leaky capacitive sensors

The relaxation oscillator circuit shown in figure 9.3 are not suitable for sensors with parasitic capacitance. The basic architecture of a relaxation oscillator circuit with parasitic capacitance, C_x, is shown in figure 9.5.

Capacitive sensors are considered to be ideal and represented using a single capacitor/resistor in many sensing applications. However, this is not always the case. Some sensors, especially ones with planar electrode geometry, are leaky, which allows the flow of continuous static current through the sensor. A timing diagram of a circuit with R_x and C_x is shown in figure 9.6. The expression for the voltage, $V_R(t)$, for the leaky capacitive sensor can be written as follows:

$$V_R(t) = -\frac{V_p}{R_i\ C_x}\ e^{-\frac{t}{R_x\ C_x}}, \tag{9.5}$$

At $t = T/2$;

$$V_R(t) = -\frac{V_p}{R_i\ C_x}\ e^{-\frac{T}{2;R_x\ C_x}} = -2\ V_p\ \frac{R_2}{R_1}. \tag{9.6}$$

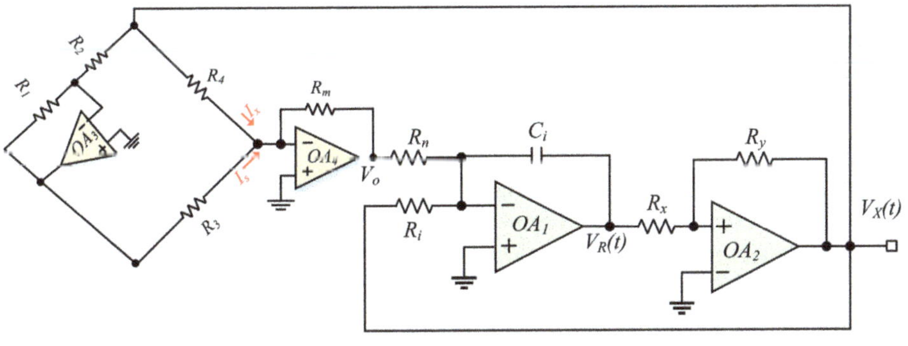

Figure 9.4. Schematic diagram of a relaxation oscillator circuit with active bridge topology to linearize the output voltage of the conventional Wheatstone bridge.

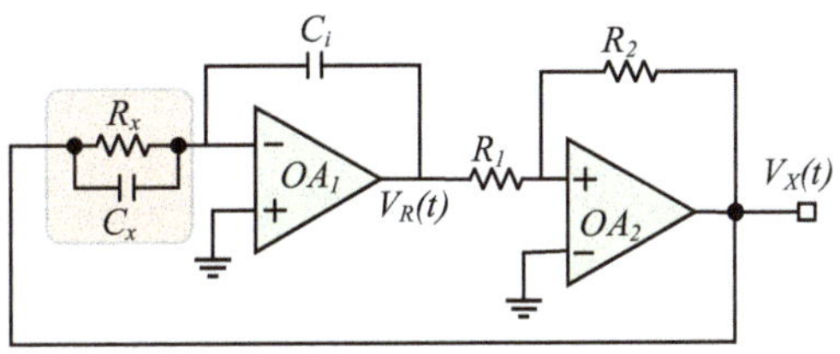

Figure 9.5. Schematic diagram of a relaxation oscillator circuit with parasitic capacitance, C_x, and sensor resistance, R_x, measurement.

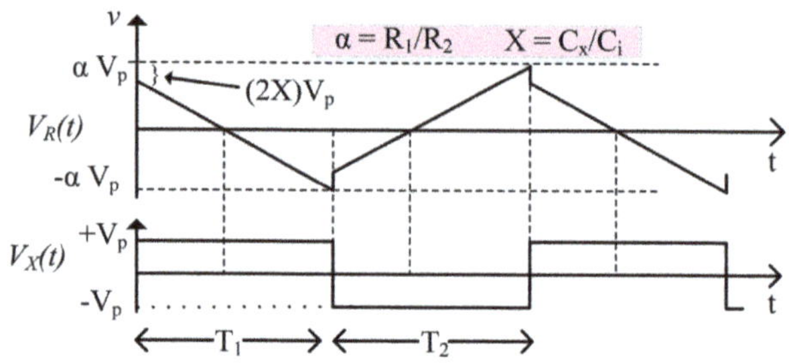

Figure 9.6. Timing diagram of a relaxation oscillator circuit with R_x and C_x.

Therefore, the expression of the time period T of the square-wave output $V_X(t)$ can be derived as follows:

$$T = 2\ R_x\ C_x\ \ln\left[2\ R_i\ C_x\left(\frac{R_2}{R_1}\right)\right]. \tag{9.7}$$

The relaxation oscillator provides a square-wave output signal, $V_X(t)$. These circuits provide linear variations in the period/frequency with respect to the variation in sensor resistance R_x. The expression of the period (T) of the square-wave signal ($V_X(t)$) can be written as follows:

$$T = 4\ R_x\ C_i\ \frac{R_1}{R_2}. \tag{9.8}$$

The period of the square-wave signal with parasitic capacitance C_x can be written as follows:

$$T = 4\ R_x\ C_i\left(\frac{R_1}{R_2} - \frac{C_x}{C_i}\right). \tag{9.9}$$

The parasitic capacitance is not constant and affects the period T of the oscillator signal. This affects the measurement of sensor resistance R_x. A relaxation oscillator circuit for resistive sensors with parasitic capacitance, as shown in figure 9.7, is presented in [31]. The circuit presented in [31] utilizes an inverting amplifier, a comparator and an XOR gate to separate the period into a pulse-width-modulated output. The charge transfer between the capacitors C_x and C_i results in a share transition at voltage $V_R(t)$, as shown in the waveform of figure 9.6. A zero-crossing detector (ZCD), implemented by OA_2 in figure 9.7, is used convert the amount of charge transferred between C_x and C_i into an equivalent time period. The values of the sensor's resistance and capacitance are calculated using the duty cycle of the XOR output [31].

To measure both parameters R_x and C_x, a ZCD (implemented using OA_3) and an XOR gate is added to separate the sensor components (R_x and C_x) into an equivalent pulse width. The ZCD separates the ramp signal into two parts: the first part is proportional to the charge transfer, and the second part is proportional to the sensor resistance, as represented by T_{p1} and T_{p2} in figure 9.8. The expressions for T_{p1} and T_{p3} can be derived as follows:

$$T_{p1} = R_x\ C_i\left(\frac{R_2}{R_1} - \frac{2\ C_x}{C_i}\right), \tag{9.10}$$

$$T_{p2} = R_x\ C_i\ \frac{R_2}{R_1}. \tag{9.11}$$

The expression of unknown impedance (R_x, C_x) can be estimated from T_{p1} and T_{p2} as follows:

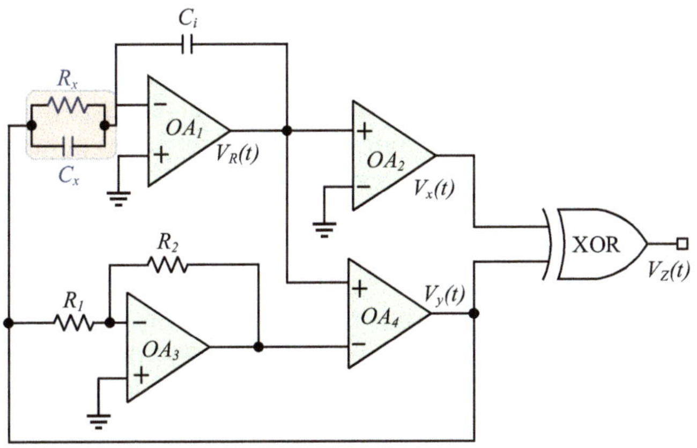

Figure 9.7. Schematic diagram of a relaxation oscillator circuit for resistive sensors with parasitic capacitance (reported in [31]).

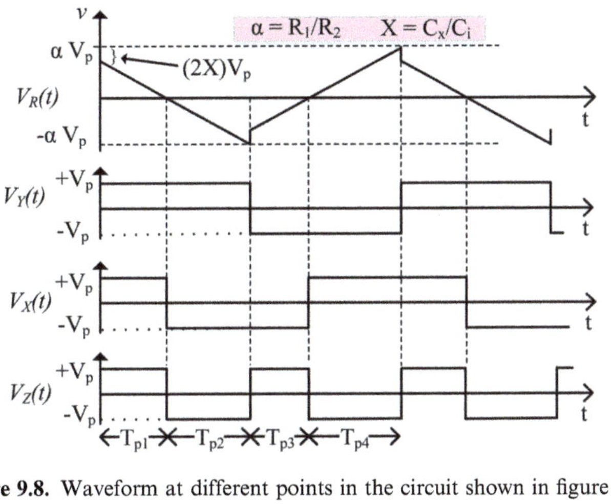

Figure 9.8. Waveform at different points in the circuit shown in figure 9.7.

$$C_x = \frac{R_2}{R_1}\frac{C_i}{2}\left(1 - \frac{T_{p1}}{T_{p2}}\right), \tag{9.12}$$

$$R_x = \frac{R_1}{C_i\ R_2}\ T_{p2}. \tag{9.13}$$

9.2.5 Relaxation oscillator circuit for differential capacitive sensors

Differential capacitive sensors consist of two capacitors connected in a differential configuration. They are widely used for detection and measurement of pressure difference, rotational angle and linear displacement. The principle of detection in differential capacitive sensors is based on a complementary change in the capacitance between two sensors based on the measurand. Accurate and linear measurement of the capacitance can be obtained by dividing the capacitance difference between two capacitors and their sum. Differential capacitive sensors provide inherent advantages of temperature and other environmental parameter compensations. Several interface circuits, including capacitance-to-voltage converters, capacitance-to-frequency converters and capacitance-to-digital converters for differential capacitive sensors have been reported.

The electrical equivalent of a differential capacitive sensor is represented by the pink block in figure 9.9 [32]. Two capacitors, C_1 and C_2, are connected together with a common electrode. In a linear displacement encoder, the area of the electrodes changes linearly with the displacement x. The capacitance of C_1 and C_2 can be expressed as follows:

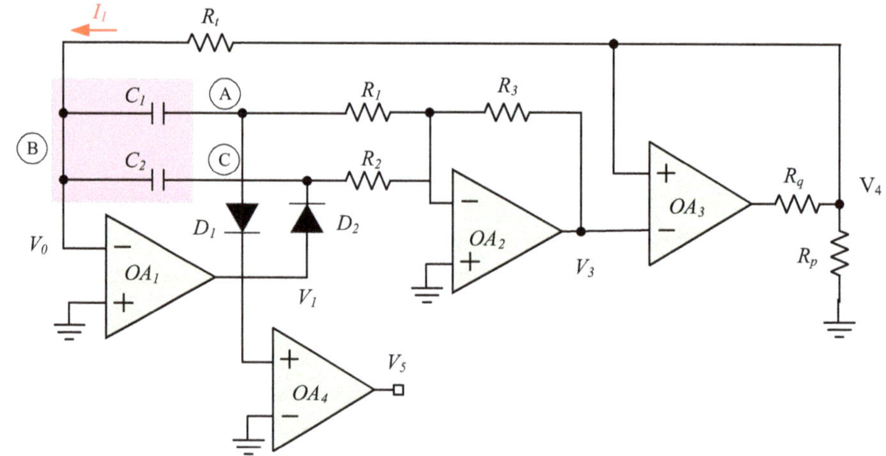

Figure 9.9. Relaxation-oscillator-based circuit for differential capacitive sensors.

$$C_{1,2} = \frac{C_o}{2}(1 \pm x), \tag{9.14}$$

where C_o is the total capacitance of the sensor.

On the other hand, in differential capacitive sensors, the spacing between the electrodes of the capacitive sensor changes linearly with respect to the pressure difference x. Therefore the capacitance of C_1 and C_2 can be expressed as follows:

$$C_{2,1} = \frac{C_o}{2}\frac{1}{1 \pm x}. \tag{9.15}$$

The measurand, x, of the sensor can be obtained independently whether the capacitance change is linear or hyperbolic and can be written as follows:

$$x = \frac{C_1 - C_2}{C_1 + C_2} = \frac{2C_1}{C_1 + C_2} - 1. \tag{9.16}$$

A circuit diagram of the relaxation-oscillator-based interface is shown in figure 9.9. The circuit performs a ratiometric operation to measure the change in the capacitance. The interface consists of an integrator followed by a comparator as the core of the relaxation oscillator. A differential capacitive sensor (C_1 and C_2) is multiplexed by diodes to form the integrator. The multiplexing is made automatic depending on the outputs and on the integrator and the comparator, respectively.

For the moment, diodes are assumed to be ideal switches. Assume the peak-to-peak value of voltage V_4 is V_u. The initial voltage across capacitor C_2 is V_d. Based on the voltage V_4 and the position of diodes D_1 and D_2, the total time period can be divided into four parts. During period T_1 the voltage V_4 is high and diode D_2 is forward biased. The period T_1 lasts until D_2 turns off and can be written as follows:

$$T_1 = C_2 R_t V_d / V_u. \tag{9.17}$$

Diode D_2 is now turned off and D_1 is turned on subsequently in period T_2 The period T_2 lasts until the voltage V_1 reaches $-V_u$ and can be written as follows:

$$T_2 = C_1 R_t. \tag{9.18}$$

During period T_3, the voltage V_1 increases from $-V_u$ to zero. The period T_3 can be written as follows:

$$T_3 = C_2 R_t V_u / V_d. \tag{9.19}$$

Similarly, period T_4 can be written as follows [27]:

$$T_4 = C_2 R_t. \tag{9.20}$$

The high and low periods at the output of V_4 can be written as follows:

$$T_H = T_1 + T_4 = C_2 R_t (V_d + V_u) / V_u, \tag{9.21}$$

$$T_L = T_2 + T_3 = C_2 R_t (V_d + V_u) / V_d. \tag{9.22}$$

The duty cycle can be obtained from the high and low periods of the oscillator as follows:

$$D = \frac{T_H}{T_L + T_H} = \frac{C_2}{C_1 + C_2}(1 + \epsilon_v), \tag{9.23}$$

where

$$\epsilon_v = \frac{C_1\left(1 - \frac{V_u}{V_d}\right)}{C_1\frac{V_u}{V_d} + C_2}. \tag{9.24}$$

If the factor ϵ_v is zero, then the duty ratio corresponds exactly to the capacitance ratio. This can be possible when $V_u = V_d$. This way, the relaxation-oscillator-based circuit is used as an interface for differential capacitive sensors.

9.3 Performance analysis with component non-idealities

For a complete performance analysis of a conventional relaxation oscillator, we introduce some non-idealities of the op-amp used in the integrator and the comparator. Such non-idealities of the op-amp are finite gain-bandwidth (GBW), the slew rate (SR) and the offset voltage. Consider C_p as the parasitic capacitance at the op-amp inverting terminal of the integrator. The op-amp transfer function, $A(s)$, is described in terms of its finite GBW as follows:

$$A(s) = \frac{A_0\omega_0}{s + \omega_0} \approx \frac{A_0\omega_0}{s}, \tag{9.25}$$

where A_0 is the DC gain, ω_0 is the dominant pole frequency and $A_0\Omega_0$ is the GBW product of the op-amp. For the analysis, the approximate expression for $A(s)$, given by equation (9.25), is used since the integrator operation at frequencies much higher than ω_0 is analyzed. Referring to figure 9.1, the integrator output voltage is derived as follows:

$$C_x s\left(V_R(s) + \frac{V_R(s)}{A(s)}\right) = -\frac{V_0(s)}{A(s)}C_p s\left(-\frac{V_R(s)}{A(s)} - V_x\right)G_x, \tag{9.26}$$

$$V_R(s)\left[s\left(C_x + \frac{G_x}{A_0\omega_0}\right) + \frac{s^2}{A_0\omega_0}(C_p + C_i)\right] = -V_x G_x, \tag{9.27}$$

$$V_R(s) = \frac{-V_x A_0\omega_0}{s^2 R_x(C_x + C_p)}\left(\frac{s}{s + \left(\frac{A_o\omega_0 R_x C_x + 1}{R_x(C_x + C_p)}\right)}\right),$$

when driven by a unit step voltage, V_p, $V_R(s)$ can be simplified as

$$V_R(s) = \frac{-V_p A_0 \omega_0}{s^2 R_x(C_x + C_p)} \left(\frac{1}{s + \left(\frac{A_o \omega_0 R_x C_x + 1}{R_x(C_x + C_p)} \right)} \right). \tag{9.28}$$

The time domain voltage, $V_R(t)$ $V_R(t)$, is obtained by transforming equation (9.28) into the time domain and taking the initial value of $V_R(t)$ as αV_P, where $\alpha = R_2/R_1$:

$$s V_R(s) = \frac{-V_p A_0 \omega_0}{s R_x(C_x + C_p)} \left(\frac{1}{s + \left(\frac{A_o \omega_0 R_x C_x + 1}{R_x(C_x + C_p)} \right)} \right)$$

$$s V_R(s) = \frac{-V_p A_0 \omega_0}{R_x(C_x + C_p)} \left[\frac{1}{s} - \frac{1}{s + \left(\frac{A_o \omega_0 R_x C_x + 1}{R_x(C_x + C_p)} \right)} \right]$$

$$\frac{d}{dt}(V_R(t)) = \frac{-V_p A_0 \omega_0}{R_x(C_x + C_p)} \left[1 - e^{-\left(\frac{A_o \omega_0 R_x C_x + 1}{R_x(C_x + C_p)} \right) t} \right].$$

Integrating both sides and applying initial conditions:

$$V_R(t) + K = \frac{-V_p A_0 \omega_0}{R_x(C_x + C_p)} \left[t + \frac{e^{-\left(\frac{A_o \omega_0 R_x C_x + 1}{R_x(C_x + C_p)} \right) t}}{\frac{A_o \omega_0 R_x C_x + 1}{R_x(C_x + C_p)}} \right].$$

Using the initial conditions at $t = 0$, the constant K is obtained, which is substituted to obtain the final $V_R(t)$:

$$K = \frac{-V_p A_0 \omega_0}{R_x(C_x + C_p)} - \alpha V_p,$$

$$V_R(t) = \alpha V_p - \frac{V_p A_0 \omega_0}{R_x C_x} \left[\frac{1}{(1 + \frac{C_p}{C_x})} \left(\frac{t}{\left(\frac{A_o \omega_0 R_x C_x + 1}{R_x(C_x + C_p)} \right)} - \left(1 - \frac{e^{-\left(\frac{A_o \omega_0 R_x C_x + 1}{R_x(C_x + C_p)} \right) t}}{\left(\frac{A_o \omega_0 R_x C_x + 1}{R_x(C_x + C_p)} \right)^2} \right) \right) \right],$$

$$V_R(t) = \alpha V_p - \frac{V_p}{R_x C_x}\left\{\frac{t}{1 + \frac{1}{A_o \omega_0 R_x C_x}} - \frac{1}{\left(1 + \frac{1}{A_o \omega_0 R_x C_x}\right)^2}\left(\frac{1 + C_p/C_x}{A_o \omega_0}\right)\left(1 - e^{-\left(\frac{A_o \omega_0 R_x C_x + 1}{R_x (C_x + C_p)}\right)t}\right)\right\}. \tag{9.29}$$

As is evident from equation (9.29), the integrator output voltage contains additional terms due to the op-amp and comparator non-idealities. Referring to equation (9.29), the first term indicates that the finite GBW product reduces the slope of the linear ramp voltage at the integrator output by a factor $1 + 1/(A_o \omega_0 R_x C_x)$. The second term offers insight into the offset voltage introduced at the integrator output due to the finite GBW product.

From a large signal point of view, the slope of the integrator output voltage is limited by the SR of the integrator op-amp. Considering a finite SR, and based on equation (9.29), the slope (SL) of the triangular waveform at the integrator output can be represented as

$$\mathrm{SL} = \min\left\{\mathrm{SR}, \frac{V_p \pm V_{\mathrm{os}}}{R_x C_x\left(1 + \frac{1}{A_o \omega_0 R_x C_x}\right)}\right\}, \tag{9.30}$$

where V_{os} is the offset voltage of the op-amp. Denoting T_{H} and T_{L} as the oscillator high time and low time at the output $V_x(t)$, using the slope we can represent the oscillator time as

$$T_{\mathrm{H}} = \frac{V_p(2\alpha + \gamma)}{\mathrm{SL}^-} + \tau_{\mathrm{c}},\ T_{\mathrm{L}} = \frac{V_p(2\alpha + \gamma)}{\mathrm{SL}^+} + \tau_{\mathrm{c}}, \tag{9.31}$$

where τ_{c} is the comparator delay, which is assumed to be the same from low-to-high and high-to-low for simplicity, and γ is the offset arising due to the finite GBW of the op-amp, as per equation (9.29), and is given by

$$\gamma = \left(1 + \frac{C_p}{C_x}\right)\frac{A_o \omega_0 R_x C_x}{(1 + A_o \omega_0 R_x C_x)^2}. \tag{9.32}$$

We now have all the ingredients to arrive at the final oscillation time period, $T = T_{\mathrm{H}} + T_{\mathrm{L}}$:

$$T = max\left\{(2\alpha + \gamma) R_x C_x\left(1 + \frac{1}{A_o \omega_0 R_x C_x}\right)\left(\frac{V_p}{V_p - V_{\mathrm{os}}} + \frac{V_p}{V_p + V_{\mathrm{os}}}\right), \frac{4\alpha V_p}{\mathrm{SR}}\right\} + 2\tau_{\mathrm{c}}, \tag{9.33}$$

$$T = max\left\{2(2\alpha + \gamma)R_xC_x\left(1 + \frac{1}{A_o\omega_0R_xC_x}\right)\left(\frac{1}{1 - (V_{os}/V_p)^2}\right), \frac{4\alpha V_p}{\text{SR}}\right\} + 2\tau_c. \quad (9.34)$$

It is evident from equation (9.34) that the oscillation period is affected by the finite GBW, parasitic input capacitance, SR and offset voltage of the integrator op-amp as well as the finite rise/fall time of the comparator. In the next section, we will discuss the design criteria for the oscillator based on the discussion detailed so far.

9.4 Design criteria

A relaxation-oscillator-based interface can be utilized to interface a resistive (R_x) or capacitive (C_x) sensor. In either cases, the error in measurement will be directly proportional to the error in the oscillation period, T, represented in equation (9.34). If ϵ is the relative error in the oscillation time measurement, assuming that the op-amp is not limited by the SR, the error can be represented as

$$\epsilon = \frac{\Delta T}{T}$$

$$\epsilon = \left(1 + \frac{1}{A_o\omega_0R_xC_x} + \frac{\gamma}{2\alpha}\left(1 + \frac{1}{A_o\omega_0R_xC_x}\right)\right)\left(\frac{1}{1 - (V_{os}/V_p)^2}\right) - 1 + \frac{\tau_c}{2\alpha R_xC_x},$$

provided

$$\text{SR} > \frac{V_p}{R_xC_x}. \quad (9.35)$$

Further, for a given supply voltage, if an op-amp is chosen with an offset-voltage relation:

$$V_{os} \ll V_p, \quad (9.36)$$

the error equation can be re-evaluated and simplified as follows:

$$\epsilon = \frac{1}{A_o\omega_0R_xC_x}\left(1 + \frac{1}{2\alpha}(1 + \frac{C_p}{C_x})\right) + \frac{\tau_c}{2\alpha R_xC_x}. \quad (9.37)$$

Hence, for the target application with an error limit of ϵ, an op-amp must be chosen with the SR condition in equation (9.35), offset-voltage condition in equation (9.36) and GBW product and parasitic capacitance condition:

$$\text{GBW} > \frac{1}{\epsilon R_xC_x}\left(1 + \frac{1}{2\alpha}(1 + \frac{C_p}{C_x})\right). \quad (9.38)$$

Further, the comparator must be chosen with a transient rise/fall time 9.1 condition:

$$\tau_c < 2\epsilon\alpha R_xC_x. \quad (9.39)$$

Table 9.1 lists the GBW, SR, parasitic capacitance, offset-voltage and transient delay time for a few of the commonly used op-amps. As a design example, consider a

Table 9.1. Specifications of the utilized operational amplifiers.

Op-Amp IC	GBW/(2π) (MHz)	SR (Vμs^{-1})	C_p (pF)	V_{os} (mV)	τ (ns)
741	1.5	0.7	—	5	300
AD844	60	2000	2	0.3	300
TL071	5.25	29	2	4	310
OPA177	0.6	0.3	—	0.6	—
LT1049	0.8	0.8	—	0.01	—

relaxation oscillator with a 3.3 V supply voltage, for a capacitive sensor with a measurement range of 10 pF–100 pF. Further, with an α of 0.5 and R_x of 1 MΩ, the oscillation frequency is limited to between 5 kHz and 50 kHz. Based on equation (9.38), equation (9.39) and equation (9.35), for 5% accuracy, the GBW of the op-amp must be more than 2.6 MHz, the SR greater than 0.35 V μs^{-1} and the propagation delay less than 500 ns. Specifically, the accuracy for this oscillator is limited to 5.3% with the 741, 3.05% with the AD844, 3.7% with the TL071, 8.75% with the OPA177, and 7.3% with the LT1049 op-amps.

References

[1] Kishore K, Malik S, Baghini M S and Akbar S A 2020 A dual-differential subtractor-based auto-nulling signal conditioning circuit for wide-range resistive sensors *IEEE Sens. J.* **20** 3047–56

[2] Baxter L 1997 *Capacitive Sensors Design and Application* (New Delhi: Wiley)

[3] Marcellis A D and Ferri G 2011 *Analog Circuit and System for Voltage-mode and Current-mode Sensor Interfacing Application* (New Delhi: Springer)

[4] Malik S, Castellví Q, Becerra-Fajardo L, Tudela-Pi M, García-Moreno A, Baghini M S and Ivorra A 2020 Injectable sensors based on passive rectification of volume-conducted currents *IEEE Trans. Biomed. Circuits Syst.* **14** 867–78

[5] Malik S, Somappa L, Ahmad M and Baghini M S 2020 AN-C2V: an auto-nulling bridge-based signal conditioning circuit for leaky capacitive sensors *IEEE Sens. J.* **20** 6432–40

[6] Malik S, Ahmad M, Somappa L, Islam T and Baghini M S 2020 AN-Z2V: auto-nulling based multi-mode signal conditioning circuit for R-C sensors *IEEE Trans. Instrum. Meas.* **69** 8763–72

[7] Ignjatovic Z and Bocko M F 2005 An interface circuit for measuring capacitance changes based upon capacitance-to-duty cycle (CDC) converter *IEEE Sens. J.* **5** 403–10

[8] Cichocki A and Unbehauen R 1990 A switched-capacitor interface for capacitive sensors based on relaxation oscillators *IEEE Trans. Instrum. Meas.* **39** 797–9

[9] Reverter F, Jordana J, Gasulla M and Pallàs-Areny R 2005 Accuracy and resolution of direct resistive sensor-to-microcontroller interfaces *Sensors Actuators* A **121** 78–87

[10] Malik S, Kishore K, Islam T, Zargar Z and Akbar S 2015 A time domain bridge-based impedance measurement technique for wide-range lossy capacitive sensors *Sensors Actuators* A **234** 248–62

[11] Malik S, Ahmad M, Somappa L, Islam T and Baghini M S 2019 Impedance-to-time converter circuit for leaky capacitive sensors with small offset capacitance *IEEE Sens. Lett.* **3** 1–4

[12] Ahmad M, Malik S, Somappa L, Sonkusale S and Baghini M S 2020 A flexible dry ECG patch for heart rate variability monitoring *2020 IEEE Int. Conf. on Flexible and Printable Sensors and Systems (FLEPS)* pp 1–4

[13] Malik S, Somappa L, Ahmad M, Sonkusale S and Baghini M S 2020 A fringing field based screen-printed flexible capacitive moisture and water level sensor *2020 IEEE Int. Conf. on Flexible and Printable Sensors and Systems (FLEPS)* pp 1–4

[14] Ahmad M, Malik S, Dewan S, Bose A K, Maddipatla D, Narakathu B B, Atashbar M Z and Baghini M S 2020 An auto-calibrated resistive measurement system with low noise instrumentation ASIC *IEEE J. Solid-State Circuits* **55** 3036–50

[15] Somappa L, Malik S, Aeron S, Sonkusale S and Baghini M S 2021 High resolution frequency measurement techniques for relaxation oscillator based capacitive sensors *IEEE Sens. J.* **21** 13394–404

[16] Marcellis A D, Depari A, Ferri G, Flammini A, Marioli D, Stornelli V and Taroni A 2008 A CMOS integrable oscillator-based front end for high-dynamic-range resistive sensors *IEEE Trans. Instrum. Meas.* **57** 1596–604

[17] Kumar S, Kumar L, Islam T and Raina K K 2020 Condition monitoring of transformer breather using a capacitive moisture sensor *IEEE Trans. Ind. Electron.* **67** 9779–89

[18] Ferrari V, Ghidini C, Marioli D and Taroni A 1998 Oscillator-based signal conditioning with improved linearity for resistive sensors *IEEE Trans. Instrum. Meas.* **47** 293–8

[19] Islam T, Kumar L, Uddin Z and Ganguly A 2013 Relaxation oscillator-based active bridge circuit for linearly converting resistance to frequency of resistive sensor *IEEE Sens. J.* **13** 1507–13

[20] Li X and Meijer G C M 2000 Elimination of shunting conductance effects in a low-cost capacitive-sensor interface *IEEE Trans. Instrum. Meas.* **49** 531–4

[21] Elangovan K and Sreekantan Anoop C 2020 Simple and efficient relaxation-oscillator-based digital techniques for resistive sensors—design and performance evaluation *IEEE Trans. Instrum. Meas.* **69** 6070–9

[22] Kim H, Chung W, Kim H and Son S 2009 A resistance deviation-to-pulsewidth converter for resistive sensors *IEEE Trans. Instrum. Meas.* **58** 397–400

[23] Kaliyugavaradan S 2000 A linear resistance-to-time converter with high resolution *IEEE Trans. Instrum. Meas.* **49** 151–3

[24] Sreenath V, Semeerali K and George B 2017 A resistive sensor readout circuit with intrinsic insensitivity to circuit parameters and its evaluation *IEEE Trans. Instrum. Meas.* **66** 1719–27

[25] Tyagi T and Sumathi P 2017 Capacitive and resistive sensing based on compensated relaxation oscillator *2017 6th Int. Conf. on Computer Applications In Electrical Engineering-Recent Advances (CERA)* pp 98–102

[26] Mohan N M, George B and Kumar V J 2008 A sigma-delta resistance to digital converter suitable for differential resistive sensors *2008 IEEE Instrumentation and Measurement Technology Conf.* pp 1159–61

[27] Mochizuki K and Watanabe K 1996 A high-resolution, linear resistance-to-frequency converter *IEEE Trans. Instrum. Meas.* **45** 761–4

[28] Johnson C D and Richeh H A 1986 Highly accurate resistance deviation to frequency converter with programmable sensitivity and resolution *IEEE Trans. Instrum. Meas.* **IM-35** 178–81

[29] Ferrari V, Ghisla A, Vajna Z K, Marioli D and Taroni A 2007 ASIC front-end interface with frequency and duty cycle output for resistive-bridge sensors *Sensors Actuators* A **138** 112–9

[30] Islam T and Zia-Ur-Rahman M 2016 Investigation of the electrical characteristics on measurement frequency of a thin-film ceramic humidity sensor *IEEE Trans. Instrum. Meas.* **65** 694–702

[31] De Marcellis A, Depari A, Ferri G, Flammini A, Marioli D, Stornelli V and Taroni A 2008 A CMOS integrable oscillator-based front end for high-dynamic-range resistive sensors *IEEE Trans. Instrum. Meas.* **57** 1596–604

[32] Ferrari V, Ghidini C, Marioli D and Taroni A 1998 Oscillator-based signal conditioning with improved linearity for resistive sensors *IEEE Trans. Instrum. Meas.* **47** 293–8

IOP Publishing

Systems for Printed Flexible Sensors
Design and implementation
Tarikul Islam, Subhas Chandra Mukhopadhyay and Boby George

Chapter 10

Sensor tags for wireless body-centric communication: challenges and opportunities

Mohammad Hashmi and Dinesh Rano

10.1 Introduction

A network of wearable, implantable and body-worn devices is collectively known as a wireless body area network (WBAN). Several bands, among them medical body area networks (MBANs) and the industrial, scientific and medical band (ISM), operate as WBANs. It provides a flexible platform with which to connect sensor nodes wirelessly for the measurement and recording of the physiological parameters of the body. Recent advances in integrated technology have brought a revolution in the miniaturization of sensor design technologies, and this has resulted in the development of numerous body-worn, ultra-low-power sensors [1–5]. The overall idea behind the use of such devices is to monitor patient healthcare seamlessly and to increase patients' mobility substantially [6]. In this context, these devices could be further categorized as body-worn devices or implanted devices [5–8]. Body temperature sensors, blood pressure sensors, hearing aids and smartwatches are a few examples of body-worn devices; whereas pacemakers, retinal prostheses and capsule endoscopy are examples of implanted devices [7, 8]. A generic depiction of a WBAN can be seen in figure 10.1.

There has been a recent emphasis on wireless health monitoring owing to the aging world population [9]. Its importance can be inferred from the fact that the Federal Communications Commission (FCC) has allocated a separate low-power band for personal communication, known as MBAN, covering a band of 2.36–2.4 GHz [10]. This band is frequently being utilized in applications such as monitoring of vital parameters like temperature, blood pressure, ligament pressure (round ligament), and many more. The need for ultra-low-power devices emanates from the fact that all these sensors and devices rely on batteries for operation. Therefore, to establish a link between an on-body node and an external node with limited power, the performance of the antenna has to be optimum in terms of gain and efficiency.

doi:10.1088/978-0-7503-3935-3ch10

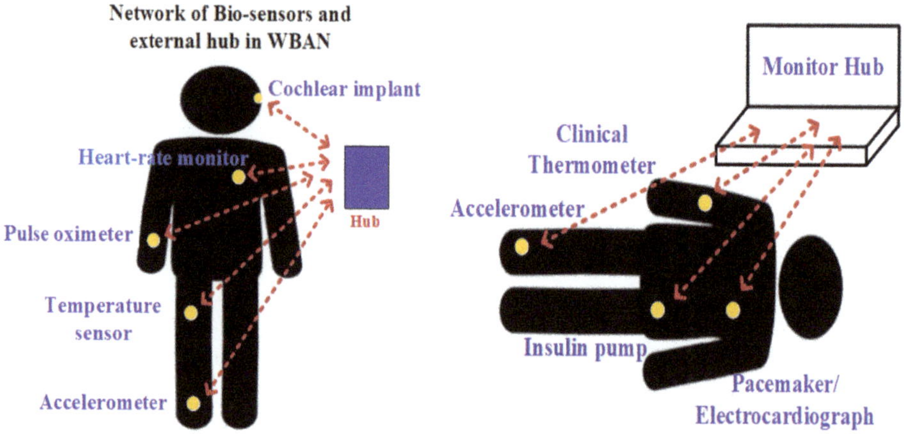

Figure 10.1. Network of sensors, known as a WBAN, and deployment of these sensors on the body for a moving body (left) and a patient resting on a bed (right).

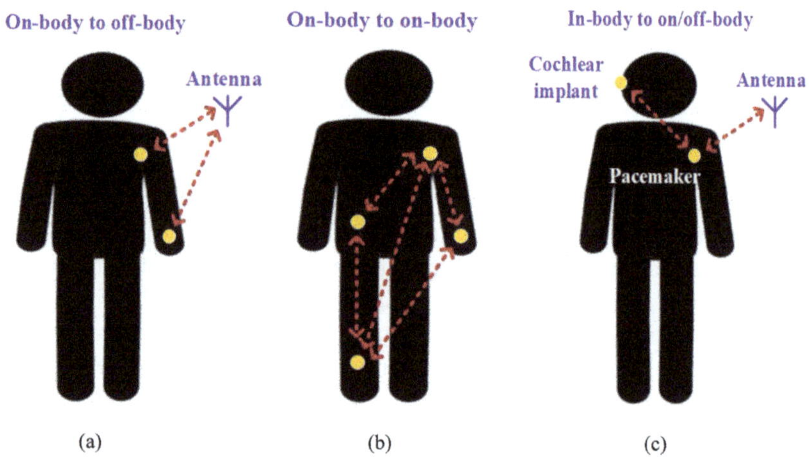

Figure 10.2. Communication scenarios in a typical WBAN: (a) on-body to off-body, (b) on-body to on-body, and (c) in-body to on/off-body.

Furthermore, power emitted in the form of electromagnetic (EM) radiation tends to heat dielectric (body) present nearby and therefore requires these wearable and implanted devices to operate within a safe level. The FCC has restricted equivalent radiated isotropic power norms for MBAN at 1 mW and medical implants at 25 μW [10–12].

As a general categorization, the successful realization of holistic WBAN includes three types of scenarios: (a) on-body to off-body; (b) on-body to on-body; and (c) in-body to on/off-body communication and vice versa.

On-body to off-body communication consists of sensors located on the body communicating with external devices, commonly known as a base station. A typical on-body to off-body communication scenario is shown in figure 10.2(a), where a wireless sensor network (WSN) located on the shoulders and waist communicates with an external device. In this scenario, the antenna should radiate away from the body and, further, its polarization should be vertical, considering that the losses for the horizontally polarized antenna are higher [13]. It, however, leads to protrusion from the surface of the antenna, which is not desirable from the WBAN perspective. Therefore, a low-profile patch antenna with horizontal polarization is a viable option [14, 15]. Typically, the range for this type of communication is from 20–30 m (approximately the size of a standard room).

On-body to on-body communication consists of sensors or antennas communicating with each other located on the body itself. Figure 10.2(b) depicts such a scenario in which a WSN located on the legs, waist and shoulders of an athlete communicates with itself. These sensors, in turn, send physiological data to an external node for further monitoring and processing. Antenna for this scenario should radiate towards the body, and minimum radiation should take place away from the body [16]. Since communication among the sensors happens mostly through the body, which has much higher attenuation than free space, the design of the antenna for this scenario is very critical.

Finally, in-body to on-/off-body communication consists of medical implants communicating with an external device or sensor placed on the body [7, 8]. For example, figure 10.2(c) shows communication between a pacemaker and an external or on-body node. Similarly, an implant in wireless capsule endoscopy when swallowed requires an external device to be placed on the body to receive images of the inside of the body [7, 8]. Such a scenario is an example of in-body to the on-body communication. Since the communication in this scenario happens through lossy tissues, the attenuation of the EM wave is higher than in the on-body to on-body case. Moreover, the implant devices operate on shallow power, are compact and, in some cases, conformal as well. This places a strict requirement on the size and performance of an implanted antenna [17].

In all these cases, the on-body to off-body communications are considered to be critical for realizing the WBAN schemes. The comprehensive study of various aspects of this scenario shows that there is a need to address several aspects to incorporate technological advancements in such a system. These are briefly explained below.

10.1.1 Antenna design

The antenna is an integral part of WBAN devices and plays a vital role in establishing a link between a wearable and base station (i.e. remote location away from the body). This area has been extensively studied in the past few years, yet still issues, such as large antenna size at the MBAN and Wi-Fi bands, significant radiation towards the body, non-conformability and reduction in performance in the presence of the body, remain and require further investigation [13–16]. The patch antennas used in WBAN applications are often limited by its operational bandwidth, size, reduced efficiency, strong coupling with the body and high specific absorption rate (SAR). Electromagnetic band gap (EBG) surfaces in this regard have shown a significant improvement in radiation properties and miniaturization of antennas used in WBAN [18–22]. The EBG is an artificially engineered structure that exhibits in-phase reflection for an incident EM wave (identified through a *reflection phase* diagram) and band-gap property (observed through a *dispersion* diagram) [23–25]. The band-gap characteristics of an EBG's surface suppress the surface wave propagation in that region established by the dispersion curve and transmission property [23–25]. It can lead to the rejection of surface currents propagating on the metal surface and cancel diffraction at the edges of the ground plane. As a consequence, this reduces the back-lobe radiation, improves gain, reduces SAR and improves the front-to-back ratio of an antenna [25]. It implies that a sound isolation can be achieved between an EBG-incorporated antenna and the human body in the absence of surface current throughout the band-gap region [25].

Furthermore, EBG surfaces also exhibit in-phase reflection for an incident electric field rather than the out-of-phase response from a typical perfect electric conductor (PEC) surface [20–25]. This leads to constructive interference of the reflected and direct electric fields from a horizontally polarized monopole antenna, which improves the radiation properties [25]. Nevertheless, EBG surfaces have a disadvantage in that only periodic repetition of the unit cells can contribute to the band-gap and in-phase reflection characteristics [23, 24]. Therefore, antennas with such an arrangement result in increased overall size and become incompatible for use in numerous WBAN applications such as smartwatches.

10.1.2 Conformability of body-worn antenna

An essential feature of wearable devices is conformability for the comfort of users. It necessitates the design of conformal antennas suitable for placement on shoulders, biceps, wrists and legs [13–16]. The fundamental challenge in the design of conformal EBG structures is that bending leads to alteration of the in-phase property and hence changes the performance of the conformed antenna [21, 22]. Therefore, either a unit cell of larger width is required so that even after bending the desired band is covered [22], or appropriate prior analysis is carried out for phase-compensated design [26, 27]. However, increased cell width is often not feasible considering that wearable applications require miniaturization of the antenna sensors.

Determination of the in-phase property in simulation is a complex process as the meshing of the conformal EBG unit cell and EBG array is critical. A discontinuity in the edges might get introduced while conforming a unit cell or array, and this necessitates the number of mesh cells to smoothen the computational area [28]. As a result, the simulation time and space requirement increases. Furthermore, the existing analytical solutions to determine the in-phase property of a conformal EBG unit cell/array are sophisticated and exist only for higher frequencies, e.g. 5.8 and 10 GHz [26, 27]. These solutions are applicable for large-array EBG unit cells and may not be appropriate for WBAN.

10.1.3 Communication through multiple textile layers

Another essential facet of WBAN is reliable communication between the on-body transmitter and receiver located externally, such that essential physiological parameters are not lost. However, the reliability is challenged by antenna covered beneath multiple layers (either in the form of *textiles* or plastic cover) for protection from environmental hazards. Such an arrangement of an antenna covered by single or multiple layers of dielectric (in the form of textiles) shifts the resonant frequency as the effective dielectric constant of the whole structure gets changed [29–31]. A high percentage of shift is not desirable as the patch antennas used in WBAN themselves have *a narrow fractional* bandwidth of 2%–3% [32]. The existing literature reports a more than 3% shift in resonant frequency for a superstrate of height 1 mm with a dielectric constant lower than 2.32 [29–31]. However, antennas for WBAN applications are commonly designed on substrates such as polyamide lossy (Nylon-6), Kapton (Dupont), Latex, polyester sheets and RO4350B, which have dielectric constants of 3.6, 3.5, 3.31, 2.8 and 3.66, respectively. The conformal mapping (CM) approach for the determination of the effective dielectric constant of an antenna covered with multilayer dielectrics can considerably mitigate such issues and therefore can significantly minimize the shift occurring due to covering of antenna by said multiple textile layers.

10.2 Existing challenges

A survey of the literature related to the antenna for WBAN (on-body to off-body) conveys that such antennas suffer from severe limitations in most performance metrics, namely compactness, desired gain and ease of conformability [20–22, 33, 34]. With the compact size of the antenna and low-profile nature, the gain and efficiency drop considerably. The gain of an antenna is directly proportional to the effective aperture and therefore smaller aperture leads to lower gain [32]. Improvement in radiation and compactness can be achieved by incorporating EBG surfaces with standalone antennas. However, the design of such surfaces in the MBAN and Wi-Fi bands is still a challenge. For example, the in-phase reflection frequency of the EBG unit cell is susceptible to the mode of polarization [21, 22]. The polarization sensitivity of the unit cells to obliquely incident tranverse electric (TE) and transverse magnetic (TM) waves also poses a challenge. Moreover, the compactness (at MBAN band) and ease of conformability of the EBG unit cells is another big challenge [13–15, 21, 22]. Current state-of-the-art designs achieve conformability in design. However, they suffer from large footprint, sensitivity to incident TE and TM modes, either normally or obliquely, and ease of conformability as well.

Measurement of SAR on the body is another challenge faced in WBAN. Instead of human testing, which involves the risk of exposing parts of the body to persistent EM radiation, two other ways, i.e. physical phantom and numerical phantom (simulation), have become very popular [21, 22, 35, 36]. A physical phantom is an accurate model of the body. However, development and testing of SAR requires a costly setup, which is not readily available. Furthermore, a numerical phantom is a simulated biological body, which consists of homogeneous tissues of average electrical property [35]. It can also be a more complex and accurate phantom with different tissues of a human body such as Emma, Gustav, Hugo and Laura in a computer simulation tool (CST) [21, 28]. Numerical phantoms are indeed accurate but require considerable simulation time and memory for the generation of results. Most of the existing solutions rely on simulated values of SAR for either 1 g/10 g of averaging mass [20–22]. Moreover, the simulated SAR values are obtained only for a finite duration until a steady state is achieved. However, this might not be the case for a typical wearable device worn on the body for a 24/7 basis.

Another, most significant, challenge faced in this direction is the shift in the resonant frequency of an on-body antenna covered by multiple textile layers. For communication taking place through multiple layers of textiles depending on terrain, it becomes essential to develop accurate formulas to determine the resonant frequency of the antenna and effective dielectric constant so that the communication between the on-body sensors with an externally located receiver does not go out of the band. Existing solutions predict a resonant frequency with a relative percentage error of more than 3% between the theoretical and measured values when the substrate and superstrate have a permittivity smaller than 2.32 [29–31]. For a patch antenna, this amount of shift in resonant frequency may lead to out-of-band communication with the receiver due to its narrow fractional bandwidth [32]. Moreover, the existing methods fail to predict the resonant frequency and effective dielectric constant of an antenna covered by multiple layers of dielectric or textiles as well.

10.3 Some solutions

10.3.1 Electromagnetic-band-gap-incorporated antenna design

A number of reports exist in this regard [20–25]. Here, design, development and analysis of two major types of EBG surfaces are described within the context of WBAN. The first is the via-*less* and compact interdigital-EBG (IDE) surface, ideally suited for smartwatch applications. The second is the compact, conformal and via-less *blocking-type EBG*. Subsequently, analyses of the *polarization stability* of the EBG cells under oblique incident TE and TM waves are included to show the effectiveness of these designs. Furthermore, the *surface wave rejection* ability of all the EBGs is also analyzed with the help of dispersion and transmission diagrams. Performance metrics of these designs are shown to demonstrate the status of the state-of-the-art EBGs. Apparently, there is a significant enhancement in performance gain (in body and air), bandwidth, efficiency and SAR of a monopole antenna incorporated with these EBG surfaces. This clearly shows that EBG-incorporated antennas are attractive for WBAN applications.

10.3.2 Determination of effective dielectric constant

In order to address the situation arising out of antenna sensors being covered with textiles and other materials, determination of effective permittivity ($\varepsilon_{\mathrm{eff}}$) a priori is important. In this context, empirical relations based on CM and equivalent capacitance between the metallic plates, partially covered by dielectric layers, to determine the ε_{eff} value of a microstrip patch antenna (MPA) covered with a superstrate layer become extremely beneficial. Furthermore, determination of the ε_{eff} and f_r of an MPA covered with either single or multiple layers of dielectric is also important. The method presented in this chapter uses modified relations in each iteration to determine the ε_{eff} of the overall geometry. It is unlike all previous approaches wherein ε_{eff} was evaluated by substituting the total thickness of the superstrate in the expression [29–31]. The model can also be seamlessly integrated with any technical computing software for rapid design and performance analysis of body-worn antenna and sensors.

10.3.3 Conformability analysis of conformal electromagnetic band gap surfaces

A planar antenna is bound to have variations in its performance. This can be addressed using a simple geometrical approach wherein the *shift in the in-phase reflection frequency* of conformal EBG surfaces can be determined easily. This technique can be considered generic and is shown to work well on three distinct conformal EBG surfaces operating at relatively low frequencies, i.e. in the ISM and MBAN bands. For this purpose, planar unit cells are drawn on a circle of radius (***r***) along the x- and y-axes to geometrically determine the shift in the in-phase property. The extra path delay due to reflected waves together with the incident wave results in the maximum phase lag an EM wave experiences while interacting with a conformal EBG surface.

10.4 Compact electromagnetic-band-gap-based antenna for smartwatch applications

In this context, several antennas for smartwatch applications have been proposed. For example, a compact and low-profile inverted F antenna has been designed at 2 GHz to study the SAR performance on a flat and wrist phantom filled with homogenous equivalent body tissue [37]. The SAR value on the flat phantom is more than 1 W kg^{-1} and, because the antenna resonates at 2 GHz, the SAR value must increase at higher frequencies such as in Wi-Fi and MBAN. A compact, ultra-wideband diversity antenna, reported in [38], shows promising results in terms of isolation among the antenna elements. However, the design has a disadvantage in that the return loss and gain have been simulated for antennas placed at a height of 15 mm, that too in the air, from the human-equivalent flat phantom of the wrist. This is impractical in general application scenarios. Copper tapes on the rubber belt of a smartwatch used to design an antenna for cellular communication have a large footprint, and the radiating surface is exposed to the environment, which makes it prone to change in performance with external force [39]. Moreover, a radiation efficiency of a mere 23% can be achieved for on-body simulation with such an arrangement. An antenna integrated into the metal frame has a compact size and low profile due to a longer resonant path at 2.4 GHz [40]. Nevertheless, the gain and efficiency of the antenna degrade substantially to −0.89 dBi and 26%, respectively, on the flat wrist phantom. It shows that the detuning effect of the body is prominent in this design and cannot be employed in practice. A monopole antenna for smartwatch has also been reported to study SAR distribution on flat and wrist phantoms [41]. The SAR value at these locations at 2.45 GHz is high enough to damage the adjacent cells in general. It is apparent from the literature survey that certain compact and low-profile characteristics can be achieved by monopole antennas in order to fit them into smartwatches. However, a strong detuning effect in the presence of the body worsens radiation efficiency, gain and SAR values. Therefore, sound isolation is needed to overcome body effects for the proper functioning of the antenna in smartwatches. These parameters can be improved significantly by incorporating EBG surfaces in the ground plane of the antenna [23–25].

10.4.1 Introduction to electromagnetic band gaps

EBG structures are a particular type of material that is non-existent but can be formed by a periodic arrangement of unit cells along with 1D, 2D or 3D axes [23]. The periodicity of the unit cell must be very small as compared to the operating wavelength. These unique materials at resonance conditions show an infinite impedance high impedance surface (HIS). The surface impedance in terms of the electric and magnetic fields is expressed by (10.1). Here, Z_S is the surface impedance of the EBG, E_x is the electric field along the x-axis and H_y is the magnetic field along the y-axis. Moreover, the magnetic field can be expressed in terms of current density J_x (10.2). The current density can also be expressed by Ohm's law, as given by (10.3). Here, σ is the conductivity of the surface. Moreover, the electric field on a surface can be given by (10.4). Here, α is the attenuation constant of the medium.

$$Z_s = \frac{E_x}{H_y}, \tag{10.1}$$

$$\nabla \times H_y = J_x, \tag{10.2}$$

$$J_x(z) = \sigma E_x(z), \tag{10.3}$$

$$J_x(z) = \sigma E_o e^{-\alpha z}. \tag{10.4}$$

For a HIS, Z_s reaches infinity, and this implies that the magnetic field given by (10.1) should be zero. Its substitution in (10.2) results in a surface current of zero value. Hence, the EBG prohibits the flow of surface current along with the metal, and this further stops diffraction at the discontinuity of the patch of an MPA and also on the sides of the printed circuit boards (PCBs). Consequently, it reduces the back lobes, enhances gain and improves efficiency. A typical EBG unit cell, commonly known as a mushroom cell, is shown in figure 10.3. It has dimensions of $W \times W$ along the x- and y-axes. A via of radius r connects the patch at the top to the ground plane. The periodic repetition of the unit cells along the x- and y-axes with a gap of g among the adjacent cells leads to the generation of an EBG surface. In general, EBG surfaces are characterized by an *in-phase reflection frequency* and a *band-gap* property.

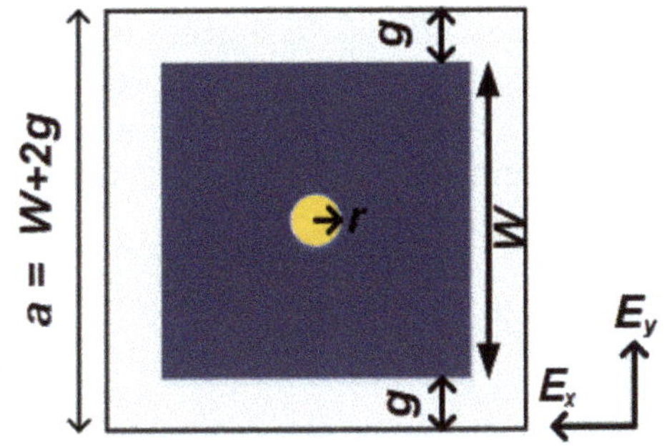

Figure 10.3. Mushroom EBG unit cell (*W*: width of the patch, *g*: gap on either side of the patch, *r*: radius of the via, *a*: period of the unit cell).

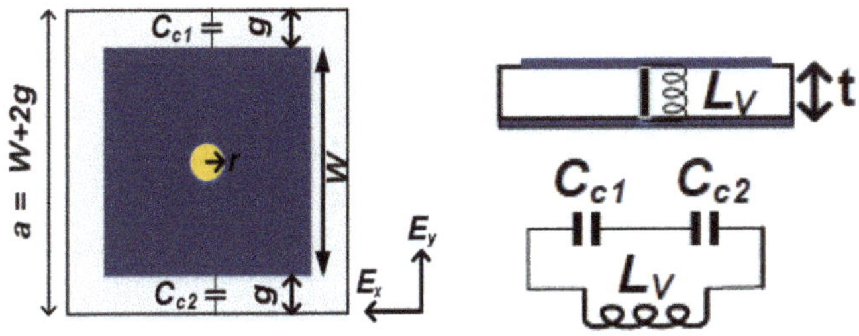

Figure 10.4. Left: Mushroom EBG unit cell with edge coupling capacitances C_{c1} and C_{c2} for E_y polarized wave. Right: Equivalent circuit model (right) (*L*: inductance of the via).

The in-phase reflection is demonstrated by a reflection phase diagram and a lumped circuit model. The band-gap property is assessed with the help of a dispersion diagram of the unit cell and the transmission characteristics of an array of EBG unit cells.

10.4.2 Lumped-element circuit model

For an EBG unit cell, if the operating wavelength is more significant than the periodicity, then the operating mechanism can be explained with the help of an equivalent lumped-element circuit model; $W+g$ is the periodicity of the EBG cell in figure 10.4. The equivalent circuit model of this mushroom cell for a *y*-polarized EM wave is also shown in figure 10.4. The capacitor C_{C1} and C_{C2} represent the capacitance developed between the adjacent patches due to the charge accumulated. Whereas, the inductor L_v is due to the via that connects the top patch with the ground plane. For an *x*-polarized EM wave, the circuit model remains the same. However, the capacitance appears along the *x*-axis.

Now, the net impedance of the parallel combination of two lumped components can be expressed by (10.5). Here, *L* is the net inductance and capacitance, which is L_v, C_{C1} and C_{C2} for the present case. The impedance can therefore be further simplified and expressed by (10.6). Since this is a parallel resonant circuit, the impedance at resonance becomes infinity. Equating the denominator of (10.6) to zero results in the resonant frequency of the circuit given in (10.7). The resonant frequency suggests that, at resonance, the circuit behaves as a stop band and thus does not allow any surface current to flow. Hence, this type of surface is also known as a HIS.

$$Z_S = j\omega L \left|\right| \frac{1}{j\omega C}, \tag{10.5}$$

$$Z_S = \frac{j\omega L}{1 - \omega^2 LC}, \tag{10.6}$$

$$f_r = \frac{1}{2\pi\sqrt{LC}}. \tag{10.7}$$

To elaborate on this concept, a fork-shaped EBG cell, shown in figure 10.5(a) with repetition along the x- and y-axes [42], is considered. For a fair comparison, the original dimensions of the unit cell are kept fixed throughout this chapter. For a y-polarized EM wave incident normally, the unit cell is bisected along the x-axis, as shown in figure 10.5(b). The edge coupling capacitances (ECCs) develop along the y-axis between the adjacent metallic strips, as depicted in figure 10.5(b) and expressed by (10.8). Here, C_c is the coupling capacitance, W is the width of the strip and g is the gap between the adjacent cells, whereas ε_o and ε_r are the respective permittivities of the air and substrate. This results in the formation of three ECCs at the top and the bottom. Due to the different widths of the strip the values of the ECCs vary, as can be seen in figure 10.5(b).

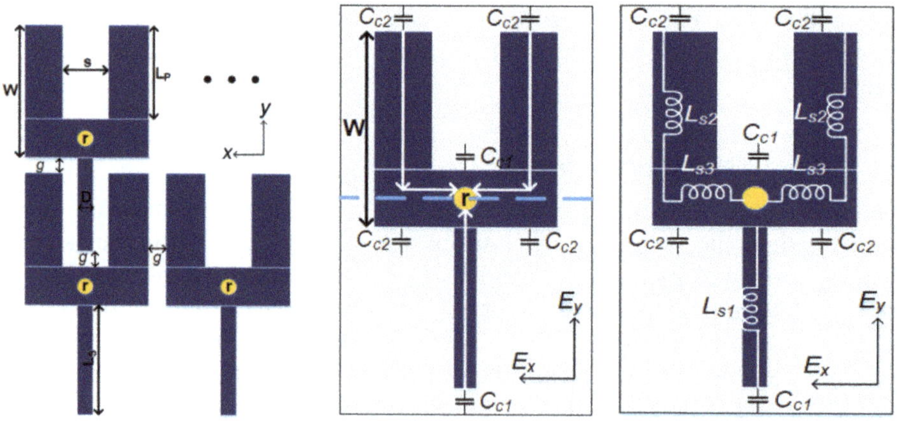

(a) (b) (c)

Figure 10.5. (a) Fork-shaped EBG cell (W: width of the patch, g: gap on either side of the patch, r: radius of the via, s: slot width, L_s and L_p: length of strips at the top and bottom, D: width of the bottom strip). (b) Surface current flow through the cell for an EM wave polarized along the y-axis (C_{c1} and C_{c2}: edge coupling capacitances). (c) Formation of inductances along the path of surface current flow (L_{s1}, L_{s2}, and L_{s3} are strip inductances).

According to Maxwell's equation for an electric field incident on a PEC/metal plate, surface current (AC) gets induced in it. This induced surface current on the patch grounds through the via taking the path as shown in figure 10.5(b). The metal surface through which the surface current flows forms the inductor. The value of the grid/strip inductance, L_g, is given by (10.9) [24], and it assumes that the length of the strip, l_s, is the dimension along the electric field while the width of the strip, w_s, is the dimension orthogonal to the electric field. For example, the dimensions L_s and D in figure 10.5(a) form the length (l_s) and width (w_s) for the bottom strip. Here, k_g is the correction factor in the presence of a full ground plane given by (10.10) [24]. The distribution of all strip inductances and ECCs in the unit cell can then be visualized as in figure 10.5(c).

$$C_c = \frac{17 \times W \times \varepsilon_o \times (\varepsilon_r + 1)}{\pi} \times \cosh^{-1}\left(\frac{W + g}{g}\right), \tag{10.8}$$

$$L_g = 0.25 \times 10^{-7} \times l_s\left(\ln \frac{l_s}{w_s} + 1.193 + 0.25\frac{w_s}{l_s}\right)k_g, \tag{10.9}$$

$$k_g = 0.57 - 0.145 \ln \frac{w_{s1}}{t}. \tag{10.10}$$

On the same note, for an x-polarized EM wave, the unit cell is bisected along the y-axis, as shown in figure 10.6(a). Once again, ECCs develop along the x-axis between the adjacent strips. The flow of induced surface current through the unit cell is similar to the approach mentioned above. However, due to the orthogonal polarization of the EM wave in the present case, the calculation of grid inductance by (10.9) results in swapping of the values of length and width. The distribution of inductances and ECCs on the unit cell are depicted in figure 10.6(b).

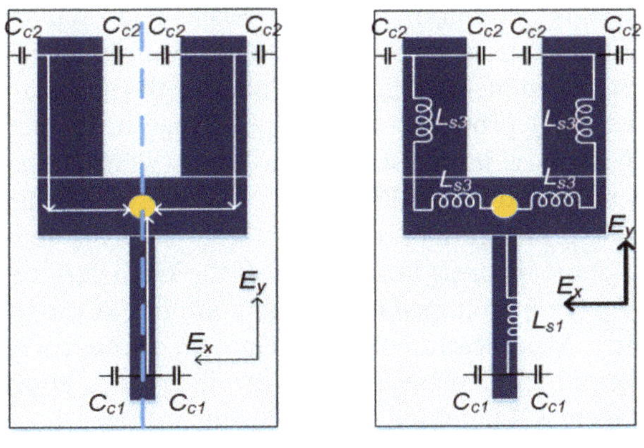

Figure 10.6. (a) Surface current flow through the cell and grounded through via for an EM wave polarized along the x-axis (C_{c1} and C_{c2}: edge coupling capacitances). (b) Formation of inductances along the path.

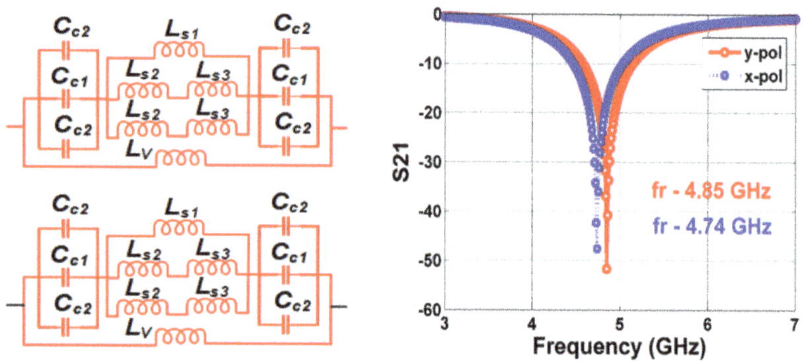

(a) (b)

Figure 10.7. (a) Simulation model for *y*- and *x*-polarized EM wave. (b) Transmission parameter from a simulation model of a fork-shaped EBG unit cell. © 2018 IEEE. Reprinted, with permission, from [42].

Circuit models of the unit cell for both polarizations are extracted from figures 10.5(c) and 10.6(b) and are depicted in figure 10.7(a). It can be inferred that the model is simply a parallel resonant circuit. L_v in figure 10.7(a) can be expressed in terms of r and t, as expressed in (10.11) [24]. Then, a circuit-level simulation in advanced design systems or other similar simulators can be performed to obtain the resonant/stop-band frequency, shown in figure 10.7(b). The effectiveness of this model can be validated by comparing these results to the one obtained using full-wave analysis [25].

$$L_v = 2 \times 10^{-7} t\left[\ln\frac{4t}{2r} + 0.5\left(\frac{2r}{t}\right) - 0.75\right] \tag{10.11}$$

10.4.2.1 Reflection phase

The reflection phase is another property of EBG cells that can be utilized by radiating structures like antennas to strengthen the electric field intensity at the far-field [25]. In general, for an incident EM wave on an EBG surface, the reflected wave exhibits an in-phase (0°) property. This implies that both the incident and the reflected electric fields are in phase. To further explain it mathematically, the expression of the reflection coefficient of an EM wave at the interface of two mediums, given in (10.12), can be utilized. Here, Γ is the reflection coefficient, whereas η_2 and η_1 are the intrinsic impedances of the EBG surface and the air. Since EBG surfaces show very high impedance, ideally infinity at the resonant frequency, therefore η_2 is infinity. Modification of (10.12) in such a scenario results in $\Gamma = 1$. On the contrary, a normal PEC plane (copper or silver as a ground plane in most antennas) exhibits $\eta_2 = 0$. Therefore, (10.12) results in $\Gamma = -1$:

$$\Gamma = \frac{\eta_2 - \eta_1}{\eta_2 - \eta_1}. \tag{10.12}$$

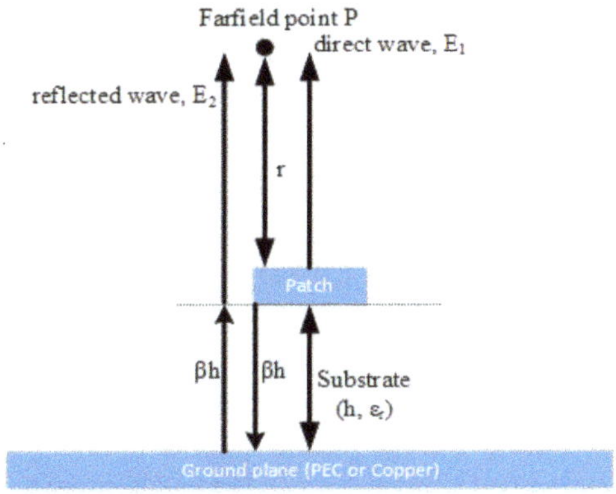

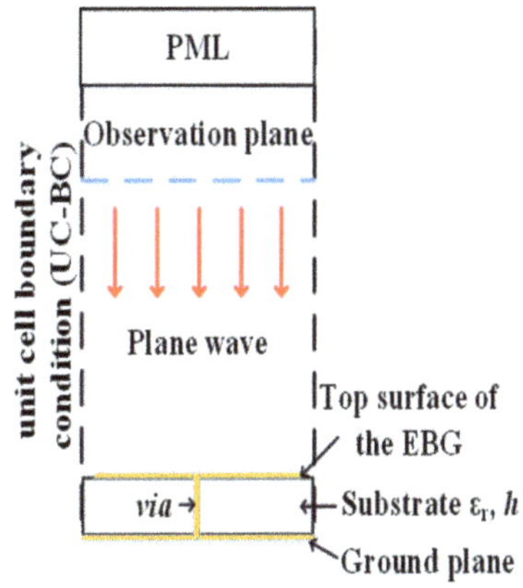

Figure 10.8. Left: Electric field from patch antenna resolved into a direct wave and reflected wave due to ground plane (βh is the electrical length from the patch to the ground plane, E_1 and E_2 represent direct and reflected electric fields). Right: FDTD model to get a dispersion diagram of a unit mushroom cell.

This has further implications on the improvement of the far-field property of a monopole patch antenna. Figure 10.8 shows a patch of an MPA separated from a ground plane by a substrate of known parameters. At the far-field, the electric field is a combination of a direct wave, E_1, and a reflected wave from the ground plane, E_2, given by (10.13). Moreover, E_1 and E_2 are expressed by (10.14–10.15). Here, r is the radial distance from the patch to the observation point P, k is the phase constant and E_0 is constant. The initial phase difference between E_1 and E_2 should be zero as both originate from the same source (patch). However, the reflected wave experiences an additional phase shift of $2\beta h$ due to reflection at the ground plane. This fact can be used in conjunction with (10.13–10.15) to get a simplified expression, given in (10.16). Next, the substitution of $\Gamma = 1$ in (10.16) results in constructive interference of both the electric fields, whereas $\Gamma = -1$ results in destructive interference. Therefore, it can be concluded that an EBG surface assists in enhancing the radiation properties of an MPA rather than a PEC plane. This property is widely used by antenna designers to improve the performance of antennas intended for use in the WBAN scenarios.

$$E_{net} = E_1 + E_2, \tag{10.13}$$

$$E_1 = E_o e^{-j(kr)}, \tag{10.14}$$

$$E_2 = \Gamma E_1, \tag{10.15}$$

$$E_{net} = E_o e^{-j(kr)} + \Gamma E_o e^{-j(kr+2\beta h)}. \tag{10.16}$$

A finite difference time domain (FDTD) setup to get the reflection phase of an EBG cell is shown in figure 10.8. The boundaries of the EBG cell are terminated with a unit cell boundary condition (UC-BC), and the TE/TM wave is incident on the surface of the cell. The reflected wave is collected in the observation plane and

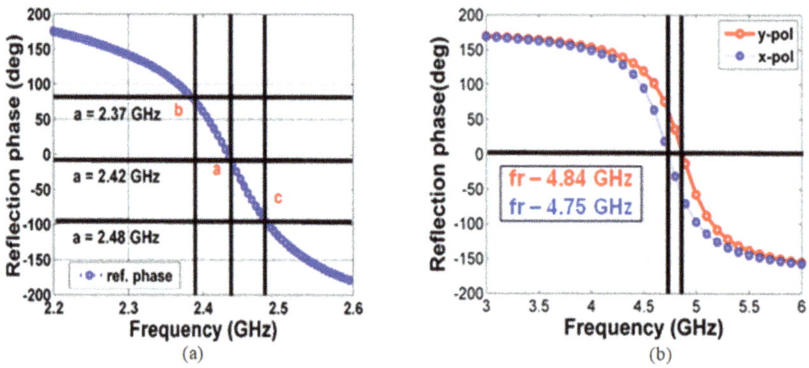

Figure 10.9. Reflection phase diagrams: (a) mushroom EBG cell, (b) fork-shaped EBG cell. © 2018 IEEE. Reprinted, with permission, from [42].

plotted against frequency to obtain the reflection phase. Here, the excitation plane has been kept at a height which is equal to the separation distance of the EBG surface and an antenna. For the mushroom cell in figure 10.3, dimensions of $14.8 \times 14.8\ mm^2$ are considered on a substrate RO4350B of $h = 1.524$ mm and $\varepsilon_r = 3.66$. The reflection phase plot is then obtained, as shown in figure 10.9(a). The point 'a' in the plot signifies the in-phase reflection frequency. In general, the bandwidth of an EBG unit cell in a reflection phase diagram is the range of frequency covered between ±90°, represented by points 'b' and 'c'. Moreover, figure 10.9(b) shows reflection phase plots for the EBG unit cell in figure 10.5(a) under x- and y-polarized EM waves. It is evident from the plots in figures 10.7(b) and 9(b) that they possess remarkable resemblance. It proves the efficacy of the technique based on the lumped model concept. It can also be concluded from the lumped model and reflection phase plot that the in-phase frequency and stop-band frequencies are equivalent.

10.4.2.2 Dispersion model

A graphical plot between the wave number in a periodic structure and frequency is known as a dispersion model [24, 25]. The dispersion diagram displays a band-gap of a periodic structure. In general, the FDTD model shown in figure 10.10(a) is used to extract the dispersion diagram. The only difference is that, instead of UC-BC, a periodic boundary condition (PBC) is used in such a case. Besides, the computation domain is truncated by periodic boundaries, given by (10.17). Here k_{xn} is the propagation constant along the x-axis in a periodic structure, k_x is the propagation constant of the unit cell and p is the periodicity of the unit cell:

$$k_{xn} = k_x + n\frac{2\pi}{p}. \tag{10.17}$$

Instead of simulating a vast number of EBG cells to display band-gap periodic boundaries along the x-and y-axes, these can be achieved through Brillouin's irreducible triangle [25], shown in figure 10.10(a). It has a significant implication as it reduces the time consumed and also space requirements primarily.

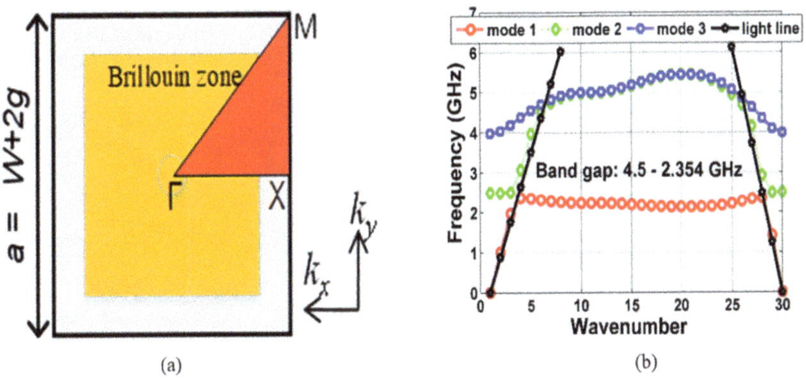

Figure 10.10. (a) Irreducible Brillouin triangle in a mushroom unit cell with propagation constants k_x and k_y. (b) Dispersion diagram of a mushroom cell. © 2018 IEEE. Reprinted, with permission, from [43].

The computation domain is truncated by perfectly matched layer boundaries such that no waves are scattered and alter the result. Figure 10.10(b) depicts a dispersion diagram for a mushroom cell for randomly selected dimensions and substrate parameters. Apparently, the band gap for the mushroom exists at 2.35–4.5 GHz.

10.4.2.3 Transmission characteristics

There is another, more practical method to obtain the band gap of an EBG surface. The physical realization of the band gap of an EBG unit cell can be done by designing an array at the bottom and a transmission line of 50 Ω at the top. The band-gap region is exhibited by a transmission characteristic, S_{21}, of the fabricated prototypes. For this, the device under test (i.e. the DUT, an array of EBG unit cells) is connected to two ports of a vector network analyzer (VNA). For an arbitrary power fed to any of the ports, the measured S_{21} at the other port results in the band gap of an EBG array. If the measured and the simulated S_{21} of the EBG array matches with the band gap of the unit cell, it justifies the design. Moreover, the band gap in the desired range also suggests that no, or very little, propagation of surface waves exists along the substrate. It implies less diffraction at the edges of a PCB and reduced back lobes. This can be assessed by considering a prototyped array of 3×5 mushroom unit cells, shown in figure 10.11(a). The back of the PCB is printed with a 50 Ω transmission line and soldered with two SMA connectors. The simulation setup is identical to the measurement setup, with the only difference being the waveguide port used to feed the array. The simulated and measured results of the array are shown in figure 10.11(b). In general, a suppression level of −20 dB is considered very good for eradicating surface waves. It is apparent from the plots in figures 10.10(b) and 10.11(b) that the band gap of the array is smaller than the individual unit cell. This is due to the fact that the band gap obtained in the previous section assumed PBC of the unit cell til infinity. However, validation of the band gap is performed with a limited number of cells (here 3×5) and this leads to an eventual reduction in the band gap.

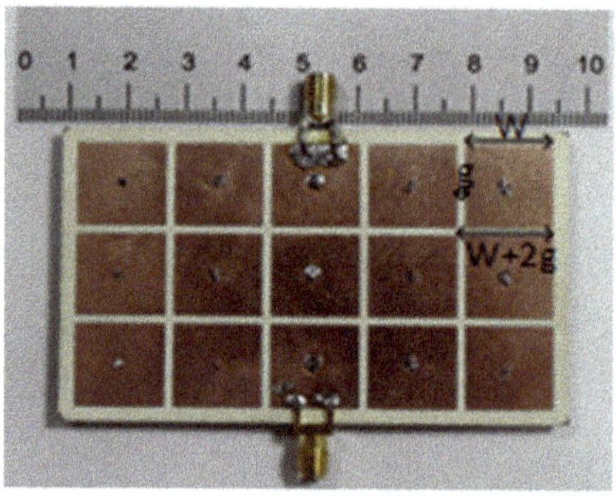

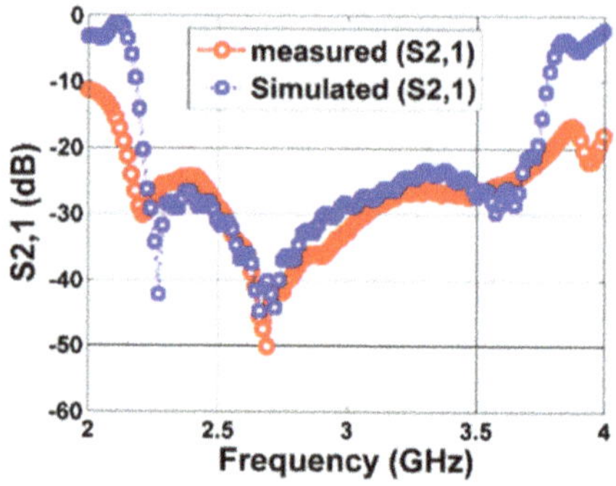

(a) (b)

Figure 10.11. (a) 3×5 EBG unit cells prototyped on 1.524 mm-thick Roger (© 2018 IEEE, reprinted, with permission, from [43]). (b) A 50 Ω transmission line at the bottom (© 2020 IEEE, reprinted, with permission, from [44]). (c) Measured and simulated S_{21} of mushroom array.

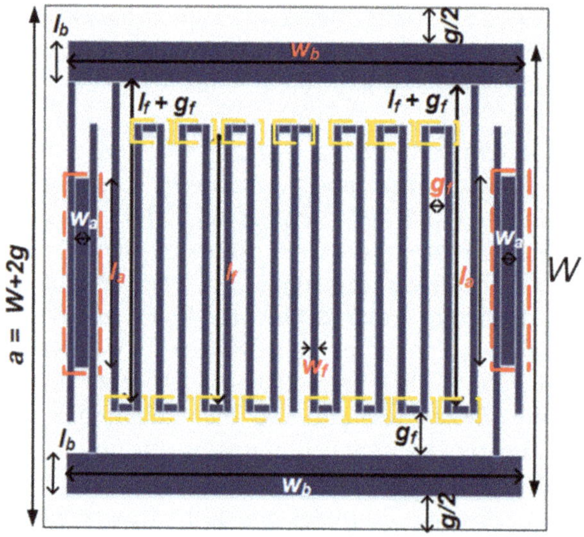

Figure 10.12. The geometry of the IDE unit cell with interconnects shown in colored boxes.

10.4.3 Interdigital electrodes unit cell

In this section, the design of a compact and via-less interdigital electrode (IDE), shown in figure 10.12, for wearable applications in the MBAN band covering 2.36–2.40 GHz is explained. Unlike conventional interdigital geometry, it consists of interconnection among adjacent fingers, marked by yellow and red boxes in figure 10.12. The cell has adjacent fingers connected in such a way that the net inductance contributed by the metallic strips increases, which aids in the reduction of the resonant frequency. The dimensional parameters of the unit cell are as follows. *W*: width of the unit cell; *g*/2:

gap on either side of the patch; l_f, and w_f: length and width of the fingers in the middle; g_f: gap among the fingers and also gap among the fingers and base strip = 0.4 mm; $l_f +$ g_f: length of the finger from the base strip; l_b, and w_b: length and width of the base strip; l_a, and w_a : length and width of the additional strip for interconnection; l_{bf}: length of the strip from the base to the bottom of the strip for interconnection on either side. A gap, g_f, of 0.4 mm can be kept to account for fabrication-related tolerances.

10.4.3.1 Lumped circuit modeling

The circuit modeling approach adopted here makes use of the surface waves generated on a metal surface for an incident EM wave (polarized along either *x*- or *y*-axes), as discussed in section 10.4.2. Surface waves are surface currents (AC) at high frequencies and flow through the metallic strip. Therefore, the strips through which the surface current flows imitate an inductor in the lumped circuit model. The gap among the adjacent cells along the *x*- and *y*-axes forms ECCs dependent on the polarization of the EM wave. For example, the flow of surface current along with the fingers of the IDE unit cell for a *y*-polarized incident EM wave is shown in figure 10.13(a). Due to the symmetrical nature of the unit cell about the *y*-axis, the surface current divides equally at the base strip and flows until reaching the strip in the middle. The surface current flows in series and this results in an increase in the overall inductance of the unit cell. This is due to the interconnection of the adjacent strips and last fingers on either side.

In the circuit modeling, the dimension parallel to the electric field forms the length, i.e. l_{s1}, l_{s2}, ..., l_{s7}, and the dimension orthogonal to the electric field form width, i.e. w_{s1}, w_{s2}, ..., w_{s6}. Then the distribution of strip inductances L_{s1}, L_{s2}, ..., L_{s7} distributed all along the fingers of the unit cell is shown in figure 10.13(b). Equation (10.18) relates L_{s1}, formed by the metal surface of a respective length and width of l_{s1} and w_{s1} [25]. Here, k_g is the correction factor given by (10.19) [25]. h is the height of the substrate. Subsequently, inductances L_{s2}, ..., L_{s7} can be evaluated by substituting the corresponding lengths and widths of the strips in (10.18) and (10.19).

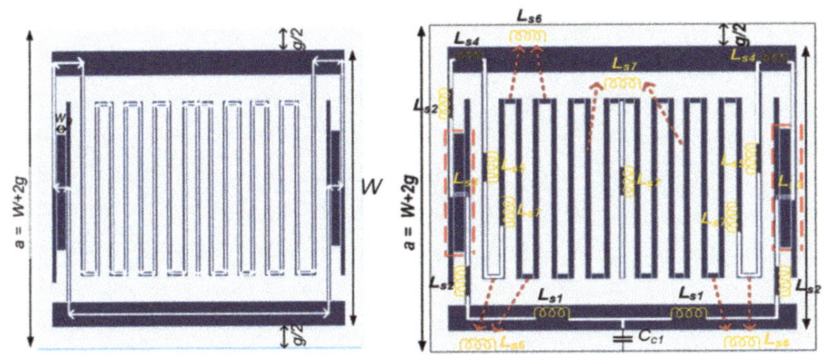

Figure 10.13. (a) Flow of surface current along the strips for a *y*-polarized EM wave (C_{c1} and C_{c2} are two edge coupling capacitances). (b) Formation of inductances and coupling capacitances in the unit cell.

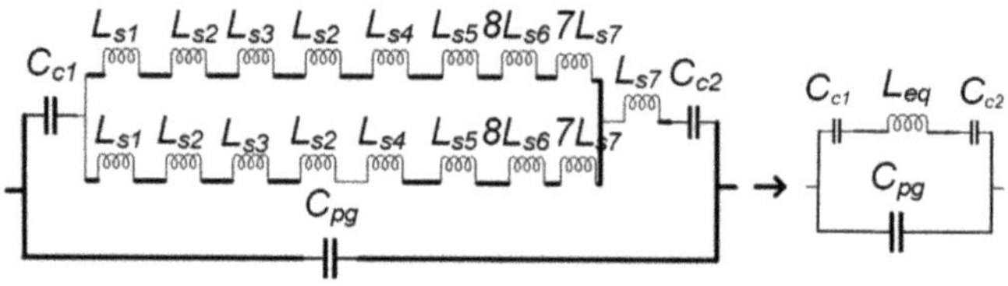

Figure 10.14. Equivalent circuit model of the unit cell consisting of strip inductances connected in series with two ECCs (L_{eq} is the net inductance of the circuit). [34] John Wiley & Sons. © The Institution of Engineering and Technology.

The whole arrangement of inductors is connected in series with ECCs to complete the equivalent circuit model of the unit cell. Here, the continuity in surface current through the strips discards the formation of static capacitances. The inductances formed by red dotted lines, L_{s6} and L_{s7}, add eight and seven times, respectively, as shown in figure 10.13 (b). The resulting inductors are connected in series with coupling capacitors, as shown in figure 10.14. Moreover, the top (patch) and bottom surfaces (ground) of the EBG cell are metal separated by a dielectric layer and thus form capacitance, C_{pg}, as shown in the circuit model. Subsequently, the total inductance, L_{eq}, in figure 10.14 can be used to express the net impedance of the equivalent circuit, Z_{eq}, in (10.20). Then, the stop band or resonant frequency of the equivalent circuit, in (10.21), can be deduced by equating the denominator of (10.20) to 0. It is important to note that only C_{c1} is mentioned in the expressions because both the capacitances are equal, and the use of any of them is acceptable. Furthermore, the capacitance C_{pg} between patch and ground can be given by (10.22). Here, h and ε_r are the height and dielectric constant of the substrate, and A is the area of the unit cell.

$$L_{s1} = 0.25 \times 10^{-7} \times l_{s1}\left(\ln \frac{l_{s1}}{w_{s1}} + 1.193 + 0.25\frac{w_{s1}}{l_{s1}}\right)k_g, \tag{10.18}$$

$$k_g = 0.57 - 0.145 \ln \frac{w_{s1}}{h}, \tag{10.19}$$

$$Z_{eq} = \frac{2 - \omega^2 L_{eq} C_{c1}}{j\omega\left(2C_{pg} + C_{c1} - \omega^2 L_{eq} C_{c1} C_{pg}\right)}, \tag{10.20}$$

$$f_r = \frac{1}{2\pi}\sqrt{\frac{\left(2C_{pg} + C_{c1}\right)}{L_{eq} C_{c1} C_{pg}}}, \tag{10.21}$$

$$C_{pg} = \varepsilon_o \varepsilon_r \frac{A}{h}. \tag{10.22}$$

As a design example, an IDE unit cell is designed on Roger 4350B with a dielectric constant of 3.66 and a height of 1.524 mm. The optimized dimensions of the unit cell to

Table 10.1. IDE unit cell parameters.

Parameter	Value (mm)	Parameter	Value
l_{s1}	1.05	L_{s1}	0.01 nH
w_{s1}	6.10	L_{s2}	1.00 nH
l_{s2}	10.9	L_{s3}	0.75 nH
w_{s2}	0.30	L_{s4}	0.02 nH
l_{s3}	10.0	L_{s5}	1.28 nH
w_{s3}	0.40	L_{s6}	3.8e–3 nH
l_{s4}	1.05	L_{s7}	1.00 nH
w_{s4}	1.70	$C_{c1} = C_{c2}$ (g = 0.4 mm)	1.63 pF
l_{s5}	11.7	$C_{c1} = C_{c2}$ (g = 0.5 mm)	1.53 pF
w_{s5}	0.30	$C_{c1} = C_{c2}$ (g = 0.6 mm)	1.47 pF
l_{s6}	0.30	$C_{c1} = C_{c2}$ (g = 0.7 mm)	1.40 pF
w_{s6}	0.40	f_0 (g = 0.4 mm)	2.40 GHz
l_{s7}	10.9	f_0 (g = 0.5 mm)	2.46 GHz
w_{s7}	0.30	f_0 (g = 0.6 mm)	2.51 GHz
		f_0 (g = 0.6 mm)	2.55 GHz

get the desired resonant frequency at 2.4 GHz is 14.3 mm × 14.5 mm. The ECCs play a crucial role in controlling the resonant frequency of the unit cell and is given by (10.8) [34]. Table 10.1 lists all the circuit parameters and their corresponding strip inductances and coupling capacitances, calculated using (10.8), (10.18) and (10.19), for gaps of 0.4 mm. Moreover, expression (10.22) results in a capacitance C_{pg} of 4.4 pF. However, as evident from (10.8), for a fixed width of the unit cell, the ECC depends only on g between the adjacent cells; therefore, three other widths, 0.5, 0.6 and 0.7 mm, have also been used to obtain the resonant frequency of the unit cell, keeping all other parameters intact.

The listed values of dimensions and L and C in table 10.1 are optimized for achieving resonance at 2.4 GHz. These values might change slightly during full-wave analysis, and, if they do not, then this proves the accuracy of the equivalent circuit model. The plots in figure 10.15 depict resonant/stop-band frequencies of the unit cell for the varying gaps listed in table 10.1. Apparently, in this case, only a gap width of 0.4 mm among the unit cells is capable of achieving the desired resonance in the MBAN band. It is clear that the influence of the gap among the interdigital elements on resonant frequency is negligible. However, it does vary substantially because the number of fingers varies if the gap between the fingers is varied for a fixed dimension of the unit cell. For example, in the present case, if the gap is greater than 0.4 mm, say 0.55 mm, the number of fingers to be accommodated on either side changes to six ($7L_{s7}$, $6L_{s7}$). While most of the circuit parameters in the equivalent circuit remain the same, the L_{s4} and L_{s6} vary slightly.

The plot in figure 10.15 also shows the significance of interconnection and the number of adjacent fingers on the resonant frequency. It can be inferred that the

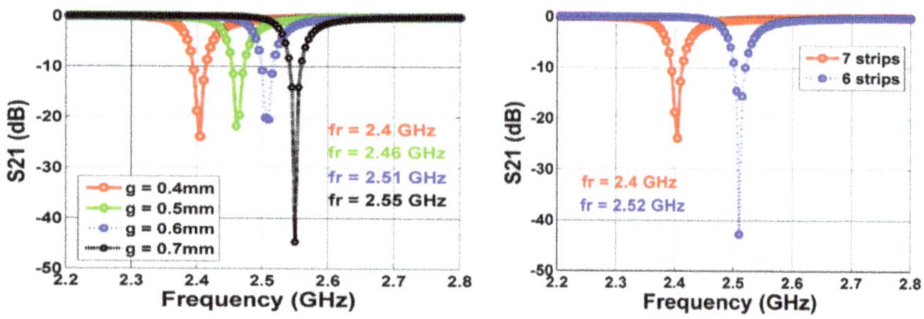

Figure 10.15. Left: The resonant frequencies of the IDE unit cell for the equivalent circuit model (given in figure 10.14) and the dimensions in table 10.1. Right: The resonant frequencies of the unit cell for seven and six connected strips/fingers. [34] John Wiley & Sons. © The Institution of Engineering and Technology.

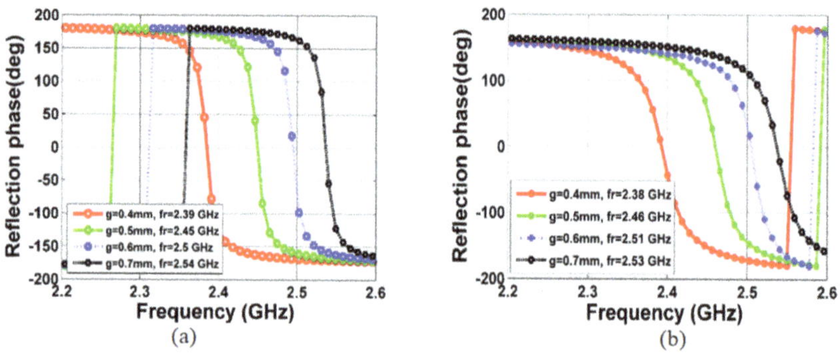

Figure 10.16. Reflection phase results of the unit cell: (a) obtained from the equivalent circuit model, (b) obtained from CST transient solver. [34] John Wiley & Sons. © The Institution of Engineering and Technology.

resonance frequency increases by 120 MHz with the reduction of finger counts from seven to six. Furthermore, an opposite effect can be expected if the g between the fingers is reduced. This explains the selection of an optimum number of fingers for any dimension to obtain a stop band in the MBAN range. It is essential to note that g has been kept at 0.4 mm to account for the fabrication facility tolerance.

10.4.3.2 Reflection phase, band gap and polarization stability

The reflection phase, ϕ, in terms of Z_{eq} and the intrinsic impedance of space, η, are related by (10.23) [24]. The reflection phase diagram of the IDE unit cell for the parameters in table 10.1 is shown in figure 10.16(a). Furthermore, the resonant frequencies for different gap widths in table 10.1 are plotted by full-wave analysis using a CST and are depicted in figure 10.16(b). The unit cell's dimensions and substrate details are identical to that in section 10.4.3.1.

$$\phi = \text{Im}\left\{\ln\left(\frac{Z_{eq} - \eta}{Z_{eq} + \eta}\right)\right\}. \tag{10.23}$$

It is apparent from the plots in figures 10.16(a) and (b) that they possess good resemblance, and this proves the appropriateness of this circuit model. Once again, only g = 0.4 mm satisfies the desired stop-band frequency at 2.4 GHz. It is also important to mention that the excitation plane for the plane wave incidence is set at a height of 1.524 mm above the unit cell's surface to record the reflection phase data. This is because the height of the antenna above the EBG base is 1.524 mm, as discussed in section 10.4.2.1. For the effective operation of an EBG surface, the bandwidth is measured between ± 90° of the reflection phase plot [25]. Evidently from the plots in figures 10.16(a) and (b), g = 0.4 mm covers the MBAN region. It is also apparent from figure 10.16(b) that the bandwidth of the IDE unit cell obtained from the full-wave analysis is smaller than that reported in [22]. This is because the bandwidth is inversely proportional to the capacitance developed between the patch and the ground plane. The capacitance, in turn, depends on the PCB thickness, which is 1.524 mm for the current case and 5 mm in [22]. Therefore, due to the smaller thickness, the capacitance of this IDE cell is more and, as a consequence, the bandwidth is smaller.

Another point to note: the uniformity in the dimensions of the unit cell, while achieving the desired result, has been maintained but still the symmetricity is slightly compromised along the longitudinal dimension to adjust the resonant frequency. This introduces a slight asymmetry in the design of the unit cell. Hence to test the polarization stability of the IDE unit cell, it is excited by an x- and y-polarized tranverse electromagnetic (TEM) wave; the reflection phase data are recorded and plotted in figure 10.17 while keeping all other parameters intact for g = 0.4 mm. It is clear from the plots that the unit cell shows excellent resistance to phase change for polarization variation even in the presence of asymmetry.

Furthermore, the polarization sensitivity of the unit cell to obliquely incident TM and TE waves can be checked. To understand this aspect, TM and TE waves at 0°, 15°, 30°, 45° and 60° are incident on the IDE unit cell, and reflection phase data are

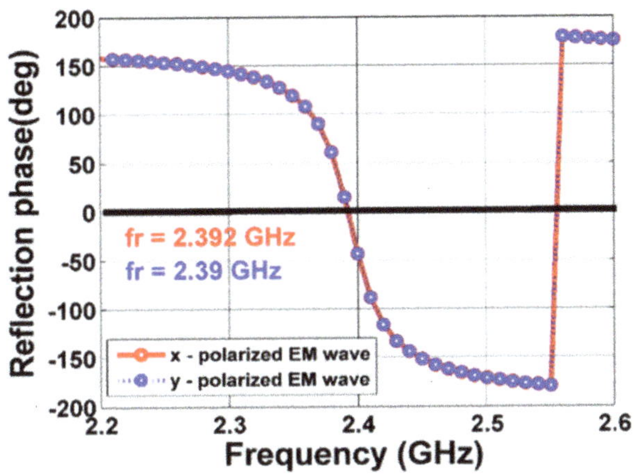

Figure 10.17. Reflection phases of the IDE unit cell for an x- and y-polarized EM wave. [34] John Wiley & Sons. © The Institution of Engineering and Technology.

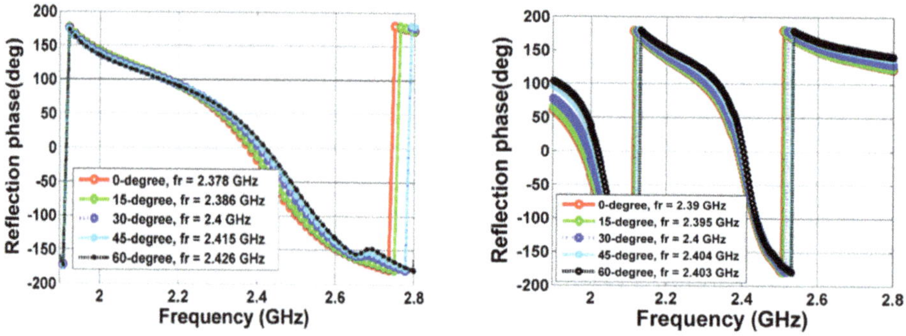

Figure 10.18. Left: Reflection phases of the unit cell for TM wave incidence at 0°, 15°, 30°, 45° and 60°. Right: Reflection phases of the unit cell for TE wave incidence at 0°, 15°, 30°, 45° and 60°. [34] John Wiley & Sons. © The Institution of Engineering and Technology.

obtained and plotted in figure 10.18. The need for such an investigation is because the TM waves are the fundamental mode of propagation, and because the distance by which the antenna is separated from the EBG base is small and might induce TE modes, as well. Moreover, the 1.524 mm separation of the EBG base from the antenna does not lie in the far-field region. This implies that the angles at which the TM and TE waves fall on the EBG base, originating from a patch antenna, may not be 0°. It is pertinent to mention that the g among the unit cells is kept the same, at 0.4 mm, for all the incidence angles. The results in figure 10.18 convey that the IDE unit cell demonstrates a stable phase relationship for all the incident angles of TM and TE wave with complete MBAN coverage. An important aspect to note is the seemingly distinct look of the reflection phase plots in figures 10.16(b) and 10.18 even though they are for the same unit cell. The difference is natural as the former is for a transient domain solver (TDS) whereas the latter is for a frequency-domain solver (FDS), and the boundary conditions for both techniques are different. For example, the TDS uses magnetic and electric boundaries along the x- and y-axes, whereas under FDS the unit cell is simulated under unit cell conditions along the x- and y-axes. The bandwidths, the frequency range between ±90° in the reflection phase plot, in both plots are different. The FDS gives a bandwidth for an infinite series of unit cells whereas TDS gives a bandwidth of a finite dimension unit cell.

To qualify the proposed interdigital structure as a band-gap surface, a dispersion diagram depicted in figure 10.19 for the given dimensions and g of the unit cell, shown in table 10.1, is obtained through a CST solver. Three consecutive modes, namely mode 1, mode 2 and mode 3, along with the corresponding band gap suggest that there is a specific region of surface wave suppression between the fundamental (mode 1) and next to fundamental mode (mode 2) effectively covering the MBAN. In the absence of surface waves along the substrate, diffraction at the edges is minimal. Therefore, it reduces the back lobes of an antenna and improves the front-to-back ratio. It is pertinent to mention that this is the key property of an IDE unit cell that helps in the significant enhancement of antenna performance when these are incorporated in antennas.

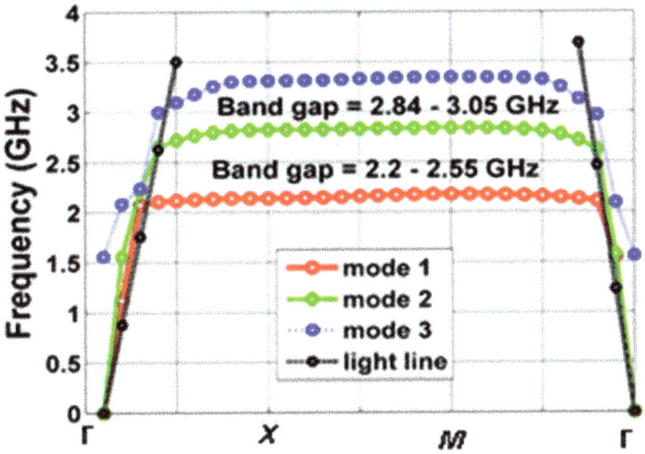

Figure 10.19. Band-gap diagram of the IDE unit cell (Γ, X and M are points in the Brillouin zone). [34] John Wiley & Sons. © The Institution of Engineering and Technology.

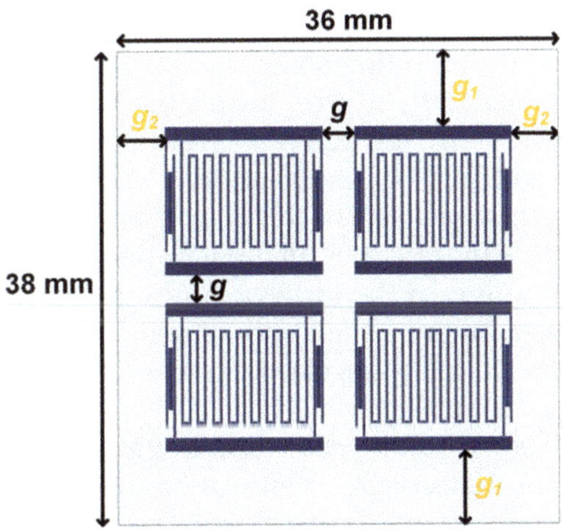

Figure 10.20. Array of 2×2 EBG unit cells (g_1 = 4.3 mm, g_2 = 3.5 mm, g = 0.4 mm).

10.4.4 Array performance enhancement

10.4.4.1 Array of 2×2 unit cells

An array of 2×2 IDE unit cells, shown in figure 10.20, designed to act as a reflector for antenna performance enhancement, is discussed. The unit cells are arranged periodically in a way such that the desired reflection phase and surface wave rejection can be achieved in the MBAN frequency range. The reason for limiting the array size is that the overall size of the incorporated antenna has been chosen to be a small size of 38 × 38 mm^2 [41]. The unit cells are placed symmetrically over the substrate while keeping an identical distance from the substrate's edges, while maintaining a spacing (g) among the unit cells of 0.4 mm. A parametric study can be

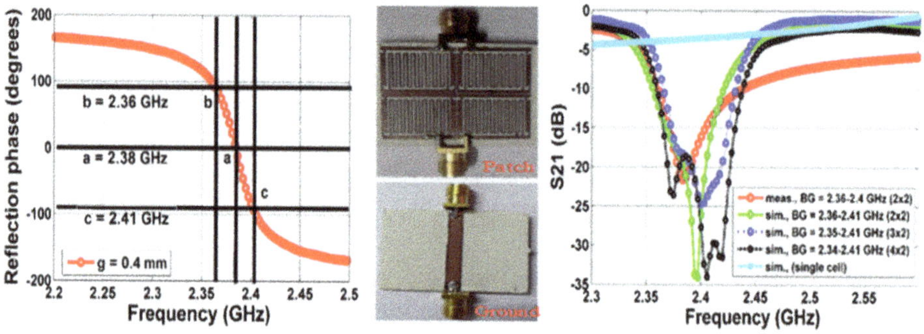

Figure 10.21. Reflection phase results of 2×2 array (left), prototyped IDE array (center), measured and simulated transmission result of the IDE array (right). [34] John Wiley & Sons. © The Institution of Engineering and Technology.

performed on gap g_2 while keeping gap g_1 fixed to 4.3 mm to obtain a reflection phase diagram covering the entire MBAN. The optimum parameters of the array for the desired reflection phase are $g_1 = 4.3$ mm and $g_2 = 3.5$ mm. Therefore, the array along the x-axis can be reduced by 2 mm, and then the overall dimensions become 36×38 mm^2. For a y-polarized TEM wave incident from a height of 1.524 mm above the EBG surface, the reflection phase is plotted in figure 10.21. It can be seen that the reflection phase plot for the array covers the MBAN region for a gap, g, of 0.4 mm. The surface wave rejection ability of the array is an essential facet in terms of demonstrating the band gap. For this purpose, a 50 Ω line is printed in the ground plane without altering the top layer. In this case, the width of the 50 Ω feed comes out to be 3.2 mm for the substrate. For an input power of 0 dBm fed into any of the ports, the transmitted power, in the form of S_{21}, is recorded at the other port. A waveguide port is used in the simulation setup, for which the properties are determined through the in-built calculator in the CST. The simulated received power is shown in figure 10.21.

Subsequently the array is prototyped; this, along with the measured results, is shown in figure 10.21. A perfect agreement can be seen between the measured and the simulated results, and a surface rejection of −15 dB can be obtained in the MBAN range. This proves the utility of the unit cell in the rejection of surface waves. The band gaps obtained in figure 10.21 in both conditions are not in agreement. This is because the band-gap diagram of the unit cell with PBC along the x- and y-axes is considered in simulation, whereas the extent of the band gap is reduced substantially in practice due to the limited size of the unit cells. Simulation results for 3×2 and 4×2 array configurations are also added in figure 10.21 to highlight this aspect. Hence, it can be expected that with larger array size and more unit cells, the band gaps obtained through this phenomenon will be in perfect agreement with those of full-wave analysis results.

10.4.4.2 Integrated monopole antenna with electromagnetic band gap array

At first, a simple monopole antenna for the dimensions given in table 10.2 is considered and designed on RO4350B substrate of height 1.524 mm with a dielectric

Table 10.2. Design parameters (in mm) of the EBG-integrated monopole antenna. [34] John Wiley & Sons. © The Institution of Engineering and Technology.

L_P	W_P	L_F	W_F	L_{gnd}	L_{gap}
27.6	22	6	3	3	0.55

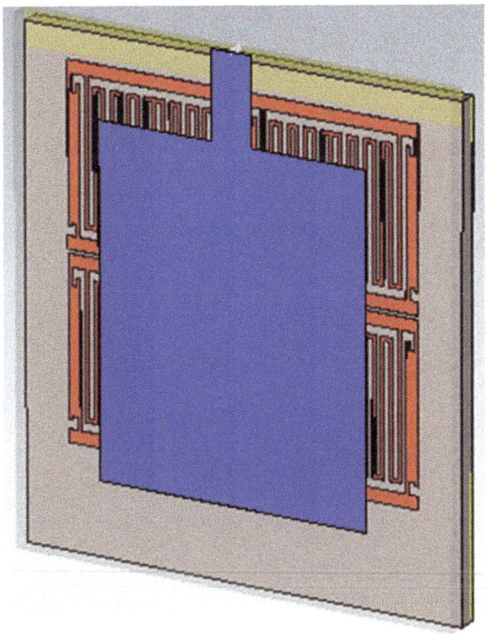

Figure 10.22. Planar monopole antenna at the top integrated with 2×2 EBG array on RO4350B substrate.

constant of 3.66, as shown in figure 10.22. The optimized design parameters, L_P, W_P, L_{gnd}, L_F and W_F, of the monopole are given in table 10.2. Its prototype, shown in figure 10.23(a), is developed and the corresponding measured and simulated performance is depicted in figure 10.24.

Subsequently, the EBG array is incorporated with the monopole antenna, as shown in figure 10.22. The prototyped EBG-incorporated antenna is shown in figures 10.23(b) and (c). In this case, there is no gap between the array and the monopole antenna. This is achieved by attaching both the individual surfaces (the monopole and the EBG array) with the help of adhesive RTV coating with a dielectric constant of 1.22. However, this leads to the formation of a gap between the layers while soldering the SMA connector during fabrication. Therefore, for suitable placement of the SMA connector to the monopole antenna, it needs to be shifted slightly upwards by 1.5 mm. A slight shift in the monopole antenna relative

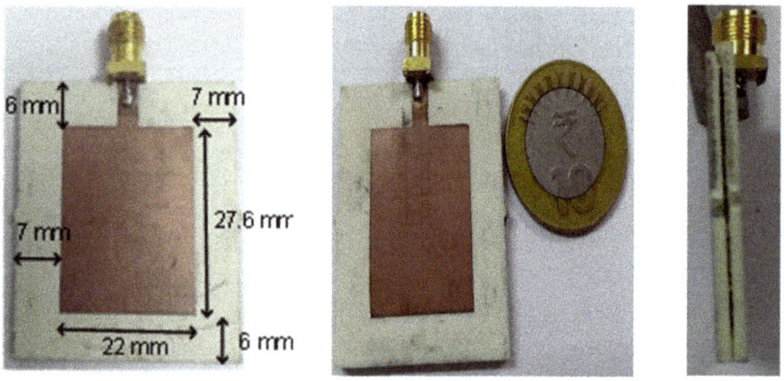

(a) (b) (c)

Figure 10.23. Fabricated IDE-incorporated antenna: (a) simple monopole, (b) incorporated antenna (front view), (c) incorporated antenna (side view). [34] John Wiley & Sons. © The Institution of Engineering and Technology.

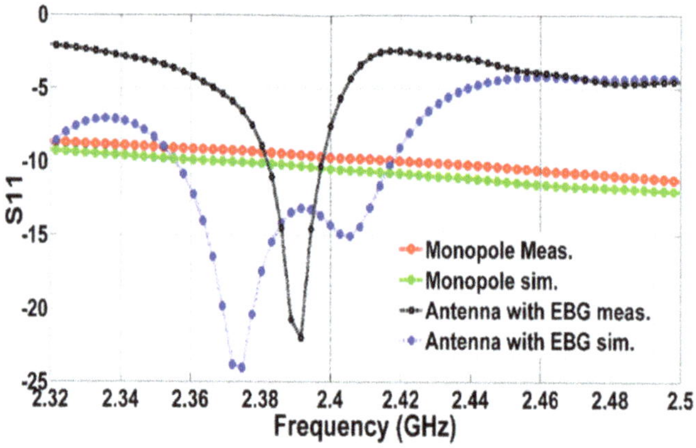

Figure 10.24. Simulated and measured return losses for simple and EBG-incorporated monopole antenna. [34] John Wiley & Sons. © The Institution of Engineering and Technology.

to the array can be observed in figure 10.23(c). The design parameter L_{gap} plays a crucial role in getting a good match in such a scenario. It was observed during the design process that the matching improves with decreasing L_{gap}, but the resonant frequency increases and vice versa. Therefore, care needs to be taken in regulating L_{gap} to get a proper match (below −10 dB) as well as the desired resonance at 2.4 GHz. The optimum L_{gap} in this situation is mentioned in table 10.2. The measured return loss of the EBG-incorporated monopole antenna, in figure 10.24, shows better results in terms of matching and desired band coverage. This is due to the reduction in the overall quality factor of the structure from the ohmic losses in the SMA connector [21].

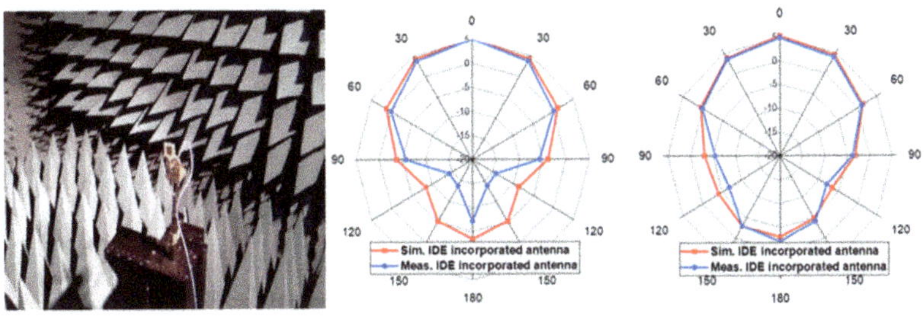

(a) (b) (c)

Figure 10.25. (a) Radiation pattern measurement setup inside an anechoic chamber and far-field results of the incorporated antenna ([34] John Wiley & Sons. © The Institution of Engineering and Technology). (b) x–z plane. (c) y–z plane.

The far-field radiation pattern of the integrated antenna is measured inside the anechoic chamber, as per the setup shown in figure 10.25(a). The simulated and measured radiation patterns of the integrated antenna along the x–z and y–z planes are given in figures 10.25(b) and (c). The simulation shows the possibility of a gain of 5.3 dB, and the measured gains are well matched with gains of 5.1 dB in both planes. The presence of back lobes in both the simulated and measured results is due to the surface waves diffracted at the edges of the substrate of the monopole antenna. This is due to the fact that the upper radiating edge of the monopole antenna is closer to the top edge by 1.5 mm when compared to the bottom edge in the overall geometry, as can be seen in figure 10.23(c). This anomaly eventually leads to enhanced diffraction at the edge and, hence, increased back-lobe radiation. Besides, the EBG array tends to suppress surface wave propagation through the substrate; however, due to the low rejection level (−15 dB), the surface waves still exist. It leads to diffraction at the edges and enhances the back lobes. Still, the accuracy of the radiation pattern measured in both planes and the gain enhancement justifies the analysis. It is imperative to note that the proposed antenna is relatively small and achieves an improvement of gain from 2.3 to 5.3 dB in free space when compared to [41].

10.4.5 Determination of effective dielectric constant

10.4.5.1 Background

An MPA covered with single or multiple dielectric layers tends to resonate at a different frequency than the original design frequency [29–31]. This is because the effective dielectric constant of the overall arrangement changes [29–31]. From the WBAN perspective, WSNs are widely used in the medical realm, as they can record and transfer vital physiological parameters of the body such as heartbeat rate (electrocardiogram, ECG), body temperature, blood pressure, respiratory rate, ligament pressure (round ligament), and many more [1–3]. Often, these sensors are placed on the body for accurate monitoring. Therefore, the communication between the on-body sensor and external BTS happens through single or multiple layers of textiles. The ZigBee, Z-wave and X-Bee are ubiquitous transmit and receive modules in WSN deployed on both on-body and off-body. These modules operate in either 800–900 MHz (Z-wave) or 2.4–2.4835 GHz (ZigBee and X-Bee) unlicensed bands. Out of the entire band (100 MHz) allocated to Z-wave, only a part of it is used in various parts of the world. For example, Z-wave in India, US, Russia, Europe and Japan operates at 862.5, 908.4, 869, 868.4–868.42, 922–926 MHz, respectively. Since the fractional bandwidth of these devices is less than 3% and any deviation from the original band (in the presence of textile layers) leads to out-of-band communication, the essential body parameters are lost. Hence, it is of utmost importance to accurately determine ε_{eff} and f_r for reliable and error-free communication between sensors. Although this analysis remains valid for multiple superstrate layers over the MPA [30, 31], only explanations for a single superstrate layer over an MPA are analyzed and discussed in this chapter.

10.4.5.2 Formulation of ε_{eff} for single superstrate over an MPA

In this section, we derive formulations to determine the ε_{eff} of an MPA covered with a single superstrate layer by using CM. For this, an MPA designed on a substrate with dielectric constant, ε_1, and thickness, h, is shown in figure 10.26. The physical parameters of the MPA, width W and length L are determined through standard expressions [32]. The MPA is covered by a superstrate layer of thickness h_2—h and dielectric constant ε_2. For $W/h > 1$ (low-profile MPA, mostly in WBAN), CM is then applied at the interface of a substrate–superstrate (ε_1–ε_2) and superstrate–air (ε_2–ε_o) using Wheeler's transformation [29]. The conformed dielectric boundaries on the flux-potential plane are shown in figure 10.27(a). Here, W_{ef} is the effective width

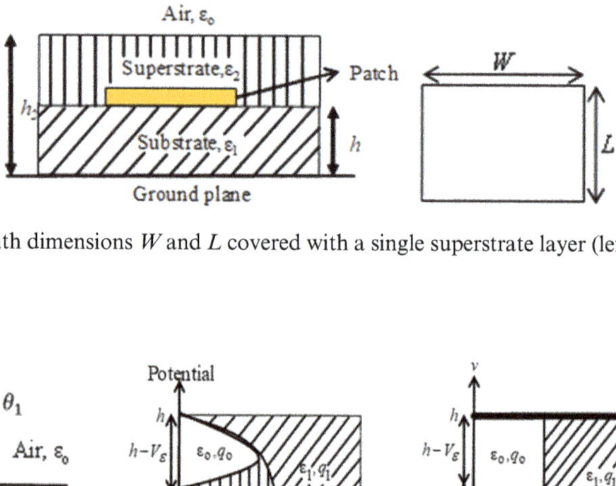

Figure 10.26. MPA with dimensions W and L covered with a single superstrate layer (left); the MPA geometry (right).

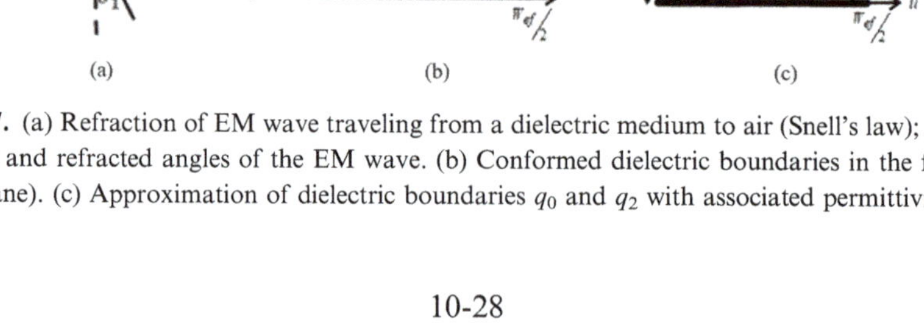

Figure 10.27. (a) Refraction of EM wave traveling from a dielectric medium to air (Snell's law); θ_1 and θ_2 are the incident and refracted angles of the EM wave. (b) Conformed dielectric boundaries in the flux-potential plane (w-plane). (c) Approximation of dielectric boundaries q_0 and q_2 with associated permittivities.

of the patch accounting for fringing at the edges, V_ε is the extent of the superstrate layer in the imaginary axis of the w-plane, and q_o, q_1 and q_2 are the effective filling fractions of the air, substrate and superstrate boundaries defined by the ratio of the area occupied by a given dielectric by total area [29].

For example, assume that the specific area occupied by air, substrate and superstrate are $S_{\varepsilon o}$, $S_{\varepsilon 1}$ and $S_{\varepsilon 2}$, respectively, whereas the total area is S. Therefore, the filling fractions q_o, q_1 and q_2 are given by (10.24). The overall filling fraction, q, is the summation of q_o, q_1 and q_2 and is expressed in (10.25). Apparently, (10.25) suggests that q reduces to unity. As a necessary condition of Wheeler's transformation, the angle of refraction at the interfaces (x–y, physical plane) must be retained while conforming the dielectric boundaries to the w plane, as shown in figures 10.27(a) and (b). This essentially means that for an EM wave traveling from substrate to superstrate or superstrate to air, the angles of refraction at these interfaces shall be retained while conforming physical planes to imaginary ones (w plane). For an MPA with width W, fringing at the edges results in W_{ef}, given in (10.26), and the point of intersection, V_ε, in figure 10.27(b) is given by (10.27). The filling fractions q_o, q_1 and q_2 can be determined from W_{ef}, V_ε, h_2 and h, and are given by (10.28–10.30). In the absence of the superstrate layer, i.e. $h_2 = h$, the terms V_ε, and q_2 must equal zero. However, it has been established that even in the absence of the superstrate layer, q_2 does not reach zero, as is evident from (10.30) [30]. This inconsistency results in a modified expression of q_2, q_{2new}, as given (10.31). Now, it can be seen that the dielectric layers in figure 10.27(c) are either in series or in parallel [30]. It implies that $S_{\varepsilon o}$ is in series with $S_{\varepsilon 2}$, and the combination of them is in parallel to $S_{\varepsilon 1}$. These dielectric regions form capacitances between the metallic layers of the MPA and ground plane, which are utilized to derive the expression of ε_{eff}.

$$q_o = \frac{S_{\varepsilon o}}{S};\ q_1 = \frac{S_{\varepsilon 1}}{S};\ q_2 = \frac{S_{\varepsilon 2}}{S}, \tag{10.24}$$

$$q = \frac{S_{\varepsilon o}}{S} + \frac{S_{\varepsilon 1}}{S} + \frac{S_{\varepsilon 2}}{S} = \frac{S_{\varepsilon 0} + S_{\varepsilon 1} + S_{\varepsilon 2}}{S}, \tag{10.25}$$

$$W_{ef} = W + \frac{2h}{\pi}\ln\left[17.08\left(\frac{W}{2h} + 0.92\right)\right], \tag{10.26}$$

$$V_\varepsilon = \frac{2h}{\pi}\tan^{-1}\left[\frac{\pi}{\frac{\pi}{2}\frac{W_{ef}}{h} - 2}\left(\frac{h_2}{h} - 1\right)\right], \tag{10.27}$$

$$q_o = \frac{1}{2} \times \frac{h - V_\varepsilon}{W_{ef}} \times \ln\left[\pi\frac{W_{ef}}{h} \times \frac{\cos\left(\frac{\pi}{h}\cdot\frac{V_\varepsilon}{2}\right)}{\pi\left(\frac{h_2}{h} - \frac{1}{2}\right) + \frac{V_\varepsilon}{2}\cdot\frac{\pi}{h}} + \sin\left(\frac{V_\varepsilon}{2} \times \frac{\pi}{h}\right)\right], \tag{10.28}$$

$$q_1 = 1 - \frac{h}{2W_{ef}} \times \ln\left(\pi\frac{W_{ef}}{h} - 1\right), \tag{10.29}$$

$$q_2 = 1 - q_1 - q_o, \tag{10.30}$$

$$q_{2\text{new}} = 1 - q_1 - q_0 - \frac{h}{2W_{ef}} \ln\left[\frac{\pi}{2} - \frac{h}{2W_{ef}}\right], \tag{10.31}$$

$$C_1 = \frac{(\varepsilon_1 q_1)}{h} C_o = \frac{(\varepsilon_o q_o)}{h_1} C_2 = \frac{(\varepsilon_2 q_{2\text{new}})}{V_\varepsilon}, \tag{10.32}$$

$$C_{\text{eqs}} = \frac{\varepsilon_2 q_{2\text{new}} \varepsilon_0 q_0}{\varepsilon_2 q_{2\text{new}} h_1 + \varepsilon_0 q_0 V_\varepsilon}, \tag{10.33}$$

$$C_{\text{eff}} = C_{\text{eqs}} + C_1, \tag{10.34}$$

$$C_{\text{eff}} = \frac{\varepsilon_2 q_{2\text{new}} \varepsilon_0 q_0}{\varepsilon_2 q_{2\text{new}} h_1 + \varepsilon_0 q_0 V_\varepsilon} + \frac{\varepsilon_1 q_1}{h}, \tag{10.35}$$

$$C_{\text{eff}} = \frac{\varepsilon_{\text{eff}} q}{h_2}, \tag{10.36}$$

$$\varepsilon_{\text{eqs}} = \frac{\varepsilon_2 q_{2\text{new}} \varepsilon_0 q_0 h_2}{\varepsilon_2 q_{2\text{new}} h_1 + \varepsilon_0 q_0 V_\varepsilon}. \tag{10.37}$$

The respective capacitances per unit area of the substrate, C_1, air, C_o, and superstrate, C_2, regions are given by (10.32). Here, $h_1 = h - V_\varepsilon$ (from figure 10.27(b)). Now, the equivalent series capacitance, C_{eqs}, between the C_o and C_2 is expressed in (10.33). Next, from figure 10.27(c), it is apparent that C_{eqs} is in parallel with C_1, and hence the effective capacitance, C_{eff}, of both of them is expressed in (10.34). After the substitution of (10.32) and (10.33) in (10.34), the resulting C_{eff} is given by (10.35). From figure 10.26, this effective capacitance in terms of the overall dielectric constant of the arrangement, ε_{eff}, the overall filling fraction, q, and the total height of the arrangement, h_2, can be deduced (10.36). All these can be simplified to achieve the expression for ε_{eqs} given in (10.37). Now, it can be said that from the known values of ε_o, ε_1, ε_2, h and h_2, and the determined values of W_{ef}, V_ε, q_o, q_1 and $q_{2\text{new}}$ that one can determine the ε_{eff} of the arrangement of layers shown in figure 10.26. Interestingly, the formula of ε_{eff} takes into account the height of the superstrate layer.

10.4.5.3 The iterative model

A unique iterative model to determine ε_{eff} and f_r at each iteration using the formulations in the last section is discussed. The model is iterative because, at the end of each step, the dielectric constant of the substrate, ε_1, gets updated with the newly computed ε_{eff}. The following can be used to realize this procedure in any technical computing software such as MATLAB. During the realization, this technique considers the optimized length of the MPA in the CST. This is done as the standard expression to determine the physical parameters of the MPA operating at some f_r assumes the ground plane to be infinite. It is, however, not the case in simulation and measurement, and this often leads to noticeable shifts in f_r from the originally obtained theoretical values. Before starting the iterative process, the design of the MPA is the first step, before a layer of superstrate is added over it. Therefore, the parameters of the MPA (width, length and impedance) are independent of the superstrate. A detailed flow chart of the model to determine ε_{eff} and f_r is shown in figure 10.28. The following steps explain the implemented model.

***Step* 1:** Choose the antenna's parameters, substrate, superstrate, such as length, width, height and dielectric constant. Here, the actual superstrate's height is $\Delta h = h_2 – h$, from figure 10.26. However, the total height of the substrate and the superstrate is taken as h_2\` rather than $h_2$ in this model, which is given by $h_2$\` $= h + n$. Therefore, h_2 in (10.27) and (10.28) is replaced by h_2\`. Divide the total superstrate layer into '*N*' smaller blocks of thickness '*n*' each. The selection of *n* is explained here with the help of an example, whereas the selection of *N* is included in step 3.

Consider an MPA designed on RO4350B substrate of dielectric constant and height 3.66 and 1.524 mm, respectively. It is covered by a superstrate layer identical to the

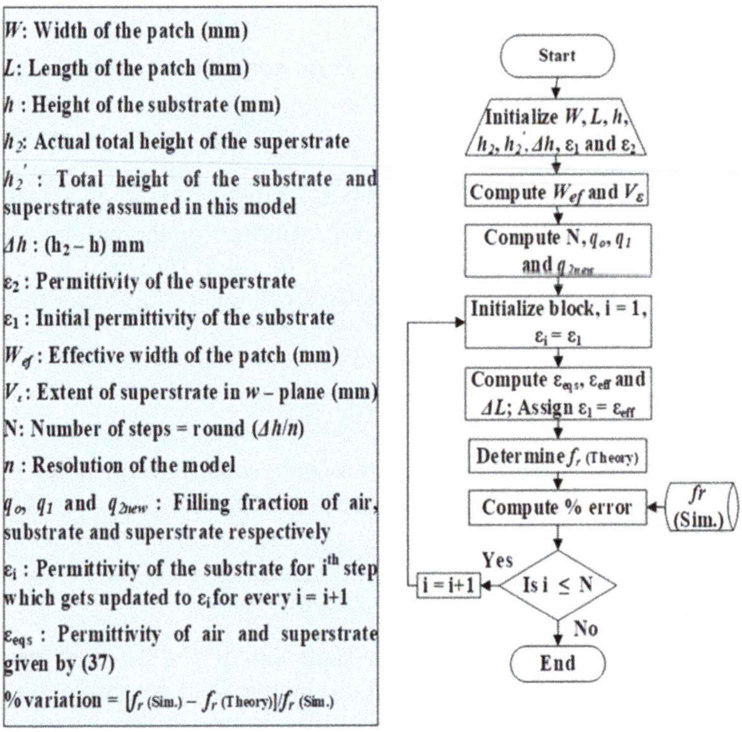

Figure 10.28. Model to determine the effective dielectric constant of an MPA covered with a single superstrate layer (the model is implemented in MATLAB).

substrate. Therefore, h_2 becomes 1.524 + n mm. Putting the value of $n = 0.1$ mm in (10.27)–(10.30), the ε_{eff} comes out to be 3.679. For any other values of n lesser than 0.1 mm, say 0.09 or 0.08 mm, ε_{eff} is smaller than the ε_1 of 3.66. Therefore, n assumes the limiting value for which ε_{eff} is just greater than ε_1. It is obvious that for larger n, say 0.11 or any higher value, the ε_{eff} becomes greater than ε_1, but due to the iterative nature of the model, the value of ε_{eff} becomes impractically larger for thicker superstrates. As a good assumption, a starting value of $n = 0.01$ can be chosen, and it ends at the value for which $\varepsilon_{\text{eff}} > \varepsilon_1$ with a step increment of 0.01.

***Step* 2:** Next, determine the values of W_{ef} and V_ε from the parameters chosen in step 1 using (10.26) and (10.27).

***Step* 3:** From the chosen and obtained parameters in steps 1 and 2, compute the filling fractions q_o, q_1 and $q_{2\text{new}}$ with the help of (10.28), (10.29) and (10.31). Subsequently, determine the number of steps, N, using round ($\Delta h/n$) in MATLAB to find the maximum iterations (no. of blocks) required to calculate ε_{eff}. The process of determination of N is explained with the help of two examples. For $\Delta h = 1.524$ mm and $n = 0.1$ mm, the computed $N = 15$, which implies that the loop in figure 10.28 runs 15 times. It also implies that the modified Δh for which the model calculates ε_{eff} now becomes $0.1 \times 15 = 1.5$ mm instead of 1.524 mm. Similarly, for $\Delta h = 1.575$ mm and $n = 0.1$ mm, N equals 16, and the modified Δh for which ε_{eff} is determined becomes 1.6 mm. It can be seen that the model may underestimate or overestimate the value of ε_{eff} based on the values of Δh. However, this difference is very small and does not have a significant impact.

***Step* 4:** Initialize a counter, i. The dielectric constant of the substrate, ε_1, gets updated in each loop. As a starting $\varepsilon_i = \varepsilon_1$.

(**Note:** ε_i for $i = 1, 2, \ldots, N$ should not be confused with the permittivities of substrates and superstrates).

***Step* 5**: Compute ε_{eqs} and ε_{eff} from (10.36) and (10.37). Then, the newly computed ε_{eff} is assigned to ε_1. Calculate the open-end extension of the patch ΔL due to the fringing field with the help of the expression in this step [32].

***Step* 6**: Next, determine the f_r of the MPA given by (10.38) using the values ε_{eff}, ΔL and L from steps 1 and 5:

$$f_r = \frac{c}{2(L + 2\Delta L)\sqrt{\varepsilon_{\text{eff}}}}. \tag{10.38}$$

***Step* 7**: Determine the percentage variation in resonant frequencies between the simulated and theoretical values with the help of (10.39):

$$\%\text{variation} = \frac{|f_r(\text{Sim}) - f_r(\text{Th})|}{f_r(\text{Sim})}. \tag{10.39}$$

***Step* 8**: Finally, use a conditional check, such that if $i \leqslant N$, then the counter is incremented by $i = i + 1$ else the loop breaks, which indicates completion of all steps.

(**Note:** Since frequency dispersion generally decreases for a wider conductor ($W/h > 1$), it has hence not been considered in this model).

10.4.5.4 Theoretical and simulation analysis

Comparison of the theoretically obtained f_r with the simulated values in the CST for an MPA covered with a single superstrate layer is included in this section. First, analysis of the formulations and model on PCB is done, and then the same for the textile layers is carried out. As a start, an MPA on a RO4350B board with dielectric constant and thickness of 3.66 and 1.524 mm is designed in the CST. The optimized physical parameters of an MPA operating at 2.38 GHz in the CST are mentioned in table 10.3. An f_r of 2.38 GHz is chosen as this is the center frequency of the band (2.36–2.4 GHz) allocated for medical applications by the FCC. Quarter-wave matching to feed the patch of the MPA with a primary feed line of 50 Ω is required. The electrical and physical parameters of the line and MPA are also given in table 10.3.

The designed MPA is then covered by superstrate layers, RO4350B, RO5880 ($\varepsilon_2 = 2.2$) and FR4 ($\varepsilon_2 = 4.7$), in such a way that all the conditions $\varepsilon_1 = \varepsilon_2$ (case 1), $\varepsilon_1 > \varepsilon_2$ (case 2) and $\varepsilon_1 < \varepsilon_2$ (case 3) are met. Tables 10.4–10.6 list the theoretical and simulated results for all the three cases up to a superstrate height of 4.5 mm. The

Table 10.3. Physical and electrical parameters of the MPA, and feed line designed on RO4350B.

Antenna para.	Value	Antenna para.	Value	Antenna para.	Value
Length, L (mm)	31.5	Patch impedance	283 Ω	Length of $\lambda/4$ line (mm)	17
Width, W (mm)	41.3	Quarter-wave transformer ($\lambda/4$) impedance	119 Ω	Width of primary feed (50 Ω) line (mm)	3.6
Effective dielectric constant	3.34	Width of $\lambda/4$ line (mm)	0.5	Length of 50 Ω line (mm)	6

Table 10.4. Comparison of simulated and theoretically obtained resonant frequencies of the MPA in this work with [29] and [30] for a superstrate of dielectric constant $\varepsilon_2 = 3.66$ (Case 1: $\varepsilon_1 = \varepsilon_2$).

Δh (mm)	ε_{eff} [29]	ε_{eff} [30]	ε_{eff} [this]	Sim. freq. (GHz) (CST)	Th. freq. (GHz) [29]	Th. freq. (GHz) [30]	Th. freq. (GHz) [This]	% var [29]	% var [30]	% var [this model]
1.0	3.60	3.50	3.83	2.317	2.401	2.434	2.338	3.61	5.05	0.91
1.5	3.62	3.51	3.92	2.305	2.392	2.429	2.312	3.80	5.40	0.30
3.0	3.69	3.54	4.13	2.284	2.372	2.421	2.268	3.86	5.95	0.70
4.5	3.74	3.56	4.30	2.270	2.358	2.414	2.230	3.88	6.34	1.76

Notes: $\Delta h = h_2 – h$, $L = 31.5$ mm, $W = 41.3$ mm, $\varepsilon_1 = 3.66$, $h_1 = 1.524$ mm at 2.38. (%var signifies % variation).

Table 10.5. (Case 2: $\varepsilon_1 > \varepsilon_2$, where $\varepsilon_1 = 3.66$ and $\varepsilon_2 = 2.2$ RO5880).

Δh (mm)	ε_{eff} [29]	ε_{eff} [30]	ε_{eff} [this]	Sim. freq. (GHz) (CST)	Th. freq. (GHz) [29]	Th. freq. (GHz) [30]	Th. freq. (GHz) [this]	% var [29]	% var [30]	% var [this model]
1.0	3.59	3.47	3.83	2.347	2.403	2.443	2.336	2.40	4.10	0.45
1.5	3.62	3.48	3.91	2.341	2.395	2.441	2.320	2.32	4.27	1.20
3.0	3.67	3.49	3.98	2.330	2.377	2.437	2.292	2.06	4.62	1.58
4.5	3.72	3.50	4.17	2.316	2.362	2.434	2.243	1.82	4.90	3.15

Table 10.6. (Case 3: $\varepsilon_1 < \varepsilon_2$, where $\varepsilon 1 = 3.66$ and $\varepsilon_2 = 4.7$ FR4).

Δh (mm)	ε_{eff} [29]	ε_{eff} [30]	ε_{eff} [this]	Sim. freq. (GHz) (CST)	Th. freq. (GHz) [29]	Th. freq. (GHz) [30]	Th. freq. (GHz) [This]	% var [29]	% var [30]	% var [this model]
1.0	3.60	3.52	3.83	2.305	2.400	2.428	2.328	4.12	5.34	1.00
1.5	3.63	3.53	3.92	2.287	2.391	2.422	2.305	4.57	5.89	0.77
3.0	3.67	3.57	4.13	2.260	2.370	2.409	2.250	4.87	6.60	0.44
4.5	3.76	3.60	4.30	2.242	2.352	2.401	2.210	4.92	7.10	1.43

tables also compares the obtained results obtained with other techniques [29, 30]. The following inferences can be made from tables 10.4–10.6.

- Even in the absence of frequency dispersion, the formulation and model performs better than the conventional techniques [29, 30].
- It is apparent from case 2 that the % variation is higher in the case of $\Delta h = 4.5$ mm w.r.t. [30]. Apparently, for a superstrate of $\varepsilon_2 \leqslant 2.2$ and smaller heights (1–3 mm), the discussed model and the formulation determine ε_{eff} and f_r with high accuracy, but accuracy decreases with more significant heights.
- It is apparent from the first and second rows of tables 10.4 and 10.6 that the % variation of the presented work goes slightly out of trend. But careful observation of (10.39) and the values in tables 10.4 and 10.6 attests to the fact that the % variation increases from a more negative value to a more positive.

Next, the presented model and the formulations on the designed MPA (on RO4350B) when covered with textile layers as the superstrate is considered. From the WBAN perspective, the antenna in the WSN has to send data through multiple layers of textiles such as cotton, rayon, polyester, wool, etc. Often there may be a

Table 10.7. Comparison of simulated and theoretically obtained resonant frequencies of the MPA covered by different textile layers as superstrate in this work with [29] and [30].

Mat.	Δh (mm)	ε_{eff} [29]	ε_{eff} [30]	ε_{eff} this	Sim. freq. (GHz) (CST)	Th. freq. (GHz) [29]	Th. freq. (GHz) [30]	Th. freq. (GHz) [this]	%var [29]	%var [30]	%var this model
1	0.3	3.56	3.45	3.71	2.372	2.408	2.442	2.3655	1.52	3.00	0.27
2	0.3	3.56	3.46	3.71	2.373	2.407	2.442	2.3620	1.50	2.91	0.46
3	0.2	3.55	3.45	3.70	2.374	2.409	2.442	2.3670	1.48	2.86	0.29
4	0.2	3.56	3.46	3.70	2.365	2.408	2.438	2.3635	1.82	3.09	0.06
5	1.5	3.60	3.45	3.91	2.363	2.397	2.442	2.3007	1.46	3.35	2.64
6	1.5	3.61	3.46	3.91	2.358	2.392	2.441	2.3005	1.45	3.54	2.44
7	1.5	3.61	3.47	3.91	2.353	2.390	2.438	2.3000	1.58	3.62	2.25

Notes: Mat. Signifies material; 1—Polyester; 2—Jeans cotton; 3—Pure cotton; 4—Rayon; 5—Felt fabric; 6—Terry wool; 7—Leather.

combination of textiles such as in a woolen jacket over cotton, felt jacket over polyester, and so on. Therefore, the analysis with textile layers can be divided into two sets. In the first set, the performance of the presented technique on textile layers which are immediately next to the MPA, such as cotton, rayon, and polyester, is assessed. In the second set, a combination of two textile layers, like wool and felt fabric (i.e. a hoodie), is considered to understand their behavior. This chapter, however, limits the analysis to an MPA covered with a single layer. Table 10.7 lists the f_r for an MPA covered with some conventional textiles obtained from the discussed model and other CM-based models in the literature [29, 30]. It can be concluded from the results that for a thinner superstrates (0.2–0.3 mm), the % variation obtained from the discussed model is superior when compared to [29, 30]. However, for the thicker superstrates (1–1.5 mm), the % variation increases when compared to [29]. Both results agree with the trend in table 10.5. It can also be inferred from tables 10.5 and 10.7 that for superstrates with $\varepsilon_2 = 2.2$ and lesser height, the % variation for the model is shallow when compared to [29, 30]. However, for superstrates with similar dielectric constant and more considerable heights, the % variation increases when compared to [29].

10.4.5.5 Experimental evaluation

In this section, the effectiveness of the discussed technique is evaluated by prototyping two MPAs operating at 2.38 GHz with the parameters given in tables 10.3 and 10.8. The prototyped MPA on the RO4350B and FR4 substrates are shown in figures 10.29(a) and (b). For PCB superstrates, dielectric layers of RO4350B, RO5880 and FR4 of identical size to the MPA are selected. Then the VNA is setup for measurement by calibrating it from 1 GHz to 3.8 GHz with a center frequency at

Table 10.8. Physical and electrical parameters of the MPA, and feed line designed on FR4. This work is licensed under a Creative Commons Attribution 4.0 License. Reprinted, with permission, from [31].

Antenna para.	Value	Antenna para.	Value	Antenna para.	Value
L, (mm)	27.7	Patch impedance	281 Ω	Length of $\lambda/4$ line (mm)	15.8
W, (mm)	38.5	Quarter-wave transformer ($\lambda/4$) impedance	118 Ω	Width of primary feed (50 Ω) line (mm)	2.8
Effective permittivity	3.946	Width of $\lambda/4$ line (mm)	0.37	Length of 50 Ω line (mm)	5.6

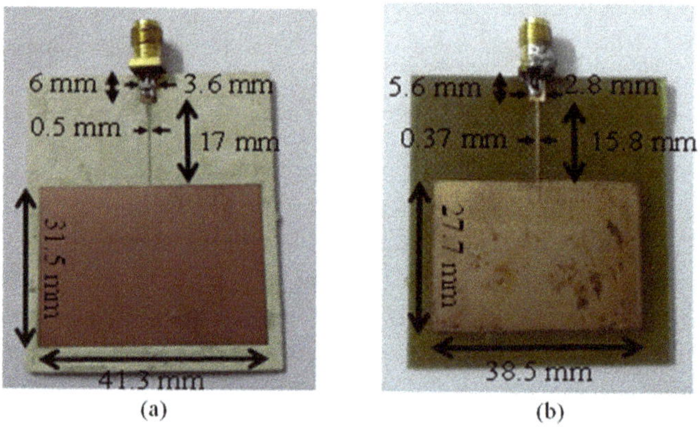

Figure 10.29. (a) Prototyped MPA on RO4350B. (b) Prototyped MPA on FR4. This work is licensed under a Creative Commons Attribution 4.0 License. Reprinted, with permission, from [31].

2.4 GHz. The successive layers of dielectrics are attached with the help of low-permittivity adhesive RTV coating ($\varepsilon_r = 1.22$). The purpose of using an adhesive of low permittivity is to nullify the effect of dielectric loading on the MPA and cause frequency shift due to change in dielectric constant. For an MPA covered with textiles, the measurement of f_r in the VNA is carried out by placing either a single layer over it or multiple layers. Proper care needs to be taken while performing measurements with textiles such that the formation of a gap does not take place between the MPA and layers.

At first, the cases given in Tables 10.4–10.6 are assessed through measurement. Figures 10.30(a) and (b) depict the measurement setup covered with a single and two layers of RO4350B of $h = 1.524$ mm and 3.048 mm. Proper care is required to apply the adhesive (RTV coating) on the dielectric layer such that no air gap is formed at the interface. Moreover, sufficient space at the feed location of the MPA is left to suitably place the dielectric layers over the MPA in the presence of the SMA

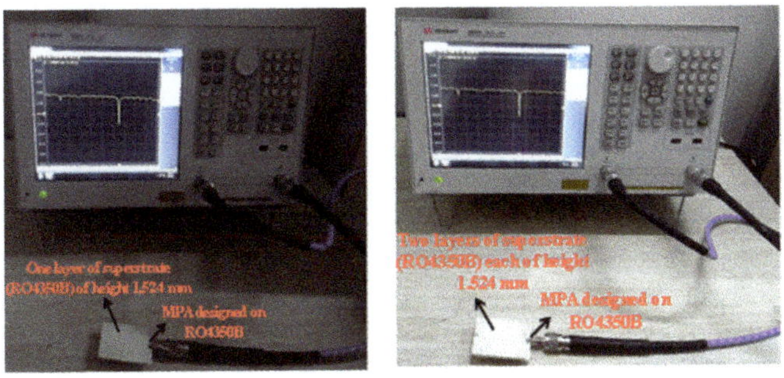

(a) (b)

Figure 10.30. Measurement setup of fabricated prototype on RO4350 substrate and superstrate for (a) Δh = 1.524 mm exhibiting f_r of 2.302 GHz, and (b) Δh = 3.048 mm exhibiting f_r of 2.29 GHz. This work is licensed under a Creative Commons Attribution 4.0 License. Reprinted, with permission, from [31].

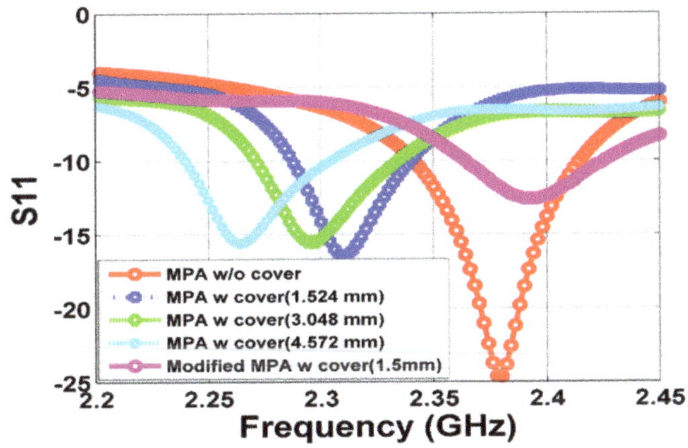

Figure 10.31. Measured return losses of an MPA on RO4350B with and without the cover of Δh = 1.524, 3.048 and 4.572 mm. This work is licensed under a Creative Commons Attribution 4.0 License. Reprinted, with permission, from [31].

connector, as can be seen in figures 10.30(a) and (b). The plot of S_{11} versus frequency in figure 10.31 depicts the measured f_r of the standalone MPA along with the MPA covered with RO4350B for all thicknesses mentioned in table 10.9. For the other two cases, the measurement results are listed in tables 10.10–10.11. It is apparent from tables 10.9 and 10.10 that the Δh, for which f_r is measured, may differ from the values used for the determination of theoretical f_r in tables 10.4 and 10.5. For example, a Δh of 1.524 mm becomes Δh = 1.5 mm following step 3 of the model in figure 10.28. Similarly, for superstrate heights Δh = 3.04, 4.572, 1.575, 3.15 and 4.725 mm, the modified Δh following step 3 becomes 3, 4.6, 1.6, 3.2 and 4.7 mm, respectively. Therefore, it can be concluded that the model tends to under/over-estimate the value of ε_{eff}. Moreover, the theoretically determined values of f_r for new Δh are also given in tables 10.9 and 10.10. Interestingly, the % variation with and

Table 10.9. Comparison of measured and theoretically obtained resonant frequencies of the MPA for superstrate of dielectric constant $\varepsilon_2 = 3.66$ (case 1: $\varepsilon_1 = \varepsilon_2$).

Δh (mm)	Δh (mm) (theoretical)	Meas. freq. (GHz)	Th. freq. (GHz) [29]	Th. freq. (GHz) [30]	Th. freq. (GHz) [This]	% var [29]	% var [30]	% var [this]
1.524	1.5	2.302	2.392	2.429	2.312	3.55	5.15	0.43
3.048	3.0	2.290	2.372	2.421	2.268	3.42	5.50	0.96
4.572	4.6 (4.5)	2.262	2.358 (2.358)	2.416 (2.414)	2.220 (2.23)	4.25 (4.24)	6.80 (6.72)	1.85 (1.41)

Table 10.10. Comparison of measured and theoretically obtained resonant frequencies of the MPA for superstrate of dielectric constant $\varepsilon_2 = 2.2$ (case 2: $\varepsilon_1 > \varepsilon_2$).

Δh (mm)	Δh (mm) (theoretical)	Meas. freq. (GHz)	Th. freq. (GHz) [29]	Th. freq. (GHz) [30]	Th. freq. (GHz) [this]	% var [29]	% var [30]	% var [this]
1.575	1.6 (1.5)	2.338	2.394 (2.395)	2.441 (2.441)	2.310 (2.320)	2.40 (2.32)	4.40 (4.27)	1.25 (1.20)
3.150	3.2 (3.0)	2.327	2.375 (2.377)	2.436 (2.437)	2.285 (2.292)	2.06 (2.06)	4.70 (4.62)	1.80 (1.60)
4.725	4.7 (4.5)	2.312	2.358 (2.362)	2.432 (2.434)	2.237 (2.243)	2.00 (1.82)	5.23 (4.90)	3.24 (31.5)

Table 10.11. Comparison of measured and theoretically obtained resonant frequencies of the MPA for superstrate of dielectric constant $\varepsilon_2 = 4.7$ (case 3: $\varepsilon_1 < \varepsilon_2$).

Δh (mm)	Meas. freq. (GHz)	Th. freq. (GHz) [29]	Th. freq. (GHz) [30]	Th. freq. (GHz) [this]	% var [29]	% var [30]	% var [this]
1.5	2.290	2.391	2.422	2.305	4.41	5.76	0.66
3.0	2.262	2.370	2.409	2.250	4.80	6.50	0.53
4.5	2.245	2.352	2.401	2.210	4.80	6.95	1.56

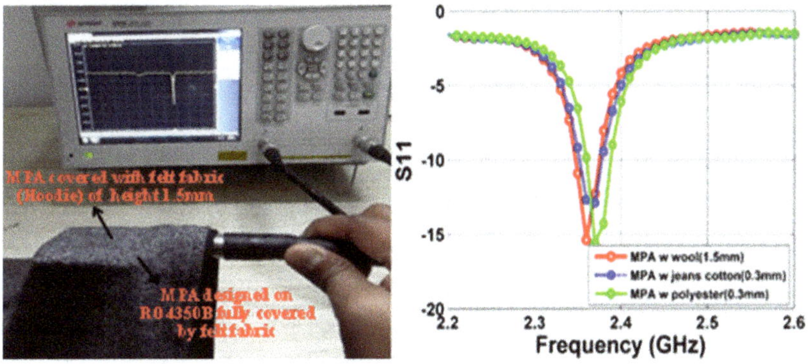

(a) (b)

Figure 10.32. (a) Measurement setup of the MPA covered with felt fabric (hoodie). (b) Return loss of the MPA covered with wool, jeans cotton and polyester with Δh = 1.5, 0.3 and 0.3 mm. This work is licensed under a Creative Commons Attribution 4.0 License. Reprinted, with permission, from [31].

without modified Δh is very small. Hence, it is safe to convey that the measured/simulated and theoretical results show excellent agreement despite the presence of inconsistency in the selection of Δh.

Next, the theoretically obtained f_r's of an MPA covered with a single textile layer with those of measured results are validated. For this purpose, the prototyped MPA in figure 10.29(a) is again used as an radio frequency source. The measurement setup of the MPA covered with a textile layer (here felt fabric) is shown in figure 10.32(a). During the measurement with textiles, proper care needs to be taken to avoid the formation of an air gap between the layer and MPA. However, due to the presence of the SMA connector at the top, a bulge appears, as is evident from figure 10.32(a). A layer of textile was pressed over the MPA with considerable force applied at the bottom to address this concern. As radiation is taking place near the patch, which is well covered by the textile layer, this air gap can be avoided. The S_{11} plot in figure 10.32(b) depicts the measured f_r of an MPA covered with wool, jeans, cotton and polyester with respective Δh of 1.5, 0.3 and 0.3 mm. Interestingly, an excellent resemblance to the shift of f_r with that of theoretical values can be seen. Table 10.12 lists the measured values of f_r for the cases listed in table 10.7. The measured results are in excellent agreement with those of the theoretical values.

10.4.5.6 Restoration of original frequency

In this subsection, the original f_r of an MPA covered by a dielectric layer is restored by altering the physical parameters based on new ε_{eff}. This analysis is critical to proving the effectiveness of the discussed model from the WBAN perspective. In the first step, we determine the ε_{eff} of an MPA covered with a dielectric layer (PCB/textile). Next, the new physical and electrical parameters of the MPA need to be recalculated by replacing the ε_1 of the substrate with a new ε_{eff}. Lastly, the modified

Table 10.12. Comparison of measured and theoretically obtained resonant frequencies of the MPA covered with single-textile superstrates.

Mat.	Δh (mm)	Meas. freq. (GHz)	Th. freq. (GHz) [29]	Th. freq. (GHz) [30]	Th. freq. (GHz) [this]	% var [29]	% var [30]	% var [present]
1	0.3	2.375	2.408	2.442	2.3655	1.46	2.84	0.40
2	0.3	2.373	2.407	2.442	2.3620	1.50	2.91	0.46
3	0.2	2.371	2.409	2.442	2.3670	1.61	3.00	0.17
4	0.2	2.367	2.408	2.438	2.3635	1.73	3.00	0.15
5	1.5	2.363	2.397	2.442	2.3007	1.46	3.35	2.64
6	1.5	2.358	2.392	2.441	2.3005	1.45	3.54	2.44
7	1.5	2.359	2.390	2.438	2.3000	1.32	3.36	2.50

Notes: Material (ε_2): 1—Polyester (1.4); Jean cotton (1.6); Pure cotton (1.7); Rayon (3.2); Felt fabric (1.22); Terry wool (1.3); Leather (1.66).

MPA is either simulated or measured to check the operation. The condition in row 3 of table 10.4 is used to elaborate on this aspect. For the given parameters in table 10.4, the ε_{eff} comes out to be 3.69, 3.54 and 4.13 for [29, 30], and for the model discussed in this chapter. To restore the original f_r of the MPA, i.e. 2.38 GHz, the modified length is obtained with the help of (10.40). Here, L^m is the modified length of the MPA:

$$f_r = \frac{c}{2(L^m + 2\Delta L)\sqrt{\varepsilon_{\text{eff}}}}. \tag{10.40}$$

The substitution of the known and determined values of $f_r = 2.38$ GHz, $c = 3 \times 10^8$, $\varepsilon_{\text{eff}} = 3.69$, 3.54 and 4.13, and $\Delta L = 0.34$ mm in (10.40) results in L^m of 32.46 mm, 33.15 mm and 29.37 mm. Consequently, for these modified values of length, the MPA resonates at 2.352, 2.34 and 2.392 GHz, respectively, for the techniques in the literature [29, 30] and the model discussed in this chapter. It becomes apparent from the calculated values that only the discussed model in this chapter predicts f_r closest to the desired resonance. The measured f_r of the modified MPA is depicted in figure 10.31. A slight shift in f_r from the original one can be attributed to the overestimation of ε_{eff} by the discussed model (Δh = 4.6 mm was used in the calculation instead of the actual thickness of 4.572 mm). It implies that, for the given case study, the obtained f_r from the discussed model is lower than the simulated value, and this results in a L^m that is smaller than the desired length to bring resonance at 2.38 GHz. Therefore, the lesser % variation leads to higher accuracy in the determination of L^m, and that, in turn, leads to better accuracy in the restoration of the original f_r even with the cover. Since only the length of the MPA is varied, keeping the width intact, this leads to the poor return loss, as is evident from figure 10.31.

10.4.6 Conformal electromagnetic band gap surfaces

An essential feature of wearable devices is conformability for the comfort of users. It necessitates the design of conformal antennas suitable for placement on shoulders, biceps, wrists and legs [20–22]. These aspects have created an increased interest in the development of conformal EBGs. Figure 10.33 shows two scenarios in which conformed antenna for WBAN applications are placed on the wrist and triceps.

The fundamental challenge in the design of conformal structures is that bending leads to alteration of the in-phase property, and hence changes occur in the performance of the conformed antenna [22]. Therefore, either a unit cell of larger width is required so that, even after bending, the desired band is covered [21, 22], or appropriate prior analysis is carried out for phase-compensated design [26]. However, increasing the cell width is often not feasible, considering that wearable applications require miniaturization. Furthermore, the determination of the in-phase property in simulation is a complex process as the meshing of a conformal EBG unit cell or EBG array is critical. A discontinuity in the edges might appear while conforming to a unit cell or array, and hence an increase in the number of mesh cells is needed to smoothen the computational area [28]. As a consequence, the simulation time is increased and, consequently, so too the memory allocation for the entire UC/array. This aspect can be addressed by appropriate analysis a priori for the conformed antenna structure.

Figure 10.34(a) shows the unit cell discussed in section 10.4.3. The conformed unit cells along the x-axis (the radius of the curvatures are 40 and 50 mm) are shown in figures 10.34(b) and (c). The reflection phase plots of the planar and conformal unit cell, when terminated by UC-BC and excited by a TE wave, are given in figure 10.35(a). Points A, B and C represent the in-phase frequency points of the planar and conformal unit cells. This is because, at these points, the reflected wave is in phase with the incident wave falling on the surface of a unit cell. The planar cell exhibits an in-phase resonance at 2.38 GHz and has a total bandwidth of

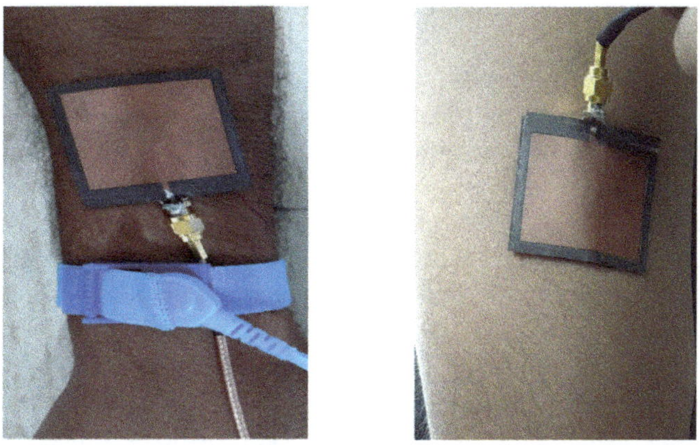

(a) (b)

Figure 10.33. (a) Conformed antenna on the wrist. (b) Conformed antenna on triceps.

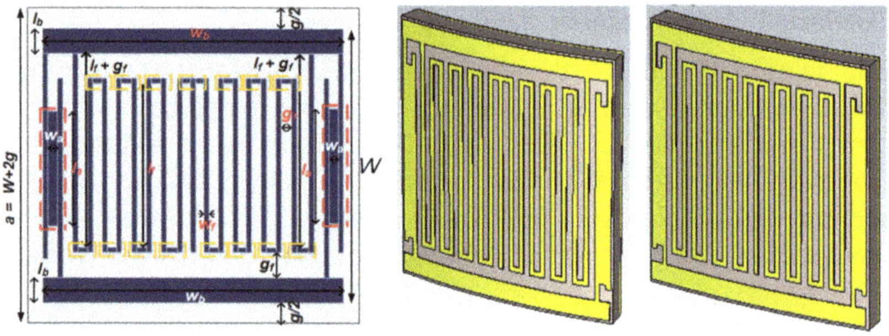

(a) (b) (c)

Figure 10.34. (a) IDE unit cell, with all dimensions given in figure 10.4. Conformed IDE unit cells along the *x*-axis: (b) $r = 40$ mm, (c) $r = 50$ mm.

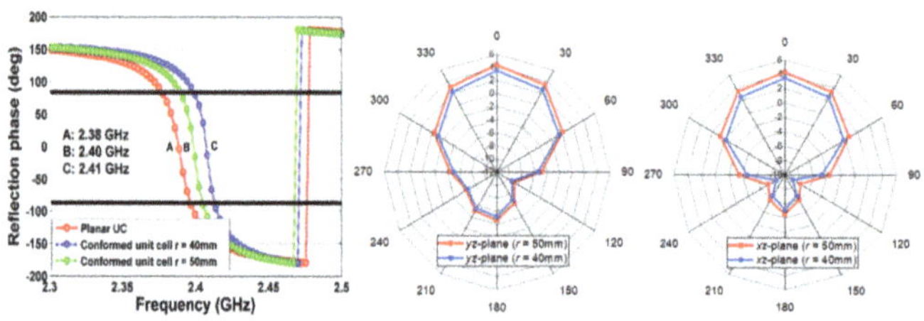

Figure 10.35. (a) Reflection phases of planar and conformal IDE cell. (b) Radiation pattern of the conformal antenna along the *y–z* plane. (c) Radiation pattern of the conformal antenna along the *x–z* plane.

approximately 40 MHz (covering the MBAN band from 2.36–2.4 GHz, frequencies between ±90°). However, the conformed unit cells resonate at 2.4 and 2.41 GHz for $r = 50$ and 40 mm, respectively. Besides, the bandwidth of the conformed unit cells, as apparent from figure 10.35(a), is out of the desired band (2.38–2.42 GHz for $r = 40$ mm and 2.4–2.44 GHz for $r = 50$ mm).

The performance of an antenna heavily relies on the band of operation of an EBG cell. Therefore, this frequency shift deteriorates the overall performance of the antenna. For example, the gain drops from 5.51 to 4.7 dBi and 3.7 dBi in the *y–z* plane for a radius of 50 and 40 mm for the planar antenna in section 10.4.4.2 when conformed along the the *x*-axis, as shown in figure 10.35(b). Similarly, the gain drops from 5.5 dBi to 4.6 dBi and 3.7 dBi in the *x–z* plane for a radius of 50 and 40 mm, as shown in figure 10.35(c). The shift in the in-phase frequency does not have an impact on the bandwidth of the EBG unit cell. Therefore, if the EBG unit cell is designed in such a way that it takes into account the shift a priori then the degradation in the performance of an antenna can be avoided. This phenomenon can be further explained with the help of another standard EBG unit cell in [22], shown in figure 10.36(a). The unit cell is designed on RO5880 with ε_r and h of 2.2 and 5 mm, respectively, and it operates in the ISM band from 2.4–2.4835 GHz. The planar and conformed geometries of the unit cells are shown in figures 10.36(b) and (c).

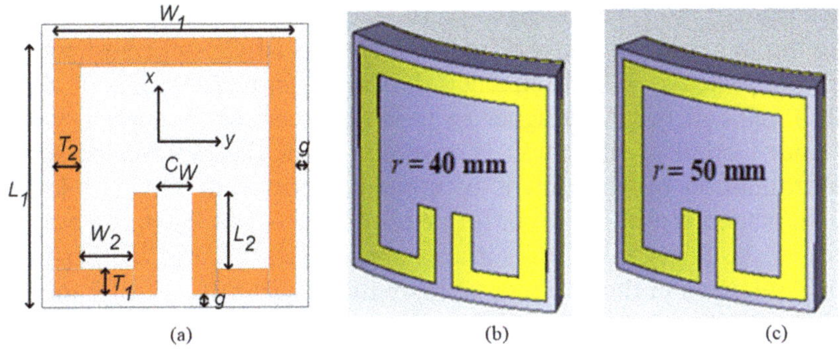

Figure 10.36. (a) Geometry of the unit cell in [15] with overall dimensions W_1 = 27.83 mm, W_2 = 6.4 mm, L_1 = 25.3 mm, L_2 = 6.32 mm, T_1 = 2.53 mm, T_2 = 3.34 mm, C_W = 2.78 mm and g = 1.25 mm. Conformed unit cells on a cylinder along the x-axis: (b) r = 40 mm, (c) r = 50 mm.

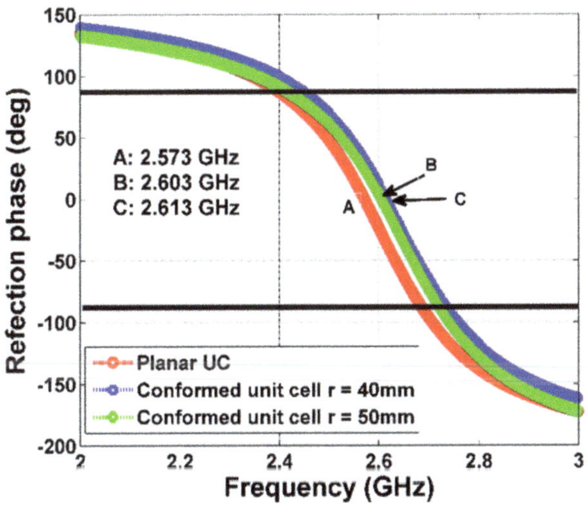

Figure 10.37. Reflection phases of the planar and conformal unit cell in figures 10.36(a)–(c).

The simulated reflection phase plots are depicted in figure 10.37. Once again, points A, B and C signify the in-phase frequency points of the planar and conformal unit cells. It is apparent from the plots that, though the conformal geometries exhibit shifts in the in-phase frequency, the intended band (i.e. the ISM band) is still covered. This is because the operational bandwidth of the planar EBG unit cell is from 2.39–2.65 GHz (approximately seven times more than the unit cell in figure 10.34). A shift of 30 MHz (for 50 mm) and 40 MHz (40 mm) in the in-phase frequency points, as apparent from figure 10.37, does not have a significant impact on the antenna's operation. It is known that the bandwidth of an EBG unit cell is inversely proportional to the capacitance developed between the top metallic patch and the ground plane. Now, the following can be inferred from this discussion.

- Determination of shift in the in-phase frequency of a conformal EBG unit cell can aid in getting better performance from an EBG-backed antenna. Having prior knowledge of the frequency shift in the in-phase points of a conformal EBG cell can help a designer to modify the circuit in such a way that the intended band of operation is covered. From the WBAN perspective, the determination of this shift is of utmost importance. This is because body-worn antennas need to be conformed for suitable placement on wrists, biceps, legs or shoulders. As EBG surfaces are used to enhance the performance of such antennas in the presence of lossy tissues [20–22], out-of-band operation might result in link failure with an external node.
- Thicker substrates can help to tackle the shift in the in-phase frequency of an EBG surface. However, from the WBAN perspective, the body-worn antennas need to be low profile and compact. Moreover, thicker PCBs are hard to bend, and this may damage the patch surface due to the application of high force during conforming.
- There are two reports in the literature that analytically compute the reflection phase of a conformal EBG surface [26, 27]. In [26], the reflection phase of a curved EBG surface is obtained from the scattered fields. The scattered fields, in turn, can be obtained from the analytical model or a full-wave solution simulator like HFSS. This technique also incorporated surface wave effects to strengthen the theory further. In [27], a phase-compensated metasurface was designed for a Fabry–Pérot cavity antenna. In this method, a group of 25 wire antennas are placed over a metallic cylinder. Then, each antenna is excited by a phase shift that compensates the phase lag from the planar surface. This technique when applied on a conformal partially reflecting surface shows improvement in gain w.r.t. an uncompensated antenna. However, both techniques have some drawbacks from the WBAN perspective. These techniques are often not suitable for miniature symmetric designs and discard many aspects such as mutual coupling between cells in the array configurations.
- The present chapter reports a conformability analysis technique for WBAN-related issues and this is appropriate both for the symmetric and symmetric structures. This is explained in brief in the next section.

10.4.6.1 Determination of phase shift

A planar EBG unit cell (P-EBG UC) and the corresponding conformal (C-EBG UC) geometry on a circle of radius $\boldsymbol{r}$ (represented by an arc) are depicted in figure 10.38 (a). The P-EBG UC is separated from the C-EBG UC and is shown on top of it. The center of the circle is represented by O, whereas A, B and C are the extreme and mid-points of the arc. The angle subtended by the arc AB about O is θ while the angle formed between the arc AC/BC and O is $\theta/2$. The incident waves XA, YB (in light brown) and C′C (in dotted blue) on the C-EBG UC, along with the excitation plane, are shown in figure 10.38(b). The term $\hat{J}$ is normal to A and B on the arc of the circle and intersects the excitation plane at X′ and Y′, respectively. To obtain the reflection phase of an EBG cell, the excitation plane is placed at a distance equivalent to the height by which an antenna is separated, as shown in

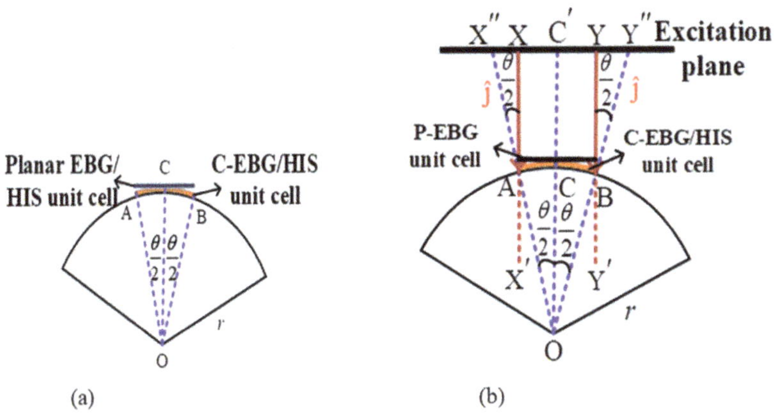

Figure 10.38. (a) Planar and conformed EBG/HIS unit cell of radius r with extreme points A, B and mid-point C (where P-and C-EBG signifies planar and conformal EBGs). (b) Arc representing a C-EBG cell with an incident EM wave making an angle $\theta/2$ at the two extreme points A and B w.r.t. normal vector $\hat{\jmath}$.

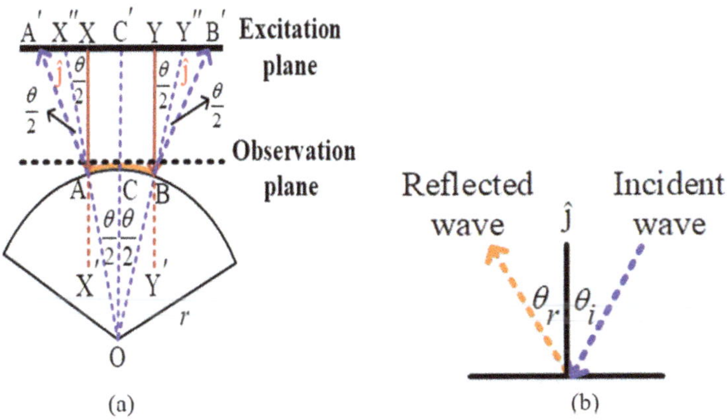

Figure 10.39. (a) Law of reflection on a planar surface (here, $\hat{\jmath}$ is normal to the interface). (b) Reflection of an EM wave on a C-EBG cell (the dotted line represents observation plane).

figure 10.38(b). Here, the height is not specified; however, it is assumed that the excitation plane is away from the C-and P-EBG UC by some length. XAX′, YBY′, and C′CO are parallel to each other as only obliquely incident waves on C-EBG UC are considered in this study. Two transversals OX′ and OY′ intersect these parallel lines, and therefore it can be deduced from elementary geometry that ∠AOC =∠X′ AX and ∠ BOC =∠Y′BY, as expressed in (10.41). For an incident EM wave, the in-phase reflection of a P-EBG UC can be obtained by positioning the observation plane at its surface, as depicted in figure 10.39(a). From the law of reflection, as shown in figure 10.39(b), *the angle of incidence* (I_i) *on an interface is equal to the angle of reflection* ($\boldsymbol{\theta_r}$). This concept is applied for the C-EBG UC in figure 10.39(a), where incident waves XA and YB make angle $\boldsymbol{\theta/2}$ about the normal $\hat{\jmath}$ (AX′ and BY′) and are reflected by an identical angle (AA′ and BB′) as expressed by (10.42).

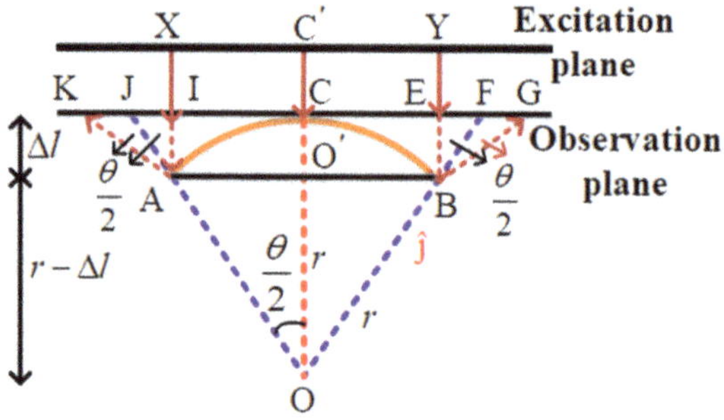

Figure 10.40. Magnified representation of reflection at the C-EBG surface along with the path lag an EM wave experiences while interacting with the C-EBG cell.

Apparently, both the extreme regions A and B of C-EBG UC behave identically for an incident EM wave.

$$\angle AOC' = \angle XAX'' = \angle BOC' = \angle YBY' = \boldsymbol{\theta}/2, \tag{10.41}$$

$$\angle XAX'' = \angle YBY'' = \angle X''AA' = \angle Y''BB' = \boldsymbol{\theta}/2, \tag{10.42}$$

$$W = \frac{\boldsymbol{\theta}}{360^o} \times 2\pi \boldsymbol{r} \Rightarrow \boldsymbol{\theta} = \frac{W \times 360^o}{2\pi \boldsymbol{r}}. \tag{10.43}$$

The angle $\boldsymbol{\theta}$ subtended by arc AB about the center O, as shown in figure 10.39(a), can be determined using (10.43). Here, W is the width of the EBG cell, $\boldsymbol{r}$ is the radius of the circle and 360° is the total angle at the center. W is identical for both the C-EBG UC and P-EBG UC because the conforming of a P-EBG cell on a conformal platform does not change its physical dimension.

It can be seen in figure 10.40 that the incident EM waves (IA/EB) while interacting with the C-EBG UC experience path lag due to reflected waves AK and BG instead of the reflection AX and BY to be seen in P-EBG UC. The path lag can also be experienced by incident waves IA and EB because IE lies on the observation plane for P-EBG UC. In contrast, AB is separated from the observation plane by $\Delta \boldsymbol{l}$ for C-EBG UC. The equivalent phase shift in terms of electrical length for the incident wave IA ($\Delta \boldsymbol{l}$) and reflected wave AK in figure 10.40 is expressed in (10.44) and (10.45). Here, β is the wave number at the operating frequency of the P-EBG UC; $\Delta \boldsymbol{l}$ and AK are the path delays due to incident and reflected waves. The use of the Pythagorean theorem in Δ IAK and Δ OO′A allows the determination of AK in terms of $\Delta \boldsymbol{l}$ given in (10.46) and $\Delta \boldsymbol{l}$ in terms of $\boldsymbol{r}$ given in (10.47). Then, the sum of individual phase lags due to incident and reflected waves gives the net shift in the in-phase reflection of the C-EBG UC. Thus the total change in radians, given by (10.48), is twice that value, considering that a similar phenomenon occurs at point B.

$$\Delta \boldsymbol{\theta}_{\text{incident}} = \beta \times (\text{IA}) = \beta \Delta \boldsymbol{l}, \tag{10.44}$$

$$\Delta \boldsymbol{\theta}_{\text{reflected}} = \beta \times \text{AK}, \tag{10.45}$$

$$\cos(\mathrm{IAK}) = \cos\boldsymbol{\theta} = \frac{\Delta \boldsymbol{l}}{\mathrm{AK}} \Rightarrow \mathrm{AK} = \frac{\Delta \boldsymbol{l}}{\cos\boldsymbol{\theta}}, \tag{10.46}$$

$$\cos\frac{\boldsymbol{\theta}}{2} = \frac{\mathrm{OO}'}{\mathrm{OA}} = \frac{\boldsymbol{r} - \Delta l}{\boldsymbol{r}} \Rightarrow \Delta \boldsymbol{l} = \boldsymbol{r} - \boldsymbol{r}\cos\frac{\boldsymbol{\theta}}{2}, \tag{10.47}$$

$$\boldsymbol{\theta}_{\text{shift}} = 2 \cdot (\Delta\boldsymbol{\theta}_{\text{incident}} + \Delta\boldsymbol{\theta}_{\text{reflected}}) = 2 \cdot \left(\beta\Delta \boldsymbol{l} + \frac{\beta\Delta \boldsymbol{l}}{\cos\boldsymbol{\theta}}\right). \tag{10.48}$$

It is apparent that there is no path lag for EM waves falling on point C of the observation plane. Here, only the maximum shift in the in-phase reflection for the unit cells is considered, and hence only the extreme points are chosen to determine the corresponding $\boldsymbol{\theta}_{\text{shift}}$ given by (10.48). This approach proves very effective, as can be seen from its application in one UC structure discussed below.

For the P-EBG cell as shown in figure 10.36(a) [22], it was identified that the P-EBG cell exhibits an in-phase reflection frequency of 2.45 and 1.413 GHz, which results in β of 51.3 and a 29.6 rad m^{-1} y- and x-polarized EM wave, respectively. Due to asymmetry in the structure, the polarization stability of the cell is low and the in-phase frequencies are far apart from each other. The polarization of the EM wave plays a critical role in the determination of the $\boldsymbol{\theta}_{\text{shift}}$ of a cell for an x- or y-bend, because for an x-bend, the plane of incidence of the EM wave coincides with the y-axis, and makes y-polarization suitable. Similarly, for a y-bend, the plane of incidence of the EM waves coincides with the x-axis and makes x-polarization convenient. The conformed EBG (C-EBG) cell along the x-axis on cylindrical platforms of radius $\boldsymbol{r}$ = 40, 50 and 60 mm are depicted in figures 10.41(a) and (b). A cylindrical platform is chosen, considering that it readily portrays the shape of body parts like a bicep, wrist, leg and many more in WBAN.

As a case study, here the $\boldsymbol{\theta}_{\text{shift}}$ of the C-EBG cell is calculated using the formulations for $\boldsymbol{r}$ = 50 mm. For this purpose, substitute $W_1 + 2g$ = 30.33 mm (from figure 10.36(a)) and $\boldsymbol{r}$ = 50 mm in (10.43), resulting in $\boldsymbol{\theta}$ = 34.75°. Then,

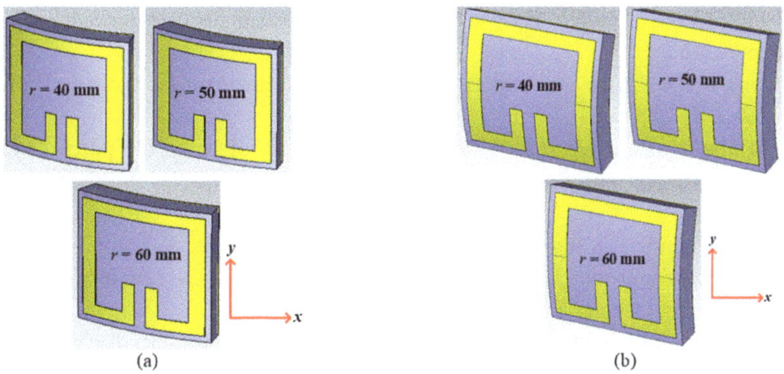

Figure 10.41. Conformed unit cells in [22] on cylinders of radius r = 40, 50 and 60 mm along (b) the x-axis and (c) the y-axis.

Table 10.13. Conformed UC-1 parameters determined by proposed method for y- and x-polarized EM waves.

Pol.	r (mm)	θ (deg)	Δl (mm)	$\Delta\theta_{\text{reflected}}$ (rad)	$\Delta\theta_{\text{incident}}$ (rad)	$\Delta\theta_{\text{shift}}$ (rad)	θ_{shift} (deg)
y–	40	43.44	2.84	0.211	0.153	0.728	41.71
	60	28.96	1.91	0.118	0.103	0.442	25.32
x–	40	39.82	2.39	0.098	0.075	0.346	19.80
	50	31.86	1.92	0.078	0.066	0.287	16.48
	60	26.54	1.60	0.056	0.050	0.212	12.17

$\Delta \boldsymbol{l} = 2.28$ mm is obtained for this value of $\boldsymbol{\theta}$ in (10.46). Placing the values of $\boldsymbol{\theta}$, $\Delta \boldsymbol{l}$ and β in (10.48) results in $\boldsymbol{\theta}_{\text{shift}}$ of 0.54 rad or 31.3°. The calculated values of $\boldsymbol{\theta}_{\text{shift}}$ for the other two radii are listed in table 10.13. It is apparent that a decrement in the radius of the curvature results in a gradual increase in $\boldsymbol{\theta}_{\text{shift}}$. This is obvious, because as $\boldsymbol{r}$ decreases (60 mm > 50 mm > 40 mm), it leads to a longer path delay, $\Delta \boldsymbol{l}$, due to incident and reflected waves, as shown in figure 10.40. Subsequently, the conformed cell along the y-axis on cylindrical platforms of radius $\boldsymbol{r}$ = 40, 50 and 60 mm are depicted in figure 10.41(b). Here, the overall width now becomes $L_1 + 2g = 27.8$ mm, and the EM wave is x-polarized. The calculated $\boldsymbol{\theta}_{\text{shift}}$'s for this case are listed in table 10.13. The value of β used in this case is 29.6 rad m^{-1}. Once again, $\boldsymbol{\theta}_{\text{shift}}$ reduces with a decrease in radii $\boldsymbol{r}$ due to longer path delay $\Delta \boldsymbol{l}$. This discussion gives rise to the idea that prior knowledge about the shift in the phase, and hence in the frequency, may enable an antenna designer to take appropriate design considerations well in advance for its appropriate operation.

References

[1] Rotariu C, Costin H, Andruseac G, Ciobotariu R and Adochiei F 2011 An integrated system for wireless monitoring of chronic patients and elderly people *Proc. 15th Int. Conf. Sys. Theory, Control and Computing* pp 1–4

[2] Wolf L and Saadaoui S 2007 Architecture concept of a wireless body area sensor network for health monitoring of elderly people *Proc. 4th IEEE Consumer Communicat and Networking Conf.* pp 722–6

[3] Armstrong S 2007 Wireless connectivity for health and sports monitoring: a review *Br. J. Sports Med.* **41** 285–9

[4] Rano D and Hashmi M S 2016 Design and analysis of wearable patch antenna array for MBAN applications *Proc. National Conf. on Communication NCC (Guwahati, India)* pp 1–6

[5] Saied I M, Chandran S and Arslan T 2019 Integrated flexible hybrid silicone-textile dual-resonant sensors and switching circuit for wearable neurodegeneration monitoring systems *IEEE Trans. Bio Circuits Syst.* **13** 1304-12

[6] Cao H, Leung V, Chow C and Chan H 2009 Enabling technologies for wireless body area networks: a survey and outlook *IEEE Comm. Mag.* **47** 84–93

[7] Chow E Y, Chlebowski A L and Irazoqui P P 2010 A Miniature-implantable RF-wireless active glaucoma intraocular pressure monitor *IEEE Trans. on Bio Circuits Syst.* **4** 340–9

[8] Thotahewa K M S, Redoutè J M and Yuce M R 2015 Propagation, power absorption, and temperature analysis of UWB wireless capsule endoscopy devices operating in the human body *IEEE Trans. on Antennas Propag* **63** 3823–33

[9] *World Population Ageing: 1950–2050* A Report, Department of Economic and Social Affairs, Population Division, United Nations

[10] 47 CFR Parts 2 and 95, [ET Docket No. 0859; FCC 1254], Medical Body Area Network, Federal Register, Rules and Regulations, **77** (176), 11 Sept 2012

[11] European Telecommunication Standards Institute website https://etsi.org/

[12] European Telecommunications Standards Institute, ETSI EN 301 839- 1 V1.3.1: Electromagnetic compatibility and Radio spectrum Matters (ERM); Short Range Devices (SRD); Ultra Low Power Active Medical Implants (ULP-AMI) and Peripherals (ULP-AMI-P) operating in the frequency range 402 MHz to 405 MHz; Part 1: Technical characteristics and test methods 2009

[13] Hertleer C, Rogier H, Vallozzi L and Van Langenhove L 2009 A textile antenna for off-body communication integrated into protective clothing for firefighters *IEEE Trans. Antennas Propagat.* **57** 919–25

[14] El Hajj W, Person C and Wiart J 2014 A novel investigation of a broadband integrated inverted-F antenna design; application for wearable antenna *IEEE Trans. Ant. Propag.* **62** 3843–6

[15] Ha S J, Jung Y B and Kim D H *et al* 2012 Textile patch antennas using double layer fabrics for wrist-wearable applications *Microw. Opt. Technol. Lett.* **54** 2697–702

[16] Lin C, Saito K, Takahashi M and Ito K 2012 A compact planar inverted-f antenna for 2.45 GHz on-body communications *IEEE Trans. Ant. Propag.* **60** 3762–5

[17] Faerber J *et al* 2018 *In vivo* characterization of a wireless telemetry module for a capsule endoscopy system utilizing a conformal antenna *IEEE Trans. Bio Circuits Systems.* **12** 95–105

[18] Rano D and Hashmi M S 2018 Interdigital based EBG: compact and Polarization stable for MBAN and Wi-Fi *Proc. EuCAP* 1–5

[19] Akaram W, Rano D, Hashmi M S and Ansari A Q 2016 Design of a new uc-planar EBG cell and its application in performance enhancement of microstrip patch antenna/antenna array *Proc. Asia Pacific Microwave Conf.* pp 1–4

[20] Rano D and Hashmi M S 2019 A modified interdigital EBG reflector for wireless body area network applications *Wiley Microw. Opt. Technol. Lett.* **61** 912–9

[21] Jiang Z H, Brocker D E and Sieber P E 2014 A compact, low-profile meta-surface-enabled antenna for wearable medical body-area network devices *IEEE Trans. Antennas Propag.* **62** 4021–30

[22] Abbasi M A B, Nikolaou S, Antoniades M A, Stevanovic M N and Vryonides P 2017 Compact EBG- backed planar monopoles for BAN wearable applications *IEEE Trans. Antennas Propag.* **65** 453–63

[23] Sievenpiper D and Yablonovitch E 1998 Circuit and method for eliminating surface currents on metals U.S. Patent 60/079953, 30

[24] Sievenpiper D F 1999 High impedance electromagnetic surfaces *PhD dissertation* University of California, Los Angeles

[25] Yang F and Rahmat-Samii Y 2009 *Electromagnetic Band Gap Structures in Antenna Engineering.* (New York: Cambridge University Press)

[26] Durgun A C, Balanis C A and Birtcher C R 2013 Reflection phase characterization of curved high impedance surfaces *IEEE Trans. Antennas Propag.* **61** 6030–8

[27] Germain D, Seetharamdoo D, Burokur S N and de Lustrac A 2013 Phase-compensated met surface for a conformal microwave antenna *Appl. Phys. Lett.* **103** 1–4

[28] CST Microwave Studio 2012 CST, Framingham, MA, USA http://cst.com/

[29] Zhong S S, Liu G and Qasim G 1994 Closed form expressions for resonant frequency of rectangular patch antennas with multi-dielectric layers *IEEE Trans. Ant. Prop.* **42** 1360–3

[30] Bernhard J T and Tousignant C J 1999 Resonant frequencies of rectangular microstrip antennas with flush and spaced dielectric superstrates *IEEE Trans. Ant. Prop.* **47** 302–8

[31] Rano D, Chaudhary M A and Hashmi M S 2020 Determination of effective permittivity and resonant frequency of patch antenna covered with same and mixed type dielectric superstrates *IEEE Access* **8** 34418–30

[32] Balanis C A 1997 *Antenna Theory Analysis and Design* 3rd edn (New York: Wiley)

[33] Ashyap A Y I, Abidin Z Z, Dahlan S H, Majid H A and Seman F C 2018 A compact wearable antenna using ebg for smart-watch applications *Proc. of IEEE Asia-Pacific Microwave Conf.* 1477–9

[34] Rano D and Hashmi M S 2019 Extremely compact EBG-backed antenna for smart-watch applications in medical body area network *IET Microw. Antennas Propag.* **13** 1031–40

[35] SPEAG Dielectric Assessment Kit http://speag.com/products/dak/

[36] Kiminami K, Iyama T, Onishi T and Uebayashi S 2008 Novel Specific Absorption Rate (SAR) estimation method based on 2-D scanned electric fields *IEEE Trans. Electromag. Compat.* **50** 828–36

[37] Gao X, Zhang Z and Chen W *et al* 2009 A novel wrist wear dual-band diversity antenna *Proc. IEEE Int. Symp. Antennas Propag. Soc. (Charleston, SC, USA)* pp 1–4

[38] Woo S, Beak J and Park H *et al* 2013 Design of a compact UWB diversity antenna for WBAN wrist-watch applications *Proc. IEEE Int. Symp. Antennas Propag.* pp 1304–6

[39] Wu C H, Wong K L and Lin Y C *et al* 2007 Conformal bluetooth antenna for the watch-type wireless communication device application *Proc. IEEE Antennas Propag. Int. Symp.* pp 4156–9

[40] Wu C H, Wong K L and Lin Y C *et al* 2007 Internal shorted monopole antenna for the watch-type wireless communication device for bluetooth operation *Microw. Opt. Technol. Lett.* **49** 942–6

[41] Chen Y S and Ku T Y 2015 A low profile wearable antenna using a miniature high impedance surface for smartwatch applications *IEEE Antenn. Wireless Propag. Lett.* **14** 1144–7

[42] Rano D and Hashmi M S 2018 A new lumped circuit modelling technique for EBG based on surface current flow *Proc. Asia Pacific Conf. on Ant. and Prop.* pp 351–2

[43] Rano D and Hashmi M 2018 A new miniaturized slot-loaded EBG cell for cancelling surface waves in PCBs *Proc. IEEE Asia-Pacific Conference on Antennas and Propagation* 2018 pp 349–50

[44] Rano D and Hashmi M 2020 Miniaturized slot-loaded mushroom EBG cell for MBAN and Wi-Fi bands *Proc. IEEE Asia-Pacific Microwave Conf.* 2020 pp 145–7

Chapter 11

The role of capacitive sensors for condition monitoring of transformer insulation

Md Manzar Nezami and Shakeb A Khan

11.1 Introduction

Oil-immersed power transformers are expensive and critical component of power systems, primarily used for altering voltage levels. The reliable and stable operation of the power sector mostly depends on the performance of these transformers [1]. Negligence in the proper maintenance of this significant electrical equipment could impact the power supply to end users. A regular health assessment of these power transformers is therefore essential to ensure their uninterrupted operation and thereby minimize any incurred financial expenses to the utilities [2].

A transformer's health heavily depends on the condition of its insulation system. The life of the transformer is mostly limited by the life of the solid insulation [3]. The solid insulation is made of cellulose paper and thereby also termed paper insulation. The cellulose is composed of a linear polymer of glucose monomers. Chemically it is represented by $[C_6H_{10}O_5]$. These monomers are linked to each other and form a chain-like structure, $[C_6H_{10}O_5]_n$. The average number (n) of glucose monomers in every chain is referred to as the degree of polymerization (DP) [4]. The DP is found to be the most reliable indicator of insulation aging, which ultimately affects the transformer's residual life. A fresh cellulose insulating paper has a DP value around 1000–1200, which decreases with the aging process. A DP value of 200 is assumed to be end of life of such paper insulation [5]. The value of DP can easily predict the current physical state of the paper insulation and thereby be applied directly to assess the transformer's health. It also establishes the degree of deterioration of the mechanical tensile strength of the insulating paper. The DP is measured through a viscometry process, following the IEEE 61198 standard, and requires original paper samples [6]. This process interrupts the transformer's operation in order to retrieve the paper from the transformer. As such, though the diagnosis of a transformer's health using the estimated DP value is very common and reliable, it is also an

doi:10.1088/978-0-7503-3935-3ch11

invasive process of a destructive nature, and therefore reduces the accessibility of insulation testing [7].

The paper insulation used in windings undergoes thermal aging and decomposes with the passage of time. The presence of moisture and oxygen inside the transformer accelerates this decomposition rapidly, which in turn affects its smooth operation [8]. Consequently, the cellulose starts to lose its dielectric strength and mechanical properties, and produces degradation byproducts which get into the transformer oil. Among several degradation byproducts of paper insulation furfurals are determined to be a direct chemical marker of cellulose aging. The furfurals are a family of six heterocyclic molecules: (i) 5 hydroxy-methyl-2-furfuraldehyde (5HMF), (ii) 2-furfuraldehyde (2-FAL), (iii) 2-acetylfuran (2ACF), (iv) 5 methyl-2-furfuraldehyde (5MEF), (v) furfuryl alcohol and (vi) 2-furoic acid; see figure 11.1 [9]. 2-furfuraldehyde, or simply 2-FAL for short, is due to its properties able to dissipate in the transformer oil in a larger proportion and, owing to its thermal stability relative to other furanic compounds, is found to be resolute and thereby used to diagnose the deterioration of solid insulation. The estimation of paper insulation's condition using concentrations of 2-FAL not only facilitates uninterrupted transformer operation but also provide the ability for regular on-line health monitoring.

This chapter comprehensively presents the alternative means of diagnosing the health of cellulose insulation by proposing and fabricating a simple, novel, less-expensive capacitive sensor design able to detect concentrations of 2-FAL in transformer oil [10, 11]. Capacitive sensors are gaining more popularity at present as they possess many inherent advantages over resistive sensors. These sensors use composite polymers to make sensing films, which exhibit capacitance changes whenever exposed

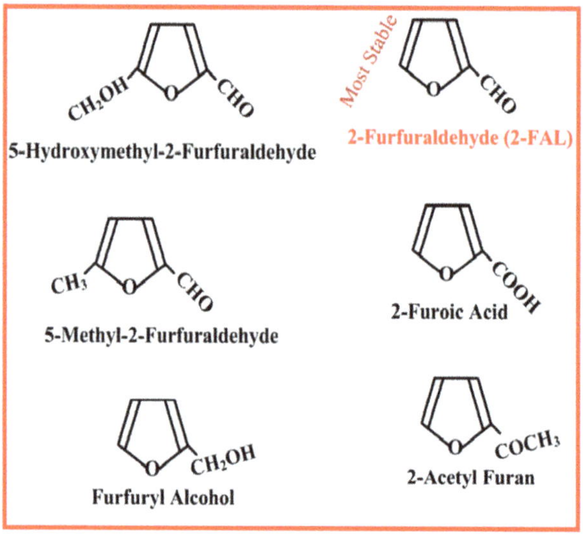

Figure 11.1. Components of furfurals produced due to paper decomposition.

in insulating oil. Capacitive sensors can be Internet-of-Things-enabled to measure the degree of deterioration of a transformer's insulation placed at remote locations.

11.2 Literature review

Incipient faults in transformers lead to severe operating conditions and can affect the uninterrupted operation of the power system. The major cause for transformer failure is attributed to insulation deterioration. The insulations used in transformers, especially the cellulose (solid) insulation, degrades at a higher rate during over-loading conditions and impacts the useful service life of the transformers. Generally, the mechanical strength of cellulose insulation weakens with its aging. Continuous accelerated aging produces cellulose degradation byproducts in the transformer, which remain dissolved in higher concentrations. The concentrations of these byproducts are very useful in order to predict the current physical state of the solid insulation directly. Close monitoring of the insulation is therefore of critical importance for utilities. Thereby, testing and analysis of the insulating oil should be undertaken and corrective action pursued in the event of any incipient conditions. This will enhance the operating life of the transformer and diminish the risk of incipient failures.

Therefore, the main focus of utilities should be pointed toward the early detection of insulation failure. The condition of the insulating oil directly reflects the state of the solid insulation and so of the power transformers. As aforementioned, the 2-FAL concentrations in the oil are more dominant and stable relative to other furanic compounds, and thereby can be used directly to assess the condition of the transformers. Different analytical and experimental approaches following the standard guidelines have been attempted to measure the contaminants of 2-FAL dissolved in oil, by many investigators. However, these endeavors have certain limitations and do not offer quantitative measurement of the oil contaminants.

Traditionally, the concentrations of 2-FAL in the oil have been measured using the technique of high-performance liquid chromatography (HPLC). This technique is based on modern instrumental technology which can measure the chemical contents of a given solution. Primarily, a HPLC arrangement has a metal capillary column packed with particles of solid material and an ultraviolet detector. An oil sample with 2-FAL molecules is injected into the column followed by pumping of an organic solvent through the column. The 2-FAL molecules in the sample are separated by the flow of organic solvent. The diverse components in the oil sample interact in differing manners with the solid material and the solvent due to their different molecular structure. Each component is eluted from the column and detected by an ultraviolet detector to record it over a chromatogram. Each component found will display a different peak over the chromatogram. The position of the peak clearly identifies the particular component and the area covered under the peak is directly proportional to the concentrations of the component in the sample [12]. However, HPLC involves costly instruments and can be only operated by skilled technicians. Also, the process of measuring 2-FAL molecules through HPLC is lengthy and time consuming and thereby not suitable for rigorous industrial applications.

In recent findings several investigations have been carried out by researchers in order to estimate the concentrations of 2-FAL in transformer oil. Some of them involve the fabrication of devices in the form of sensors and others are based on analytical approaches. The most common sensors fabricated for 2-FAL detection are optical sensors that are based on the colorimetric principle [13–17]. Additionally, many spectroscopy techniques, like Raman spectroscopy, ultraviolet spectroscopy [18–20], infrared spectroscopy [21], x-ray diffraction and dialotometry [22] have also been developed to detect molecules of 2-FAL. However, both methods have their shortcomings as they may require either interaction with toxic chemicals or a costly measurement set-up with skilled and efficient personnel able to operate on it.

In [23, 24], potentiometric and amperometric techniques were developed and implemented to detect only the presence of concentrations of the furanic derivatives, but they were unable to measure the exact quantity of their concentrations in the oil. Several methodologies using different soft-computing techniques have also been implemented in the literature to explore the feasibility of measurement of 2-FAL and other degradation products of paper insulation, and with which to predict its current physical state [25–29].

Though the existing methods and techniques prove very useful, they face many limitations in their implementation. Some of them are not very reliable while others lack the capacity for real-time health assessment of transformers. Hence, it is necessary to fabricate a device which facilitates the condition monitoring of transformers accurately and precisely, even though the transformers may be placed at remote locations.

This chapter presents the development of an electrochemical sensor. The sensor is employed with electrical techniques of a capacitive nature, which is capable of detecting changes in its capacitance values when subjected to different 2-FAL concentrations dissolved in transformer mineral oil.

11.3 Development of a capacitive sensor for condition assessment of an oil-immersed power transformer's insulation

A new, simple, less-expensive capacitive sensor employing a molecularly imprinted polymer (MIP) technique to detect molecules of 2-FAL in transformer mineral oil has been developed. The developed sensor is able to sense 2-FAL molecules in a 0–20 ppm range and thereby identify both the slow deterioration and advanced deterioration of insulating paper. The following sections describe the materials used and the process of fabrication of the developed capacitive sensor.

11.3.1 Materials required for the fabrication of the 2-FAL sensor

To fabricate the proposed 2-FAL sensor, the important materials required are (i) polydimethylsiloxane (PDMS, a silicon-based organic polymer) with its cross-linker, (ii) copper-cladded double-sided printed copper board (PCB); (iii) industrial-rated 2-FAL chemical (99% 2-FAL); and (iv) some consumables such as ethanol, acetone, isopropyl alcohol, deionized water, filer and zero-size sandpaper. PDMS is used because of its high chemical stability, hydrophobicity and ease of fabrication [30]. The PCB uses copper sheets separated by a glass epoxy of thickness 1.6 mm.

11.3.2 Process of fabrication

To fabricate the capacitive 2-FAL sensor, double-sided copper-cladded PCB sheet is used. The PCB sheet is cut to produce strips of dimensions 5 cm × 0.5 cm, shown in figure 11.2 [31]. The PCB strips are properly filed with a filer on the edges and smoothed with sandpaper. Then they are kept in an ultrasonicator filled with acetone for 15 min at 60 °C and finally cleaned using deionized water. The PCB strips are kept inside a controlled oven at 70 °C for 10 min. To take measurements, electrical connections are created by soldering lead wires at one end of the PCB strip. Now the PCB electrode is ready for coating with the prepared polymer (MIP).

An MIP sensitive to 2-FAL concentrations has been prepared using PDMS (functional monomer) and 2-FAL (template molecule). At first the PDMS monomer is mixed with its curing agent (cross-linker) followed by the 2-FAL molecule. The materials are mixed in a ratio of 10:1:4 (w/w) to form the MIP solution.

The MIP solution is stirred thoroughly and then kept inside a vacuum desiccator for degassing for 30 min and thereby produces a thick sticky MIP solution. The PCB strips are then coated with this MIP solution using a dip-coating device. The thickness of the MIP film over the PCB electrode is measured to be 60 μm. Figure 11.3 shows the cavities created in the polymer prepared using the process of polymerization.

The coated PCB electrode is thermally treated thereafter for perfect polymerization of the MIP film. It is dried inside a microcontrolled furnace at 90 °C for 3 hr. The PCB electrode is then rinsed in ethanol to take away the 2-FAL molecules from the MIP film, thereby forming complementary 2-FAL cavities in it. Finally, the capacitive sensor, capable of detecting 2-FAL concentrations, is ready for testing [32]. Figure 11.4 shows the capacitive sensor with and without the coating of the

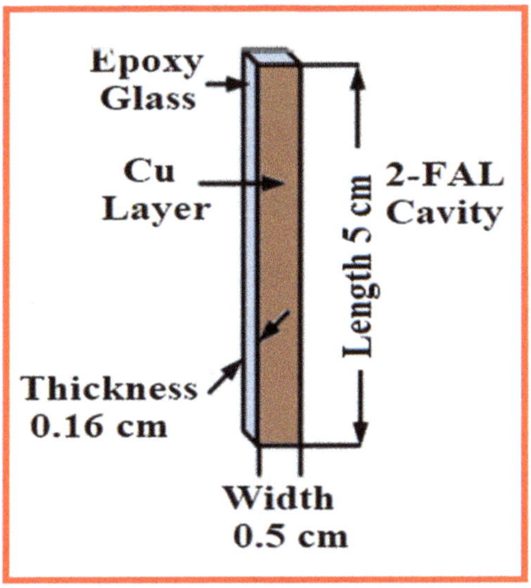

Figure 11.2. Schematic of the PCB strip.

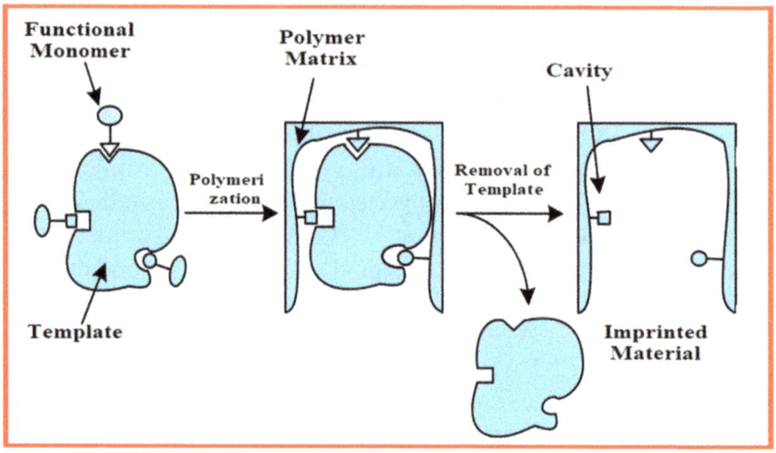

Figure 11.3. Molecularly imprinted polymer (MIP).

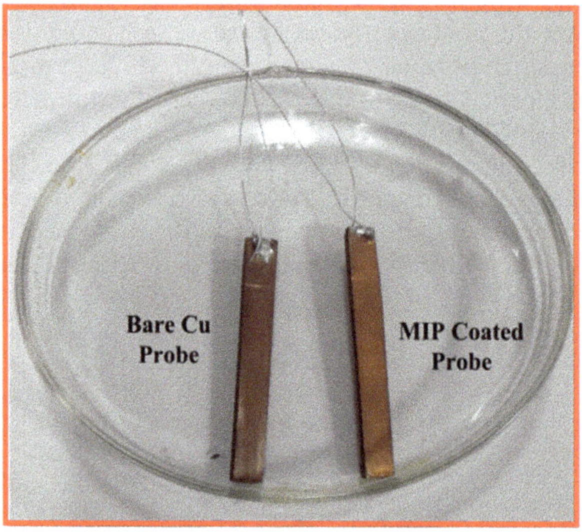

Figure 11.4. Coated and uncoated PCBs.

sensing film. The sensor is fabricated in batches and measurements are taken as an average of five readings in each case.

11.3.3 Experimental setup

To perform the experiments on the fabricated capacitive sensors, different test samples are first prepared. These test samples are composed of transformer oil properly mixed with different concentrations of 2-FAL molecules. Six test samples are prepared in a 100 mL beaker filled with 100 mL of transformer oil and varying concentrations of 2-FAL. At first, 0.1 gm of 2-FAL is poured into 100 mL of fresh transformer oil and mixed properly. This will give a stock solution of 100 ppm of 2-FAL in 100 mL of transformer oil. Then, this stock solution is diluted turn by turn with fresh transformer oil to obtain five test samples with 2-FAL concentrations of 2.5, 5, 10, 15 and 20 ppm of 2-FAL, respectively. One pure sample (pure oil, i.e. 0 ppm of 2-FAL) is also prepared in a separate beaker. The 2-FAL in the test samples ranges from 0–20 ppm. Thus, six test samples are ready for experimentation, which contain 0, 2.5, 5, 10, 15 and 20 ppm of 2-FAL, respectively. The lead of the sensor is connected to an Agilent 4294 A impedance analyzer (40 Hz–110 MHz) and dipped into the test samples one by one and the capacitance values recorded. The dip length of the sensor into every test samples is kept constant (i.e. 4 cm in this study) using a three-axis movable stage. The sensor is washed and dried properly with 2-FAL solvent before and after every reading. The data are recorded after 10 min incubation of the sensor in a test sample. It is worth noting here that all the experiments are performed at a constant temperature of 20 °C.

11.4 Electrical equivalent model of the developed sensor

11.4.1 Electrical equivalent circuit of the sensor

The electrical equivalent circuit of the sensor is derived with the help of electrical equivalent models of similar systems reported in the literature [31–33]. Since the developed sensor is fabricated with a PCB, the copper electrodes are on either side of an FR4 glass epoxy. When connected to an impedance analyzer, the sensor dipped into the test samples experiences two paths of electrical field distribution. One path is straight between the electrodes through the glass epoxy, and the other is fringing between the electrodes through the transformer oil. The straight field produces circuit elements like C_{epoxy} and R_{epoxy} while the fringing field produces C_{oil}, R_{oil}, C_{MIP} and C_T. The electric field distribution and the physical phenomenon of considering these circuit elements is shown pictorially in figure 11.5.

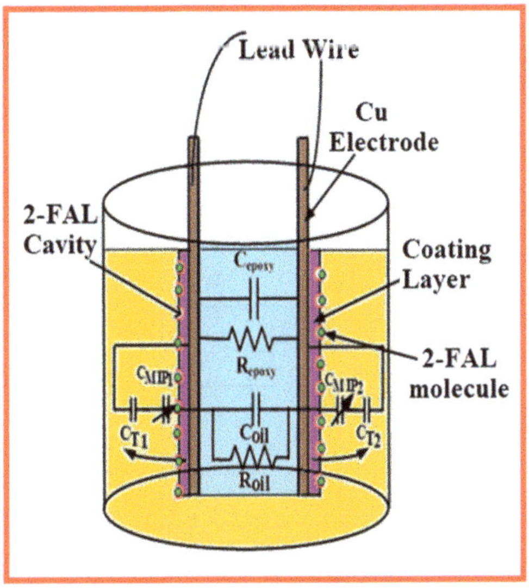

Figure 11.5. Pictorial representation of circuit parameters.

Figure 11.6(a) is the electrical equivalent circuit of the sensor when tested with the test samples. The capacitance, C_{epoxy}, along with a parallel resistance, R_{epoxy}, are produced due to the straight field between the electrodes through the glass epoxy. In the same time, the fringe field, especially at the edges of the electrode, produces a capacitance (C_{oil}) and a resistance (R_{oil}) due to oil. There also exist two coating capacitances, C_{T_1} and C_{T_2}, due to the coating layers on either side of the electrode, along with two variable capacitances, C_{MIP_1} and C_{MIP_2}, which are also produced in series with C_{T_1} and C_{T_2} due to the adsorption of 2-FAL molecules into the MIP film on both sides of the electrode.

The values of R_{epoxy} and R_{oil} are found to be very high and are thereby considered open [31], and hence the equivalent circuit can be reproduced as shown in the figure 11.6(b). The sensor is coated with the MIP film and hence produces two coating capacitances, C_{T_1} and C_{T_2}, in series on either side of the electrode. The thickness of the sensing film is fixed on both the sides of electrode, i.e. it remains unchanged irrespective of whether the sensor is dipped in any of the test samples of different 2-FAL concentrations. Here, C_T is considered as the resultant capacitance of two coating capacitances, C_{T_1} and C_{T_2}, and found constant. Since the sensing layer over the electrode has 2-FAL cavities, it is filled with 2-FAL molecules when dipped into different test samples. This produces adsorption layers of varying thickness and thereby obtains two variable capacitances, C_{MIP_1} in series with C_{MIP_2}, on either side of the electrode. C_{MIP} is the resultant capacitance of C_{MIP_1} and C_{MIP_2}. The overall equivalent circuit can be reproduced as shown in figure 11.6(c) and can be analytically analyzed with

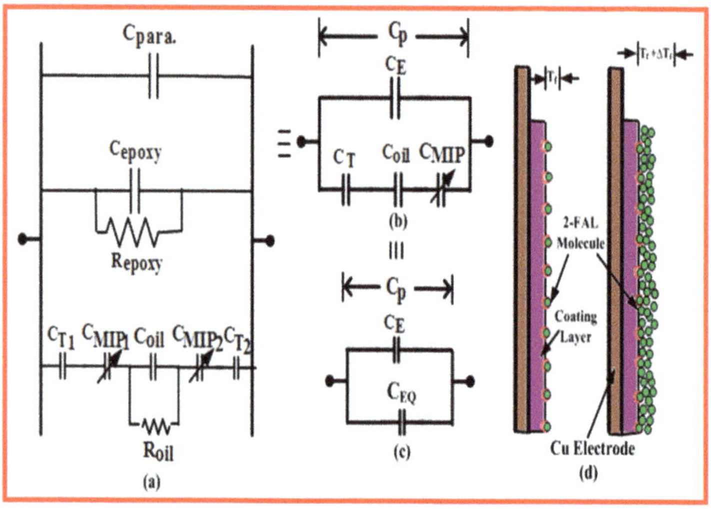

Figure 11.6. Electrical equivalent circuit and adsorption phenomenon of 2-FAL molecules over the MIP sensing film.

equations (11.1)–(11.5). It is important to note here that C_{para} is the parasitic capacitance and C_E is the resultant capacitance of the C_{para} and C_{epoxy} of the fabricated sensor in figure 11.5:

$$C_E = C_{\text{para}} + C_{\text{epoxy}}, \tag{11.1}$$

$$C_T = \frac{C_{T_1} \times C_{T_2}}{C_{T_1} + C_{T_2}} = f\left(\frac{1}{T}\right) = \text{constant}, \tag{11.2}$$

Where T is the thickness of the MIP coating on either side of the electrode:

$$C_{\text{MIP}} = \frac{C_{\text{MIP}_1} \times C_{\text{MIP}_2}}{C_{\text{MIP}_1} + C_{\text{MIP}_2}} = f\left(\frac{1}{T_f}\right), \tag{11.3}$$

where T_f is the thickness of the adsorbed 2-FAL molecule layer on the MIP's surface on either side of the electrode:

$$C_{\text{EQ}} = \frac{C_T \times C_{\text{oil}} \times C_{\text{MIP}}}{C_T \times C_{\text{oil}} + C_{\text{oil}} \times C_{\text{MIP}} + C_{\text{MIP}} \times C_T} = \frac{C_T \times C_{\text{oil}}}{C_T + C_{\text{oil}} + \frac{C_T \times C_{\text{oil}}}{C_{\text{MIP}}}}, \tag{11.4}$$

$$C_P = C_E + C_{\text{EQ}} = C_{\text{para}} + C_{\text{epoxy}} + \frac{C_T \times C_{\text{oil}}}{C_T + C_{\text{oil}} + \frac{C_T \times C_{\text{oil}}}{C_{\text{MIP}}}}$$

$$= C_{\text{para}} + C_{\text{epoxy}} + \frac{C_T \times C_{\text{oil}}}{C_T + C_{\text{oil}} + \frac{C_T \times C_{\text{oil}}}{f\left(\frac{1}{T_f}\right)}}. \tag{11.5}$$

11.4.2 Relation between the sensor's capacitance and the concentration of 2-FAL

The sensing mechanism in the fabricated sensor is established using the MIP technique. This technique has the ability to construct cavities in the shape of a desired template into a polymer. Since the sensor has to detect 2-FAL molecules in the test samples, so the 2-FAL template is used here for this purpose. This MIP is able to adsorb the 2-FAL molecules which are present in the test samples. The polymer (MIP) is prepared using the process of polymerization of the functional monomers and cross-linker along with the template. Thus, the prepared MIP film is deposited over the strips (PCB) and then thermally treated to develop the sensor. When this sensor is immersed into the test samples, 2-FAL molecules commence adsorption into the template cavity of the polymer and thereby form a 2-FAL molecules surface over the MIP, as shown in figure 11.6(d). Lower concentrations of 2-FAL in the test samples cause a thinner adsorption layer, i.e. T_f, while higher concentrations lead to a thicker layer, i.e. $T_f + \Delta T_f$. The change of thickness of the adsorption layer over the MIP causes a variable capacitance, C_{MIP_1} and C_{MIP_2}, on either side of the electrode; see figure 11.6(a). With the increase of concentrations (in ppm) of 2-FAL in the test samples, the thickness of adsorption layer increases also, and thereby leads to a decrease in the capacitance value of C_{MIP_1} and C_{MIP_2}. In the same way, when this concentration decreases, the value of the capacitances C_{MIP_1} and C_{MIP_2} increases. Therefore, the equivalent capacitance C_{MIP} changes with the concentration of 2-FAL, i.e. depends on T_f.

The equivalent capacitance of the sensor, i.e. C_P, is the resultant capacitance of C_{para}, C_{epoxy}, C_T, C_{oil} and C_{MIP}, as can clearly be seen from equation (11.5). However, C_{para}, C_{epoxy}, C_T and C_{oil} remain constant and C_{MIP} changes only as the sensor is dipped into different test samples with varying concentrations of 2-FAL (in ppm). Hence, the C_P changes as a function of C_{MIP}, which ultimately depends on the concentrations of 2-FAL in the test samples. As C_{MIP} decreases with the increase of 2-FAL concentration, the overall sensor capacitance, C_P, also decreases; see equation 11.5. Therefore, it is established that C_P is higher for pure test samples and gradually decreases with an increase in 2-FAL concentration.

11.5 Performance of the sensor

11.5.1 Response of the sensor at different frequencies

The proposed sensor is evaluated and analyzed for its response at different frequencies. For this, the sensor was connected to the impedance analyzer and immersed into different test samples with varying 2-FAL concentrations (0–20 ppm). The impedance analyzer is set to a frequency range from 1 to 200 kHz for every test sample to measure the value of the sensor's capacitance. Here, the frequencies below 1 kHz were not taken into account to overcome the noise from the supply frequency. The capacitance values are measured in the set frequency range and recorded in the log book of the impedance analyzer. The response of the sensor for each test sample is shown in the plot of figure 11.7. This plot shows how the sensor's capacitance changes at different frequencies for every test sample. The plot clearly shows a decline in capacitance with the rise of frequency and is almost consistent for higher frequencies (approx. beyond 50 kHz).

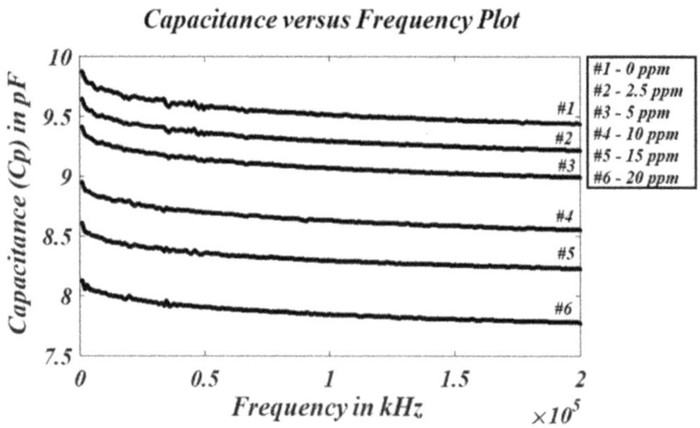

Figure 11.7. Sensor response as a function of 2-FAL change in the test samples at different frequencies.

Table 11.1. Sensor response in every test sample at different frequencies.

S. No.	Frequency (kHz)	0 ppm	2.5 ppm	5 ppm	10 ppm	15 ppm	20 ppm	Sensitivity pF ppm^{-1}
		Capacitance (C_P) pF						
1	1	9.87	9.64	9.41	8.95	8.60	8.12	0.087
2	40	9.60	9.38	9.16	8.71	8.37	7.92	0.084
3	100	9.51	9.29	9.06	8.62	8.29	7.84	0.083
4	150	9.47	9.24	9.02	8.57	8.25	7.80	0.083
5	200	9.43	9.21	8.99	8.54	8.22	7.77	0.083

The sensor gives its maximum capacitance of 9.87 pF at 1 kHz for the fresh test sample (i.e. with 0 ppm of 2-FAL). For the remaining test samples, the values of the sensor's capacitance are 9.64, 9.41, 8.95, 8.60 and 8.12 pF, respectively, at 1 kHz. The sensor response for all the test samples is also estimated and tabulated in table 11.1 at different frequencies. The last column of table 11.1 shows the average sensitivity of the sensor at frequencies of 1, 40, 100, 150 and 200 kHz, respectively, for the test samples of 0–20 ppm of 2-FAL. The sensor's average sensitivity is 0.084 pF ppm^{-1} throughout the frequencies ranging between 1 and 200 kHz.

To verify the sensor response with MIP film, experiments are also performed with sensors coated with PDMS (i.e. without a 2-FAL template) only. The capacitance is measured for all the test samples following the same procedure and a negligible capacitance change is observed. Here, the sensor gives capacitance values of 9.929, 9.91, 9.918, 9.919, 9.915 and 9.911 pF for the 0, 2.5, 5, 10, 15 and 20 ppm test samples, respectively, at 1 kHz. These results confirm that pure PDMS alone is unable to detect the presence of 2-FAL molecules in the transformer oil and thereby validates the MIP technique.

11.5.2 Response of the sensor with time

The proposed sensor is also evaluated and analyzed for its response with time. For this, the sensor is again connected to the impedance analyzer to measure the capacitance value at a fixed frequency for 1 hr, turn by turn, in every test sample. The frequency is set to 1.5 kHz and, using the time sweep function available in the impedance analyzer, the capacitance values are measured and recorded. The plots of capacitance versus time for various test samples are shown in figure 11.8.

11.5.3 Calibration of the sensor response

The response of the sensor is calibrated by establishing the relation of the sensor's capacitance with the concentration of 2-FAL in the test samples. Figure 11.9 clearly shows a decreasing trend in the value of capacitance for an increase in the 2-FAL concentrations in the test samples. The same is also demonstrated with the help of the sensor's equivalent model (figure 11.6).

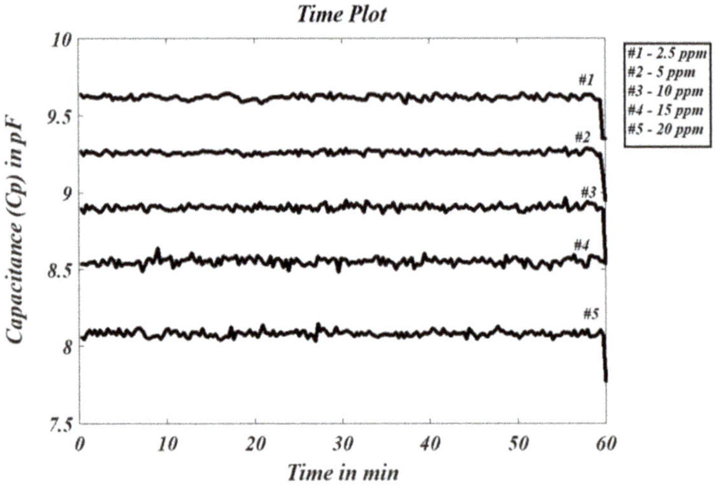

Figure 11.8. Plot of sensor response with time at a fixed frequency for different test samples.

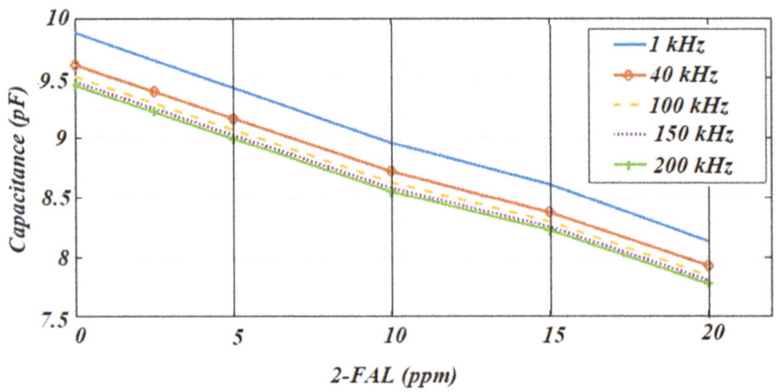

Figure 11.9. Calibration of sensor response for change in 2-FAL concentrations in test samples at various frequencies.

The graph of figure 11.9 exhibits an almost linear change in the capacitance value for the 2-FAL concentration change at 1 kHz. Similar linear change trends in the capacitance value with 2-FAL concentrations are observed for frequencies of 40 100, 150 and 200 kHz, respectively (table 11.1). For every curve in figure 11.9, a linear-fit expression and the non-linearity at various frequencies are listed in table 11.2. It is evident from table 11.2 that the sensor shows more linear characteristics at lower frequencies. The average non-linearity is measured to be 0.22%, 0.20%, 0.23%, 0.23% and 0.24% at 1, 40, 100, 150 and 200 kHz, respectively (table 11.2).

The sensitivity of the sensor varies slightly in the frequency range of 1–200 kHz. It is observed that the sensitivity is a little higher at lower frequency and becomes almost consistent at upper frequencies above 60 kHz (≈ 0.0831 pF ppm^{-1}).

Table 11.2. Non-linearity analysis of sensor at different frequencies.

Frequency (kHz)	Linear-fit expression	Maximum non-linearity (%)	Average non-linearity (%)
1	$y = -0.086x + 9.86$	0.47	0.22
40	$y = -0.083x + 9.59$	0.48	0.2
100	$y = -0.082x + 9.49$	0.51	0.23
150	$y = -0.082x + 9.44$	0.56	0.23
200	$y = -0.082x + 9.41$	0.56	0.24

Table 11.3. List of statistical parameters.

Parameters	2.5 ppm	5 ppm	10 ppm	15 ppm	20 ppm
Mean	9.614	9.258	8.902	8.538	8.081
Mode	9.612	8.950	8.894	8.531	8.077
Median	9.617	9.260	8.903	8.540	8.081
Standard deviation	0.030	0.026	0.030	0.027	0.029

11.5.4 Statistical analysis

The response of the developed sensor is also statistically analyzed by plotting normal distribution curves. A normal distribution curve is plotted using data obtained from a time sweep experiment. The experiments are conducted by dipping the sensor in all the test samples turn by turn and measuring the capacitance for 1 hr at 1.5 kHz. For every test sample, the impedance analyzer gives 201 capacitance data points at 18 s interval. The mean (μ) and the standard deviation (σ) values of the data obtained for each test sample are estimated first to plot the normal distribution curve. The mean, standard deviation, median and mode values of each measured data set for all test samples are listed in table 11.3. The curves of the measured capacitance values of each test sample are found to be normally distributed; see figure 11.10. The values of probability density for the mean of the sensor's capacitance are calculated to be 0.3, 0.356, 0.384, 0.358 and 0.354 for the test samples with 2.5 ppm, 5 ppm, 10 ppm, 15 ppm and 20 ppm of 2-FAL concentrations, respectively. It should be noted here that each curve is narrowly spread due to the least value of the standard deviation ($\sigma_{avg} \approx$ 0.028 pF). It can be seen from table 11.3 that the mean (μ), mode and median values of the measured capacitance are almost equal for every test sample. For instance, for the 2.5 ppm 2-FAL test sample the mean (μ), mode and median values are 9.614, 9.612 and 9.617 pF, respectively.

Likewise, the other test samples also witness the same trends in their mean, mode and median values, i.e. they are almost identical; see table 11.3. This therefore validates the fact that the measured values are distributed normally.

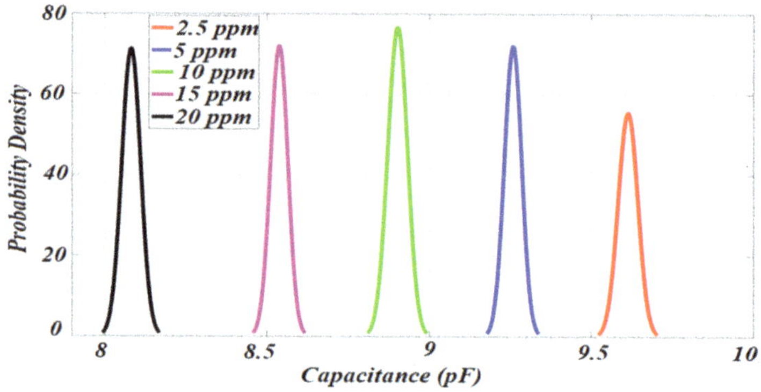

Figure 11.10. Curves of probability density for the sensor response in every test sample.

Table 11.4. Response of the sensor fabricated in the month of October 2017 at a regular interval of 2 days.

Date	0 ppm	2.5 ppm	5 ppm	10 ppm	15 ppm	20 ppm
	Capacitance (C_p) pF					
3 October 2017	9.820	9.563	9.353	8.901	8.532	8.112
5 October 2017	9.833	9.579	9.362	8.921	8.540	8.121
7 October 2017	9.845	9.586	9.362	8.936	8.543	8.131
9 October 2017	9.842	9.617	9.374	8.932	8.551	8.120
11 October 2017	9.853	9.617	9.394	8.922	8.578	8.138
13 October 2017	9.851	9.624	9.382	8.937	8.583	8.139
15 October 2017	9.854	9.621	9.397	8.945	8.583	8.141
17 October 2017	9.864	9.632	9.401	8.942	8.596	8.129
19 October 2017	9.872	9.626	9.418	8.953	8.598	8.124
21 October 2017	9.875	9.636	9.425	8.968	8.602	8.134
23 October 2017	9.863	9.642	9.421	8.955	8.612	8.152
25 October 2017	9.874	9.647	9.433	8.973	8.622	8140
27 October 2017	9.876	9.654	9.442	8.962	8.627	8.156
29 October 2017	9.887	9.654	9.437	8.973	8.638	8.144
31 October 2017	9.865	9.667	9.453	8.975	8.637	8.143

11.5.5 Repeatability and drift study of the sensor response

The developed sensor is also analyzed for its repeatability and drift performance. For its repeat performance, the sensor is properly cleaned and refreshed with 2-FAL solvent after every measurement and stored inside a vacuum desiccator thereafter. The drifts in the measured capacitance values are studied by testing the sensor continuously in every test sample over a period of 1 month at an interval of 2 days. The measured capacitance values in the month of October 2017 at a frequency of 1 kHz are listed in table 11.4. From the listed data, it can clearly be stated that the capacitance of the sensor decreases with an increase in the 2-FAL concentrations in the test sample. While with repeat testing, the drift in the capacitance values is nearly negligible for all test samples (table 11.4).

In addition, to observe the repeatable response of the developed device, it is fabricated and tested continuously over a period of 6 months from October 2017 to April 2018 in batches. As before mentioned, PDMS material is used to develop the sensing layer and it is highly hydrophobic and stable. Hence, the sensors attain a very stable performance irrespective of their month of fabrication. This is also evident from table 11.5. The measured capacitance values as listed in the table are an average of the capacitance estimated for a particular test sample throughout the month. The measured values listed in table 11.5 clearly iterate that the sensor is quite robust with repeatable performance every time.

Figure 11.11 shows a bar representation of the sensor response fabricated in different months (batches) and tested in every test sample. The average value of measured capacitance for every test sample in every month is used in preparing the bar chart. The bar representation is a clear indication of the approximately similar characteristics of the sensor response over the period of its fabrication.

Table 11.5. Sensor fabrication over a period of 6 months (in batches) and its performance in test samples.

		0 ppm	2.5 ppm	5 ppm	10 ppm	15 ppm	20 ppm
Batches	Fabrication date	Capacitance (C_P) pF					
1	**October 2017**	9.870	9.573	9.352	8.901	8.543	8.112
2	**November 2017**	9.884	9.585	9.382	8.932	8.573	8.121
3	**December 2017**	9.893	9.586	9.392	8.954	8.562	8.152
4	**January 2018**	9.883	9.627	9.425	8.948	8.580	8.135
5	**February 2018**	9.893	9.656	9.438	8.957	8.624	8.122
6	**April 2018**	9.892	9.642	9.413	8.957	8.603	8.134

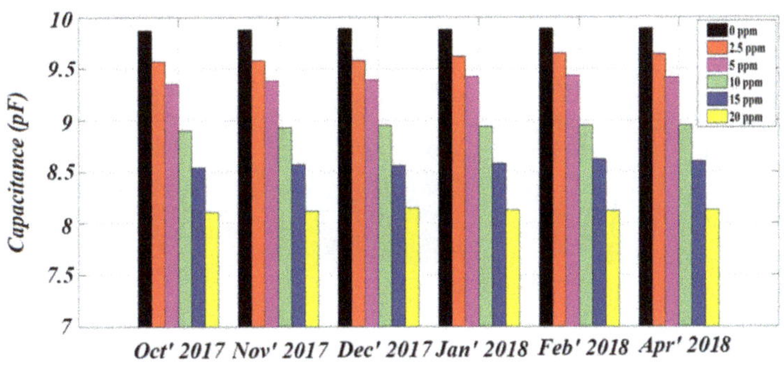

Figure 11.11. Bar chart representation of the sensor response over a period of 6 months (batches) for different test samples.

11.6 Modification of the sensor's structure to improve its sensitivity in detecting 2-FAL concentrations in transformer oil

The capacitive sensor developed for 2-FAL detection shows an average sensitivity of 0.084 pF ppm^{-1} for a range of 0–20 ppm. This 2-FAL detection sensor needs an improvement in its sensitivity. Since the developed sensor works on the principle of fringing field capacitance, especially at the edges of the electrode, a change in the geometry of the sensor's electrode is proposed. This will enhance the effective edge length of the electrode and henceforth the sensor's sensitivity. Within the available limited resources, the electrode is given a comb-like structure with five similar fingers. These fingers enhance the total edge length of the electrode where the fringe field is found to be more prominent.

The comb sensor is fabricated using PCB of dimensions 50 mm × 18mm. The PCB is cut using a computerized numeric control vertical drill machine to give the electrode a comb-like design; see figure 11.12. The fingers of the comb are of dimensions 40 mm × 2 mm and the gaps of the two adjacent fingers are 2 mm; see figure 11.12(a). The total effective length of the comb sensor's edge has now become 41 cm while it was only 8.5 cm in the previous sensor. This increase in the effective edge length of the comb sensor causes a further enhancement in the fringe, which ultimately helps to generate a sufficient change in the sensor response (capacitance) in response to 2-FAL detection. Hence, the sensitivity of the comb sensor is improved as compared to the previous sensor design. Also, the effective surface area of the comb sensor has increased to 4 cm^2 as compared to the surface area of the previous sensor, i.e. 2 cm^2. Therefore, the comb-shaped sensor has the benefit in that, even though the effective surface area of the sensor has increased by two times, the total effective length of the edges has increased five times.

The other processes in the fabrication of the comb sensor do not alter except the change in its geometry. Therefore, the same MIP technique with the same materials (i.e. PDMS, cross-linker and 2-FAL) is used to prepare the sensing film of the comb sensor. The materials are used in the same ratio and follow the same procedures to

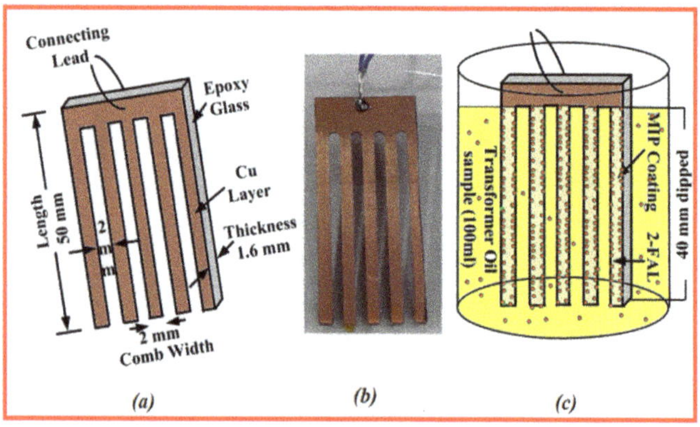

Figure 11.12. (a) Schematic of the comb sensor design. (b) Fabricated sensor in comb design and coated using MIP. (c) Schematic representation of the comb sensor dipped into test sample.

prepare and deposit the sensing film over the comb electrode. The thickness of sensing film is calculated to be 60 µm. The comb electrode is finally rinsed with the 2-FAL solvent (i.e. ethanol) to remove the template from the MIP and thereby create the 2-FAL cavities. Finally, the comb sensor is ready for experiments; see figure 11.12(b).

To perform experiments with the comb sensor, the same type of test samples is used. Six fresh test samples are prepared with 100 mL of transformer mineral oil, which contains 0 ppm, 2.5 ppm, 5 ppm, 10 ppm, 15 ppm and 20 ppm of 2-FAL concentrations. The sensor is connected through its lead wire to the impedance analyzer with operating frequencies of 40 Hz–110 MHz to measure its response. The capacitance measurements are recorded in the data logger. Measurements are carried out after 10 min of incubation of the sensor in every test sample at a temperature of 20 ± 1 °C in the laboratory; see figure 11.12(c). The recorded capacitance values are the average of five measurements in every test sample. It is worth noting here that the comb sensor is properly cleaned and refreshed after every measurement with the 2-FAL solvent to remove the 2-FAL molecules from the cavities of the MIP over the sensing layer.

11.7 Performance evaluation of the comb sensor

11.7.1 Response of the comb sensor at different frequencies

The comb sensor is also evaluated and analyzed for its response at different frequencies. To perform this experiment, the comb sensor is connected to the impedance analyzer and dipped into the test samples, turn by turn, maintaining a length of 4 cm. The frequency of the impedance analyzer is fixed in the range of 1 kHz to 200 kHz using its frequency-sweep function to measure the comb sensor's response (capacitance).

The sensor is dipped into every test sample for 10 min before recording the capacitance values. The sensor response in the selected frequency range for every test sample is shown in the graph of figure 11.13. This graph clearly shows that the value of the sensor's capacitance decreases with the increase of 2-FAL concentrations in the test samples. The capacitance slightly decreases at lower frequencies and then settles to an almost constant value at upper frequencies (approx. beyond 50 kHz) for every test sample.

At 1 kHz, the comb sensor has a maximum value of 24.76 pF for the test sample without 2-FAL molecules (i.e. 0 ppm of 2-FAL test sample). The comb sensor's capacitance value then decreases to 24.2, 23.6, 22.8, 22.01 and 21.2 pF for 2.5, 5, 10, 15 and 20 ppm test samples at 1 kHz, respectively. The responses of the comb sensor for every test sample at the other frequencies are recorded in table 11.6. The average sensitivity of the comb sensor is measured to be 0.1840, 0.1821, 0.1811, 0.1801 and 0.1792 pF ppm^{-1} at selected frequencies of 1, 50, 100, 150 and 200 kHz, respectively, in 0–20 ppm concentrations of 2-FAL in the test samples. The overall average sensitivity of the comb sensor throughout the frequency range is found to be 0.181 pF ppm^{-1}. Hence, it can be seen here that the change in the geometry of the electrode structure (the comb sensor) leads to an improvement of the sensor's sensitivity.

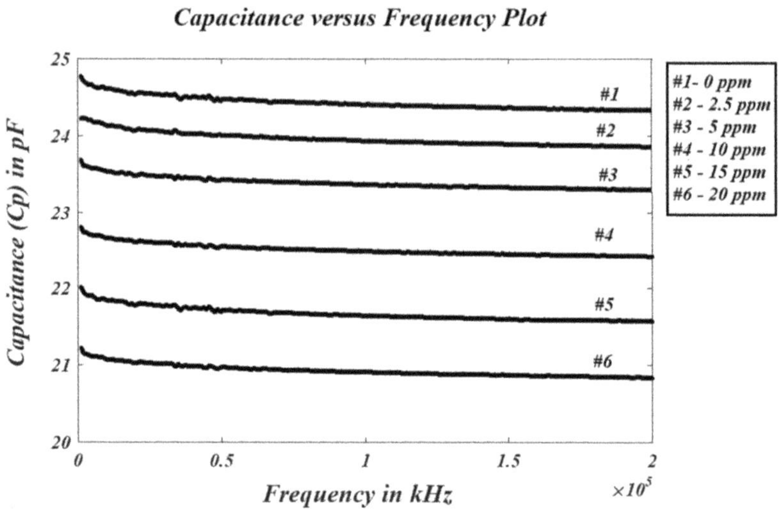

Figure 11.13. Comb sensor response as a function of 2-FAL change in the test samples at various frequencies.

Table 11.6. Comb sensor response for all the test samples at different frequencies.

S. No.	Frequency (kHz)	0 ppm	2.5 ppm	5 ppm	10 ppm	15 ppm	20 ppm	Sensitivity pF ppm^{-1}
		Capacitance (C_P) pF						
1	1	24.76	24.21	23.68	22.81	22.01	21.22	0.184
2	50	24.47	24.01	23.42	22.55	21.72	20.96	0.182
3	100	24.40	23.92	23.36	22.49	21.64	20.91	0.181
4	150	24.36	23.88	23.33	22.46	21.61	20.87	0.180
5	200	24.32	23.85	23.29	22.42	21.57	20.87	0.179

11.7.2 Response of the comb sensor with time

The response of the comb sensor at a particular frequency for a certain period of time is evaluated and analyzed in this experiment. The sensor is properly cleaned and dried before performing this experiment with the 2-FAL solvent. This aids in clearing the filled cavities over the MIP sensitive film. The comb sensor is again connected to the impedance analyzer fixed at a particular frequency of 1 kHz. The sensor is dipped into the test samples turn by turn and capacitance values are recorded for every test sample at 1 kHz for 4 hr with the help of the time sweep function provided in the impedance analyzer. It is observed that the comb sensor behaves almost consistently throughout the period of measurement; see figure 11.14. There are 201 capacitance values for every test sample recorded during time sweep measurement at 1 kHz. The capacitance values so recorded are statistically analyzed

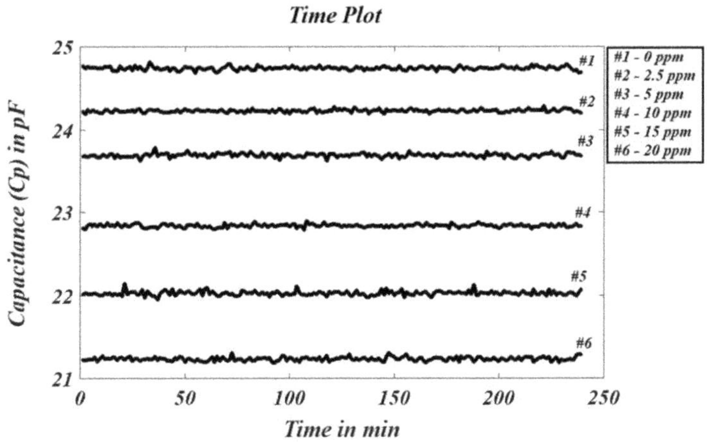

Figure 11.14. Plot of comb sensor response with time at a fixed frequency for different test samples.

Table 11.7. List of statistical parameters.

Parameters	0 ppm	2.5 ppm	5 ppm	10 ppm	15 ppm	20 ppm
Mean	24.736	24.225	23.696	22.836	22.024	21.236
Median	24.734	24.229	23.693	22.834	22.029	21.235
Mode	24.724	24.216	23.628	22.829	21.947	21.185
Standard deviation	0.022	0.016	0.024	0.018	0.027	0.024
Mean deviation (median)	0.017	0.015	0.019	0.016	0.018	0.019
Mean deviation (mean)	0.017	0.0143	0.019	0.016	0.018	0.017

to observe the sensor performance. Different statistical parameters like mean (μ), standard deviation (σ), mode, median, mean deviation about median and mean deviation about mean are calculated. The values of these parameters are listed in table 11.7. The average value of standard deviation (σ) in the 0–20 ppm range is calculated to be 0.022 pF, which evidently validates a stable pattern (a decrease) in the capacitance values as the concentrations of 2-FAL increase in the test samples.

11.7.3 Calibration of the comb sensor response

The comb sensor also undergoes calibration to establish a correlation between the sensor's response and the concentrations of 2-FAL in the test samples. Figure 11.15 shows a plot of the measured values of capacitance against change in 2-FAL concentrations in test samples at different frequencies. At all chosen frequencies, the plot clearly shows a decrease in the value of capacitance with the 2-FAL change. The same pattern is also observed in the frequency-sweep characterization of the comb sensor (figure 11.13).

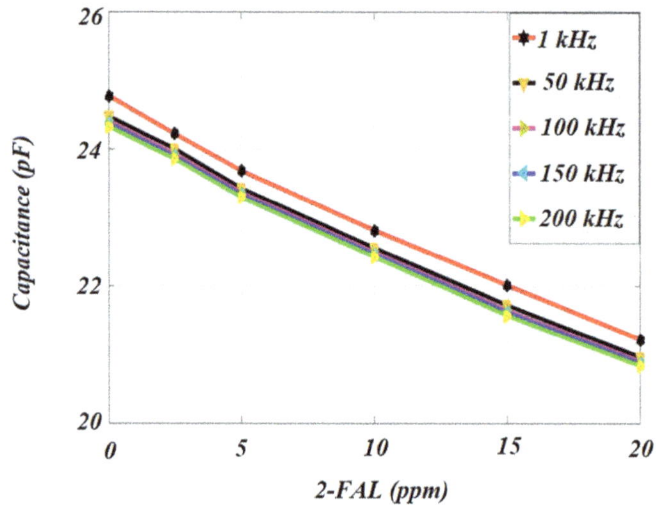

Figure 11.15. Calibration of the comb sensor response due to change in 2-FAL concentrations in test samples at different frequencies.

Table 11.8. Comb sensor non-linearity analysis.

Frequency (kHz)	Linear-fit expression	Maximum non-linearity (%)	Average non-linearity (%)
1	$y = -0.176x + 24.66$	0.37	0.27
50	$y = -0.175x + 24.44$	0.43	0.38
100	$y = -0.175x + 24.35$	0.54	0.37
150	$y = -0.176x + 24.283$	0.58	0.38
200	$y = -0.174x + 24.265$	0.76	0.33

The calibration between measured capacitance and change in 2-FAL concentrations possesses an almost linear relation at all frequencies. The linear curves so obtained are fitted with linear equations and recorded in table 11.8. The non-linearity is also calculated and recorded at different frequencies. The comb sensor displays more linear characteristics at lower frequencies, with an average non-linearity of 0.27% at 1 kHz (table 11.8). Hence, the comb sensor is found to be linear in the 0–20 ppm range of 2-FAL concentrations in the test samples.

11.7.4 Repeatability of the comb sensor response

To study the repeatability and continued performance of the comb sensor, it is continuously tested in every test sample repeatedly for 16 days at an interval of 3 days. The comb sensor capacitances are measured for all the test samples at 1 kHz. The sensor is properly refreshed by cleaning it with ethanol, followed by drying, before each new measurement. The sensor is stored inside a vacuum desiccator to save the sensing layer from environmental hazards. The measured capacitance values throughout the period of experiment (15 days) for each test sample are listed in table 11.9.

The measured capacitance values listed in table 11.9 clearly show an almost repeatable response of the comb sensor in every test sample.

The repeatability index, $\mathfrak{R}$, of the comb sensor is estimated using equation (11.6) [34]:

$$\mathfrak{R}(\%) = \pm\left[\frac{[(C_1 - C_m)^2 + (C_2 - C_m)^2 + \ldots.+(C_n - C_m)^2]^{\frac{1}{2}}}{2(n - 1)C_m}\right]. \tag{11.6}$$

Here C_1, C_2, …, C_n are the measured values of capacitances for n times (here, $n = 6$) taken every 3 days continuously for 16 days in each test sample, and C_m is calculated to be their mean values. The average $\mathfrak{R}$ in 0–20 ppm of 2-FAL is measured to be ±0.012%, and hence demonstrates the highly repeatable performance of the comb sensor. The graph in figure 11.16 shows the repeat performance of the developed sensor tested on

Table 11.9. Repeatability study of the comb sensor in each test sample.

		Capacitance (C_p) (pF)					
S. No.	Days	0 ppm	2.5 ppm	5 ppm	10 ppm	15 ppm	20 ppm
1	**First**	24.763	24.218	23.687	22.812	22.012	21.226
2	**Fourth**	24.778	24.227	23.699	22.814	22.025	21.209
3	**Seventh**	24.776	24.235	23.678	22.809	22.026	21.234
4	**Tenth**	24.767	24.207	23.688	22.805	22.037	21.216
5	**Thirteenth**	24.757	24.198	23.706	22.795	21.994	21.220
6	**Sixteenth**	24.76	24.232	23.647	22.891	22.071	21.281
$\mathfrak{R}$		±0.011%	±0.010%	±0.013%	±0.013%	±0.014%	±0.012%

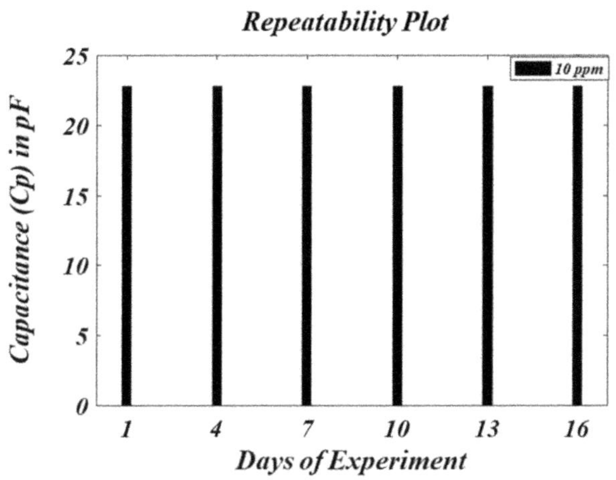

Figure 11.16. Repeatability of the comb sensor tested for 10 ppm test sample.

the 10 ppm test sample. The peak value of the capacitance at every adjacent measurement looks alike, i.e. the comb sensor shows consistency in its response every time.

11.8 Conclusion

This article presents the development of a capacitive sensor to detect concentrations of 2-FAL in transformer oil. The developed sensor is able to determine the paper insulation's deterioration condition at both incipient and advanced stages. The sensor is fabricated using a strip of PCB deposited with MIP film sensitive to detecting 2-FAL molecules. The PDMS monomer, its cross-linker and 2-FAL are used to prepare the MIP film. The fabricated sensor has been tested in six test samples, all prepared in 100 mL of fresh transformer oil and mixed with 2-FAL concentrations of 0, 2.5, 5, 10, 15 and 20 ppm, respectively. Different concentrations of 2-FAL molecules in the test samples lead to a change in the sensor's capacitance. The sensor is characterized with a linear capacitance change as the 2-FAL concentrations change in the test samples, with an average sensitivity of 0.084 pF ppm^{-1} over the 1–200 kHz frequency range. At every frequency, a linear decrease in the sensor capacitance is observed with a change of 2-FAL concentration in the test samples, as also verified by the sensor's electrical equivalent model. The sensor is fabricated in batches for different months to study its drift and repeatability performance. A negligible drift in the sensor capacitance values and high repeatability are observed during the fabrication period. The estimated capacitance value of the sensor is statistically analyzed at chosen frequencies to validate the reliable performance of the sensor. Furthermore, an additional sensor is fabricated with an improved sensitivity for 2-FAL detection. This sensor is termed a comb sensor and is developed simply by changing the PCB electrode's structure. The PCB is given a comb shape with five similar fingers to increase the total effective edge length as compared to the previous sensor. This change of edge length led to an improvement of the fringing field at the edges of the comb electrode and hence the sensor's sensitivity. The sensing film for the comb sensor is prepared with the same MIP technique and is tested in test samples containing 2-FAL concentrations in the 0–20 ppm range. The comb sensor also gives a linear change of capacitance for different 2-FAL concentrations in the 1–200 kHz frequency range, with an improved average sensitivity of 0.184 pF ppm^{-1}. The reliability of the comb sensor is also statistically analyzed with estimated capacitance values and found to be consistent in its performance. The comb sensor is highly repeatable over a period of fabrication, with an average repeatability index of ±0.012%.

References

[1] Abu-Elanien A E B, Salama M M A and Ibrahim M 2012 Calculation of a health index for oil-immersed transformers rated under 69 kV using fuzzy logic *IEEE Trans. Power Deliv.* **27** 2029–36

[2] Emsley A M and Stevens G C 1994 Review of chemical indicators of degradation of cellulosic electrical paper insulation in oil-filled transformers *Proc. Inst. Elect. Eng., Sci Meas. Technol.* **141** 324–34

[3] Prevost T A and Oommen T V 2006 Cellulose insulation in oil-filled power transformers: part I—history and development *IEEE Electr. Insul. Mag.* **22** 28–35

[4] Baird P J, Herman H, Stevens G C and Jarman P N 2006 Non-destructive measurement of the degradation of transformer insulating paper *IEEE Trans. Dielectr. Electr. Insul.* **13** 309–18

[5] dos Santos E O, Silva A M S, Fragoso W D, Pasquini C and Pimentel M F 2010 Determination of degree of polymerization of insulating paper using near infrared spectroscopy and multivariate calibration *Vib. Spectrosc.* **52** 154–7

[6] Ali M, Heywood R J and Emsley A M 2000 Degradation of cellulosic insulation in power transformers. Part 1: factors affecting the measurement of the average viscometric degree of polymerisation of new and aged electrical papers *IEEE Proc.—Sci. Meas. Technol.* **147** 86–90

[7] Lelekakis N, Guo W, Martin D, Wijaya J and Susa D 2012 A field study of aging in paper-oil insulation systems *IEEE Electr. Insul. Mag.* **28** 12–9

[8] Emsley A M, Xiao X, Heywood R J and Ali M 2000 Degradation of cellulosic insulation in power transformers Part 3: effects of oxygen and water on ageing in oil *Science, IEE Proc. Meas. Technol.* **147** 115–9

[9] Scheirs J, Camino G, Avidano M and Tumiatti W 1998 Origin of furanic compounds in thermal degradation of cellulosic insulating paper *J. Appl. Polym. Sci.* **69** 2541–7

[10] Nezami M M, Wani S A, Khan S A, Khera N and Sohail S 2018 An MIP-based novel capacitive sensor to detect 2-FAL concentration in transformer oil *IEEE Sens. J.* **18** 7924–31

[11] Nezami M M, Wani S A, Khan S A, Khera N and Sohail S 2019 A comb type capacitive 2-FAL sensor for transformer oil with improved sensitivity *IEEE Trans Instrum. Meas.* **69** 4524–32

[12] Unsworth J and Mitchell F 1990 Degradation of electrical insulating paper monitored with high performance liquid chromatography *Electr. Insul.* **25** 737–46

[13] Cennamo N, De Maria L, Chemelli C, Profumo A, Zeni L and Pesavento M 2016 Markers detection in transformer oil by plasmonic chemical sensor system based on POF and MIPs *IEEE Sensors J.* **16** 7663–70

[14] Cennamo N *et al* 2019 A novel sensing methodology to detect furfural in water, exploiting MIPs, and inkjet-printed optical waveguides *IEEE Trans. Instrum. Meas.* **68** 1582–9

[15] Blue R, Uttamchandani D G and Farish O 1998 A novel optical sensor for the measurement of furfuraldehyde in transformer oil *IEEE Trans. Instrum. Meas.* **47** 964–6

[16] Blue R, Uttamchandani D and Farish O 1997 A novel solid-state material for furfuraldehyde detection *IEEE Trans. Dielect. Elect. Insul.* **4** 341–3

[17] Yuste A, Vallejo V, Peña E, Engel T, Orellana G and Bondi M 2016 Furfural determination with disposable polymer films and smartphone-based colorimetry for beer freshness assessment *Anal. Chem.* **88** 3959–66

[18] Chen W, Gu Z, Zou J, Wan F and Xiang Y 2016 Analysis of furfural dissolved in transformer oil based on confocal laser Raman spectroscopy *IEEE Trans. Dielectr. Electr. Insul.* **23** 915–21

[19] Somekawa T, Fujita M, Izawa Y, Kasaoka M and Nagano Y 2015 Furfural analysis in transformer oils using laser raman spectroscopy *IEEE Trans. Dielectr. Electr. Insul.* **22** 229–31

[20] Iruthayarajan M W, Karthik R and Sree Renga Raja T 2013 Investigations of furanic derivatives in transformer oil using ultra-violet light absorption property *IET Electr. Power Appl.* **7** 123–30

[21] Blue R, Uttamchandani D and Farish O 1998 Infrared detection of transformer insulation degradation due to accelerated thermal aging *IEEE Trans. Dielectr. Electr. Insul.* **5** 165–8

[22] Mohan Rao U and Jarial R K 2019 Measurement of transformer solid insulation degradation using dilatometry and x-ray diffraction analysis *Measurement* **131** 701–5
[23] Shamsipur M, Miran Beigi A A, Teymouri M, Rasoolipour S and Asfari Z 2009 Highly sensitive and selective poly(vinyl chloride)-membrane potentiometric sensors based on a calix [4]arene derivative for 2-furaldehyde *Anal. Chem.* **81** 6789–96
[24] Bosworth T, Setford S, Heywood R and Saini S 2001 Pulsed amperometric detection of furan compounds in transformer oil Anal. Chem. *Acta* **450** 253–61
[25] Booth C and McDonald J 1998 The use of artificial neural networks for condition monitoring of electrical power transformers *Neurocomputing* **23** 97–109
[26] Mohamed E A, Abdelaziz A Y and Mostafa A S 2005 A neural network-based scheme for fault diagnosis of power transformers *Electr. Power Syst. Res.* **75** 29–39
[27] Barbosa F R, Almeida O M, Braga A P S, Amora M A B and Cartaxo S J M 2012 Application of an artificial neural network in the use of physicochemical properties as a low cost proxy of power transformers DGA data *IEEE Trans. Dielectr. Electr. Insul.* **19** 239–46
[28] Arshad M and Islam S M 2006 A novel fuzzy logic technique for power transformer asset management *Conf. Record of the 2006 IEEE Industry Applications Conf. Forty-First IAS Annual Meeting* 1 pp 276–86
[29] Abu-Siada A, Lai S P and Islam S M 2012 A novel fuzzy-logic approach for furan estimation in transformer oil *IEEE Trans. Power Deliv.* **27** 469–74
[30] Fujii T 2002 PDMS-based microfluidic devices for biomedical applications *Microelectric Engineering* (Amsterdam: Elsevier) **61–62** 907–14
[31] Bose A and Biswas K 2017 Performance study of urease-PMMA-based aqueous urea sensor *IEEE Sens. J.* **17** 6850–8
[32] Demori M, Ferrari V and Poesio P 2010 A microfluidic capacitance sensor for fluid discrimination and characterization *Sensors Actuators* A **172** 212–9
[33] Adhikary A, Kumar G, Banerjee S, Sen S and Biswas K 2016 Modelling and performance improvement of phase-angle–based conductivity sensor *Proc. IEEE 1st Int. Conf. Control, Meas. Instrum. (CMI) (Kolkata, India)* pp 403–7
[34] Zargar Z H and Islam T 2019 A novel cross-capacitive sensor for noncontact microdroplet detection *IEEE Trans. Ind. Electron.* **66** 4759–66

www.ingramcontent.com/pod-product-compliance
Ingram Content Group UK Ltd.
Pitfield, Milton Keynes, MK11 3LW, UK
UKHW051042120926
8556IPUK00030B/727

9 780750 339339